Heat Shock Proteins

Volume 16

Heat Shock Proteins: key mediators of Health and Disease. Heat shock proteins (HSP) are essential molecules conserved through cellular evolution required for cells to survive the stresses encountered in the environment and in the tissues of the developing and aging organism. These proteins play the essential roles in stress of preventing the initiation of programmed cell death and repairing damage to the proteome permitting resumption of normal metabolism. Loss of the HSP is lethal either in the short-term in cases of acute stress or in the long-term when exposure to stress is chronic. Cells appear to walk a fine line in terms of HSP expression. If expression falls below a certain level, cells become sensitive to oxidative damage that influences aging and protein aggregation disease. If HSP levels rise above the normal range, inflammatory and oncogenic changes occur. It is becoming clear that HSP are emerging as remarkably versatile mediators of health and disease. The aim of this series of volumes is to examine how HSP regulation and expression become altered in pathological states and how this may be remedied by pharmacological and other interventions.

More information about this series at http://www.springer.com/series/7515

Alexzander A. A. Asea • Punit Kaur
Editors

Chaperokine Activity of Heat Shock Proteins

Springer

Editors
Alexzander A. A. Asea
Department of Medicine and
Director, Precision Therapeutics
Proteogenomics Diagnostic Center
Eleanor N. Dana Cancer Center
University of Toledo College of Medicine
and Life Sciences
Toledo, OH, USA

Punit Kaur
Department of Experimental Radiation
Oncology
MD Anderson Cancer Center
Houston, TX, USA

ISSN 1877-1246 ISSN 1877-1254 (electronic)
Heat Shock Proteins
ISBN 978-3-030-02253-2 ISBN 978-3-030-02254-9 (eBook)
https://doi.org/10.1007/978-3-030-02254-9

Library of Congress Control Number: 2019930723

This Springer imprint is published by the registered company Springer Nature Switzerland AG.
The registered company address is: Gewerbestrasse 11, 6330 Cham, Switzerland

Preface

Chaperokine is a term that describes the unique function of extracellular heat shock protein (eHsp) as both chaperone and cytokine. The cellular consequence of binding and signaling of eHsp is the stimulation of a potent and long-lasting immune response. eHsp induces a plethora of immune responses including the release of bioactive mediators like cytokines, chemokines, nitric oxide, apotogenic mediator, stimulation of the innate and adaptive immune response, migration and maturation of dendritic cells (DC), and the enhancement of natural killer cell-mediated cellular cytotoxicity.

The book *Chaperokine Activity of Heat Shock Proteins* provides the most comprehensive review on contemporary knowledge on the chaperokine activity of heat shock proteins (HSP) in biology and medicine. Using an integrative approach to understanding the chaperokine activity of HSP, the contributors provide a synopsis of novel mechanisms, signal transduction pathways, and how the principles of the chaperokine activity of HSP has been harnessed for therapeutic gain.

To enhance the ease of reading and comprehension, this book has been subdivided into various sections, including Part I which reviews current progress on our understanding of immunological and inflammatory responses and Part II which evaluates the role of physiological responses.

Key basic and clinical research laboratories from major universities, academic medical hospitals, and biotechnology and pharmaceutical laboratories around the world have contributed chapters that review present research activity and importantly project the field into the future. The book is a must-read for researchers, postdoctoral fellows, and graduate students in the fields of translational medicine, clinical psychology, human physiology, zoology, botany, biotechnology, molecular medicine, infectious diseases, pathology, pharmaceutical science, and research involved in drug discovery.

Toledo, OH, USA — Alexzander A. A. Asea
Houston, TX, USA — Punit Kaur

Contents

Editors Biography

Alexzander A. A. Asea is a highly innovative and accomplished world-renowned clinical and basic research scientist and visionary executive leader who has exceptional experience spearheading clinical and basic science research, training, education, and commercialization initiatives within top-ranked academic biomedical institutes. Prof. Asea's initial findings studying the effects of Hsp72 on human monocytes lead to the proposal of a novel paradigm that Hsp72, previously known to be intracellular molecular chaperones, can be found in the extracellular milieu where it has regulatory effects on immunocompetent cells – a term now called chaperokine. Prof. Asea has authored over 255 scientific publications, including peer-reviewed articles, reviews, books, book chapters, editorials, and news headliners in a wide range of biomedical-related disciplines. Prof. Asea is the series editor of the widely successful book series *Heat Shock Proteins* (Springer Nature Publications) and is an editorial board member of 13 other scientific peer-reviewed journals. Currently, Prof. Asea is at the University of Toledo College of Medicine and Life Sciences in Toledo, USA.

Punit Kaur is an expert in onco-proteogenomics, with extensive training and experience in quantitative mass spectrometry imaging, protein chemistry, and biomarker discovery. Dr. Kaur's main research focus is on the use of heat-induced nanotechnology in combination with radiotherapy and chemotherapy in the cancer stem cell therapy. Dr. Kaur has published more than 40 scientific articles, book chapters, and reviews and currently serves as editorial board member for the *European Journal of Cancer Prevention* and the *Journal of Proteomics and Bioinformatics*. Dr. Kaur is an editor of five books in the highly successful *Heat Shock Proteins* book series by Springer Nature Publishers. Currently, Dr. Kaur is a Visiting Scientist Professor at the University of Texas MD Anderson Cancer Center in Houston, USA.

Part I
Immunological and Inflammatory Responses

Chapter 1
The Chaperokine Activity of Heat Shock Proteins

Punit Kaur and Alexzander A. A. Asea

Abstract Enhanced expression of intracellular heat shock proteins (HSP) primarily promotes protein chaperoning, transport and folding of naïve, aberrantly folded, or mutated proteins, resulting in cytoprotection during variety of stressful stimuli. In contrast, exposure of immunocompetent cells to extracellular HSP activates antigen presenting cell-mediated effectors functions; including enhanced pro-inflammatory and anti-inflammatory responses, chemokine and costimulatory molecule expression and in anti-tumor surveillance. In addition, extracellular HSP has been shown to play a role in situations of both acute psychological stress and exercise. This chapter covers recent advances in understanding the complex nature of the chaperokine activity of HSP and briefly discusses the biological significance of circulating serum HSPA1A (Hsp70) to host physiology and includes recent application of HSPA1A (Hsp70)-based immunotherapies.

Keywords Chaperokine · Heat shock proteins · Inflammatory responses · Signal transduction pathways

---

P. Kaur
Department of Experimental Radiation Oncology, MD Anderson Cancer Center, Houston, TX, USA

A. A. A. Asea (✉)
Department of Medicine and Director, Precision Therapeutics Proteogenomics Diagnostic Center, Eleanor N. Dana Cancer Center, University of Toledo College of Medicine and Life Sciences, Toledo, OH, USA
e-mail: alexzander.asea@utoledo.edu

A. A. A. Asea, P. Kaur (eds.), *Chaperokine Activity of Heat Shock Proteins*, Heat Shock Proteins 16, https://doi.org/10.1007/978-3-030-02254-9_1

Abbreviations

APC	Antigen presenting cells
CTL	Cytotoxic T lymphocytes
HSPA1A	Inducible form of the 72 kDa heat shock protein
HspA8	Constitutive form of the 73 kDa heat shock protein
IFN-γ	Interferon-gamma
IL	Interleukin
TLR	Toll-like receptors

1.1 Introduction

Many of the effects of extracellular stress proteins are mediated through cell-surface receptors including TLR2, TLR4, CD40, CD91, CCR5, and scavenger receptor family (LOX-1 and SREC-1). These receptors permits binding to a diverse range of cells and the performance of complex functions in immune cells and neurons (Calderwood et al. 2007a, b). Natural killer (NK) cells (Multhoff 2006, 2007) and APC including DC (Asea et al. 2002b), macrophages, peripheral blood monocytes (Asea et al. 2000c, f) and B cells have been shown to efficiently bind HSPA1A.

Previously demonstrated that the binding of extracellular HSPA1A to human monocytes and dendritic cells results in production of the pro- inflammatory cytokines including TNF-α, IL-1β, IL-6 and IL-12 (Asea et al. 2000f, 2002b) and IFN-γ (Fig. 1.1b). Pre-treatment of cells with 1 μg/ml *Rhodopseudomonas sphaeroides* LPS (RSLP), an endotoxic LPS antagonist, did not affect HSPA1A-induced IFN-γ expression (Fig. 1.1c), completely inhibited LPS-induced IFN-γ expression (Fig. 1.1d) and controls were cells pre-treated with control protein OVA (Fig. 1.1a). Exposure of human dendritic cells for 3–5 days with extracellular HSPA1A resulted in the significant increase in proliferation of immature dentritic cells and augmentation of co-stimulatory molecules (Asea et al. 2002b). The mechanism by which peptides within the HSPA1A-peptide binding groove get incorporated into the MHC was addressed by Binder et al. (Binder et al. 2001a, b). Free peptides introduced into the cytosol become ligands of MHC I molecules and treatment of cells with deoxyspergualin (drug known to specifically bind HSPA1A and Hsp90), abrogates the ability of cells to present antigenic peptides through MHC I molecules (Sevin et al. 2015).

Stimulation of murine and human macrophages with HSPA1A and gp96 results in induction of inducible nitric oxide (iNO) synthase and the release of NO (Panjwani et al. 2002). HSPA1A-induced NO release was unaffected by pre-treatment of cells with the LPS antagonist, RSLP, negating the possibility of endotoxin contamination. Co-incubation of macrophages with IFN-γ and HSPA1A resulted in the synergistic production of NO (Panjwani et al. 2002). *Mycobacterium tuberculosis*-derived HSPA1A (Hsp70/DnaK) stimulated the production of the CC chemokines RANTES,

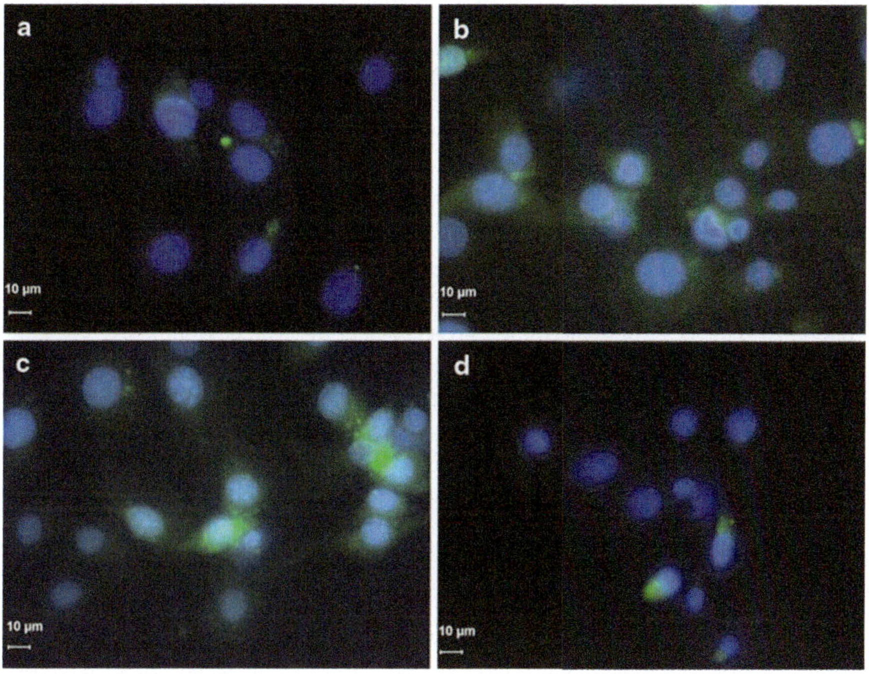

Fig. 1.1 Chaperokine-induced IFN-γ expression. Human monocytic cells THP1 were stimulated with 100 ng/ml control protein OVA (**a**), or 100 ng/ml HSPA1A alone (**b**), or pre-treated with 1 µg/ml *Rhodopseudomonas spheroids* (RSLP), a LPS antagonist, for 1 h prior to stimulation with 100 ng/ml HSPA1 A (**c**), or pre-treated with 1 µg/ml RSLP for 1 h prior to stimulation with 100 ng/ml LPS (**d**). Cells were incubated at 37 °C for 4 h in the presence of 10 µM Brefeldin A (Sigma), and simultaneously fixed and permeabilized using Cytofix/CytopermTM kit (BD Biosciences, San Diego, CA) and counter stained with anti-human IFN-γ-FITC (BD Biosciences). Results show fluorescence microscope pictograms of overlays of DAPI stain that reveal the cells nuclear morphology (blue) overlaid onto FITC stained cells which reveals the intensity of intracellular IFN-γ expression (green). Results are a representative experiment from three independently performed experiments with similar results

MIP-1α and MIP-1β from three leukocyte subsets: PBMC, $CD8^+$ and $CD4^+$ T lymphocytes and $CD14^+$ monocytes from macaques. This is significant since $CD8^+$ produce human immunodeficiency virus (HIV) and simian immunodeficiency virus (SIV) suppressor factors.

With immunocompetent cells, malignant tumor cells express high levels of surface bound HSPA1A,which correlates with an increased sensitivity to natural killer (NK)-mediated cytolysis following cytokine stimulation (Multhoff et al. 2001). Cytolytic activity of HSPA1A can be transduced by specific fragments of the HSPA1A protein with both the full-length HSPA1A protein and the C-terminal domain of HSPA1A stimulates the cytolytic activity of naive NK cells (Gross et al. 2003a, b, c). Tumor growth in mice with severe combined immunodeficiency was shown to be inhibited by HSPA1A-peptide-activated, CD94-positive natural killer cells (Moser et al. 2002). A 14 amino acid peptide sequence of the HSPA1A protein termed TKD (TKDNNLLGRFELSG, aa450–463) is the extracellular recognition site for NK cells (Multhoff et al. 2001). Granzyme B specifically binds to portions of the HSPA1A expressed on the plasma surface of tumors (Gross et al. 2003a, b, c) eliminate HSPA1A-expressing tumors in a perforin-independent, granzyme B-dependent manner and demonstrating that a $CD8^+$ CTL-independent mechanism, associated with HSP. The surface expression of HSPA1A in metastatic melanoma (Farkas et al. 2003), acute myeloid leukemia (Gehrmann et al. 2003), and head and neck cancer (Kleinjung et al. 2003) stimulates specific NK cell-mediated cytolytic functions. The development of a HSPA1A peptide which stimulates NK cell-mediated killing of leukemic blast cells (Gross et al. 2008a, b; Stangl et al. 2008) and the demonstration that NK cell-mediated targeting of membrane HSPA1A on tumors can be greatly enhanced after treatment with the cmHsp70.1 monoclonal antibody (Stangl et al. 2011a, b, c). Numerous strategies targeting HSPA1A that takes advantage of its chaperokine activity are being developed in cancer therapy (Boudesco et al. 2018).

1.1.1 Chaperokine-Induced Signal Transduction Pathways

Several cell surface proteins are described as the receptor for HSPA1A including TLR 2 and 4 (Asea et al. 2002a, b), scavenger receptor, CD36 (Delneste et al. 2002), and the co-stimulatory molecule, CD40 (Becker et al. 2002). Our group demonstrated steps involved in the HSPA1A-induced signal transduction cascade and HSPA1A binds with high affinity to the plasma membrane of APC elicits a rapid intracellular Ca^{2+} ($[Ca^{2+}]I$) flux (Asea et al. 2000a, b, c, d, e, f, g; Asea and Calderwood 2000). Endotoxin contamination might confound our results addressed by using Polymyxin B and Lipid IVa (LPS inhibitor) abrogates LPS-induced and not HSPA1A-induced, cytokine expression. HSPA1A-induced $[Ca^{2+}]i$ flux is followed by the phosphorylation of I-κBα (Asea et al. 2000a, b, c, d, e; f, g; Asea and Calderwood 2000). Activation of NF-κB is regulated by its cytoplasmic inhibitor, I-κBα, *via* phosphorylation at Serine 32 (Ser-32) and 36 (Ser-36) which targets it for degradation by the proteosome and releases NF-κB to migrate to the nucleus and activate the promoter of target genes. As early as 30 min post exposure to extracellular HSPA1A, I-κBα was phosphorylated at Serine 32 (Ser-32) and 36 (Ser-36) resulting in the release and nuclear translocation of NF-κB (Asea et al. 2000a, b, c, d, e, f, g; Asea and Calderwood 2000). Mechanistic studies using the HEK293 model system revealed that HSPA1A-induced NF-κB promoter activity is MyD88-dependent, CD14-dependent and transduced *via* both TLR2 and TLR4 (Asea et al. 2002a, b; Asea and Gunning 2002). Synergistic activation of NF-κB promoter by co-expression of both TLR2 and TLR4 is MyD88-independent, suggesting an alternative pathway by which extracellular HSPA1A stimulates cells of the immune system. Our working hypothesis is summarized in Fig. 1.2.

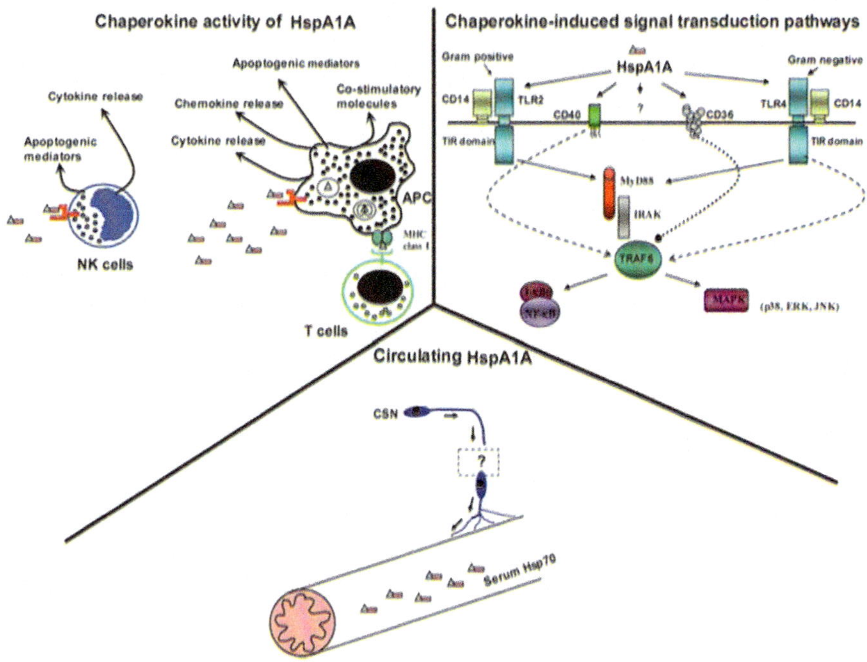

Fig. 1.2 Schematic representation of the spectrum of events mediated by exogenous HSPA1A. The HSPA1A-peptide complex binds to specific receptors on APC and stimulates co-stimulatory molecule expression and the release of pro-inflammatory cytokines, chemokines and apoptogenic mediators. Chaperokine-induced signal transduction pathways; exogenous HSPA1A binds to a yet unknown receptor (?) and/or TLR4 and TLR2 in a CD14-dependent fashion and activates the signal cascade that results in the phosphorylation of adaptor protein MyD88 IRAK TRAF6 at which the signal bifurcates to stimulate either NF-κB or MAPK signaling pathways. Exogenous HSPA1A also binds surface bound CD40 and CD36 and activates NF-κB or MAPK signaling pathways in a MyD88-independent fashion (*dotted lines*). *APC* antigen presenting cell, *CSN* central nervous system, *CTL* cytotoxic T lymphocytes, *HSPA1A* the inducible 70 kDa heat shock protein, *IRAK* IL-1 receptor-associated kinase, *MAPK* mitogen activated protein kinase, *MHC* major histocompatability complex, *MyD88* myeloid differentiation factor 88, *NF-κB* nuclear factor-κB, *TLR* toll-like receptor, *TRAF6* TNF receptor associated factor 6

CD40 is a co-stimulatory molecule expressed on APC has been demonstrated to bind HSPA1A-peptide complexes *via* its exoplasmic domain (Becker et al. 2002). The HSPA1A-CD40 interaction is mediated by the NH2-terminal ATPase domain of HSPA1A in its ADP-bound state and is strongly augmented by the presence of substrate peptides in the COOH-terminal domain of HSPA1A, however suppressed by Hip, a co-chaperone that is known to stabilize the HSPA1A ATPase domain in the ADP state (Becker et al. 2002). Specific HSPA1A-CD40 binding stimulates signal transduction *via* the phosphorylation of p38, which results in the activation of NF-κB and uptake of peptide (Becker et al. 2002).

HSPA1A has been discovered to also bind to two classes of receptor: c-type lectin receptors (CLR) and scavenger receptors (SR). HSPA1A binds LOX-1 (a member of both the CLR and SR), with the c-type lectin binding domain (CTLD), SR family members SREC-I and FEEL-1/CLEVER-1/STABILIN-1, which by contrast have arrays of EGF-like repeats in their extracellular domains (Calderwood et al. 2007a, b, c, g; Theriault et al. 2005). Members of SR family bind Hsp70-peptide complexes and mediate its efficient internalization (Theriault et al. 2006). The scavenger receptor, CD36 has been shown to bind HSPA1A (Delneste et al. 2002). Specifically, LOX-1, on human dendritic cells was shown to bind HSPA1A, and incubation of cells with a neutralizing anti-LOX-1 mAb abrogated HSPA1A binding to dendritic cells and suppressed HSPA1A-induced antigen cross-presentation (Calderwood et al. 2007a, b, c, d, e, f, g; Delneste et al. 2002; Theriault et al. 2005, 2006). In vivo studies show that pre-treatment of animals with anti-LOX-1 mAb induces antitumor immunity.

Hsp70-induced NF-κB promoter activity is MyD88-dependant, CD14-dependant and is transduced *via* both TLR2 and TLR4 synergistically stimulates Hsp70-induced cytokine production (Asea et al. 2002b). Synergistic activation of NF-κB promoter by co-expression of both TLR2 and TLR4 is MyD88-independent, suggesting an alterative pathway by which exogenous Hsp70 stimulates cells of the immune system. As early as 2–4 h post exposure of APC to exogenous Hsp70, there is significant release of TNF-α, IL-1β, IL-6 and IL-12 (Asea et al. 2000f, 2002b). The human monocytic cell line THP1, transfected with the dominant negative MyD88 plasmid or a combination of both dominant negative TLR2 and TLR4 inhibited a portion of Hsp70-induced IL-6 (Asea et al. 2002b) and IL-1β expression (Asea 2003).

1.1.2 Biological Significance of Circulating Serum HAPA1A: Role in Host Physiology

Danger theory suggests that innate immune cells are activated by danger signals that are derived from stressed or damaged self-proteins. Our working hypothesis is that indeed extracellular HSPA1A is such a danger signal further demonstrates acute psychological stress, which results in the release of extracellular HSPA1A (Campisi and Fleshner 2003; Campisi et al. 2003a, b, c). Rats exposed to inescapable tail-shock stress (IS; 100 1.6 mA tail shocks, 5 s duration, 60 s intertrial interval) or exhaustive exercise stress (EXS; treadmill running to exhaustion), but not sedentary rats, had elevated Hsp72 in dorsal vagal complex, frontal cortex, hippocampus, pituitary, adrenal, liver, spleen, mesenteric lymph nodes, and heart, as judged by blood, brain, and peripheral tissue samples collected 2 h post stress termination (Campisi et al. 2003c). Splenocytes from physically active rats responded to extracellular HSPA1A stimulation ex vivo with a greater nitric oxide as compared to sedentary rats. Studies in humans showed that physical exercise results in marked increase in circulating serum HSPA1A levels (Walsh et al. 2001). It is hypothesized that the released circulating HSPA1A serves as a chaperokine and enhances the host defense against pathogens during exercise. Indeed, we demonstrated that HSPA8 is released from neoplastic, but not normal, cells in response to cytokines including IFN-γ and IL-10, cytokines normally found in high concentrations within the tumor microenvironment, and in response to heat shock (Barreto et al. 2003). Neuroblastoma cells took up biotinylated HSPA1A and exhibited a thermotolerance phenotype when exposed to lethal heat shock (44 °C) and to staurosporine-induced apoptosis, suggesting a mechanism by which extracellular HSPA1A might affect neuronal function (Guzhova et al. 2001).

1.1.3 Application of Extracellular HSPA1A: HSPA12A-Based Immunotherapies

Due to the unique ability to: chaperone antigenic peptides, interact with APC, mediate release of chemokine and cytokines from APC and induce the maturation of dendritic cells, makes heat shock proteins an important new generation of prophylactic and therapeutic vaccine against cancers and infectious diseases (Srivastava 2004). Oncophage®, HSP/peptide-complex (HSPPC) derived from autologous liver metastases of colorectal carcinoma induces cancer-specific T-cell responses in patients rendered disease-free by surgery (Mazzaferro et al. 2003), elicits a significant increase in CTL $CD8^+$ T-cell response. Further, in this limited number of patients, 2-year overall survival and disease-free survival were significantly improved in subjects with post-vaccination anti-tumor immune response (Calderwood et al. 2016).

Malaria is one of the most devastating health problems in large parts of the world. However, the development of an effective vaccine is marred because there are significant antigenic changes associated with the malaria parasite during its various life cycles. The merozoite surface protein 1 (MSP1) fused to heat-shock cognate protein 70 (HSPA8) was constructed and its efficacy tested against the malaria parasite (Kawabata et al. 2002). Merozoite surface protein 1 (MSP1) is expressed during blood-stage in the exoerythrocytic forms of malaria. Immunization of C57BL/6 mice with the fusion protein prior to challenge infection with *Plasmodium yoelii* sporozoites, resulted in the significant delay in or complete inhibition of parasitemia. Additional ways to combat malaria have been to use inhibitors of HSP. The hypothesis is that the joint disruption of the parasite antioxidant enzymes/HSP system would interfere with parasite growth and open new perspectives for anti-malaria therapy. Akide-Ndunge and colleagues reported that mRNA expression of parasite antioxidant enzymes and HSP was co-ordinated and stage-dependent (Akide-Ndunge et al. 2009). In addition, both systems were redox-responsive and showed remarkably increased and co-ordinated expression in oxidatively-stressed parasites and in parasites growing in antioxidant blunted G6PD-deficient RBCs (Kempaiah et al. 2017).

It has been suggested that the major difficulties in developing a preventive vaccine is because of the rapid rate of mutation of HIV and the lack of induction of neutralizing antibodies to the virus (Whittall et al. 2011). In addition, the anti-HIV drug treatment is associated with severe side effects and the development of drug-resistant strains of HIV. Immunization of nonhuman primates with HSPA1A359-610 induced the production of RANTES and IL-12, and acted as an adjuvant when loaded with CC5-peptide (Bergmeier et al. 2010).

1.1.4 *Active Release Mechanism: Inflammation, Exercise, Psychological Stress, Brain and Cerebral Spinal Fluid (CSF)*

The alternative to passive release of HSP is the active release mechanism. Hightower and Guidon demonstrating that a small groups of HSP namely Hsp110, Hsp71, Hsc73 and Hsp72 are released from mammalian cells (Hightower and Guidon, 1989). In this study, supernatant from cultured rat embryo cells exposed to control conditions (37 °C) or heat stressed (45 °C) for 10 min and allowed to recover for 2 h at 37 °C. At control conditions, Hsp110, Hsp71 and Hsc73 were shown to be released and heat stress additionally induced Hsp72. Pre-treatment of cells with inhibitors of the classical protein secretory pathways including monensin (known to disrupts the Golgi apparatus and vesicular transport) and colchicine (an inhibitor of microtubule assembly) did not abrogate stress-induced HSP release. Guzhova and colleagues demonstrated that Hsp70 is released by glia cells in the absence of necrotic cell death (Guzhova et al. 2001) and Hsp70 has been shown to be released by B cells (Clayton et al. 2005), peripheral blood mononuclear cells (Hunter-Lavin et al. 2004) and tumors (Bausero et al. 2005) (Gastpar et al. 2005), under non necrotic conditions.

The extracellular molecular chaperone Hsp90 (eHsp90) has been identified as a critical effector in cancer cell invasion and metastasis by virtue of its interaction with a diverse cohort of molecules that serve as key nodal points in oncogenic pathways. Inflammatory stress results in the active release of Hsp90 from pancreatic islets (Gardiner et al. 2015). This is interesting since Hsp90 inhibitors have already shown encouraging clinical efficacy in a number of oncological diseases (Calderwood and Neckers 2016; Neckers et al. 2018).

Heat shock proteins (HSP) are families of molecular chaperones that play important homeostatic functions in the central nervous system (CNS) by preventing protein misfolding, promoting degradation of improperly folded proteins, and protecting against apoptosis and inflammatory damage especially during hyperthermia, hypoxia, or oxidative stress (Asea and Brown 2007; Lackie et al. 2017). An important, an overlooked function of some HSP is their ability to function as extracellular messengers also termed chaperokines that modulate immune responses within the CNS (Prado 2017; van Noort et al. 2017).

1.1.5 Chaperokine-Mediated Inflammatory Responses

After admixing eHsp70 to APCs resulted in the stimulation of an inflammatory response including release of cytokines TNF-α, IL-1β, IL-6 and IL-12 (Asea et al. 2000c, d, e, f; g, 2002a, b; Asea and Calderwood 2000) and GM-CSF (Srivastava 2002a, b); nitric oxide, a potent apotogenic mediator (Panjwani et al. 2002); chemokines, including MIP-1α, MCP-1β, and RANTES (Panjwani et al. 2002). We also demonstrated that both peptide-bearing and non peptide-bearing eHsp70 are capable of inducing pro-inflammatory cytokine production by APCs (Asea et al. 2000c, d, e, f, g; Asea and Calderwood 2000). eHsp72 induces the DC maturation by augmenting the surface expression of CD40, CD83, CD86 and MHC class II molecules on DC (Asea et al. 2000c, d, e, f, g; Asea and Calderwood 2000) and migration of DC (Binder et al. 2000) and NK cells (Gastpar et al. 2005).

Our group was the first to demonstrate that IFN-γ and IL-10, pro-inflammatory cytokines, known to be found in high concentrations in inflammatory foci, induce the active release of constitutive Hsc73 from tumors (Barreto et al. 2003). We further showed that whereas some Hsp72 could be found as free Hsp72, a proportion of Hsp72 was released within exosomes (Bausero et al. 2005; Gastpar et al. 2005), highly packed with immunostimulatory mediators including MHC class I and II (Zitvogel et al. 1999) and costimulatory molecules (Escola et al. 1998). Also, our group demonstrated that Hsp72 was released by a non classical protein transport pathway and that intact surface membrane lipid rafts were required for efficient stress-induced Hsp72 release (Bausero et al. 2005; Gastpar et al. 2005). Similar findings were demonstrated in B cells (Clayton et al. 2005). Newly synthesized Hsp72 protein localizes within the Golgi region of HELA cells and also concentrates on the surface of the plasma membrane and in the ruffled zone of migrating cells (Schneider et al. 2002). Patients with type 1 diabetes and rheumatoid arthritis show that treatment with HSP-derived peptides promote a switch from a pro-inflammatory cytokine-secretion profile of T lymphocytes to a regulatory anti-inflammatory cytokine-secretion profile (van Eden 2006). T regulatory (Treg) cells display CD4+CD25+ phenotype, a down-regulation of the CD4, CD3 and TCR transcripts and plasma membrane antigenic expression similar to activated T lymphocytes. The transcription factor FoxP3 has recently been shown to play an important role in Treg function (Hori et al. 2003a, b) supported by studies in knock-in transgenic mice.

In addition to T regulatory cells (Tregs), natural killer (NK) cells have been shown to respond to HSP. The transfer of NK cells from Hsp70-pc-immunized mice to recipients sensitized for EAE abolished the development of the disease. The cytolytic activity of Hsp70 can also be greatly augmented by specific fragments of the Hsp70 protein. In addition, tumor growth in mice with severe combined immunodeficiency was shown to be naive inhibited by Hsp70-peptide-activated, CD94-positive NK cells. A 14 amino acid sequence of the Hsp70 protein, termed TKD (TKDNNLLGRFELSG, aa450–463) is the extracellular recognition site for NK cells (Multhoff et al. 2001). Immunization of the peptide binding C-terminal portion of Hsp70 aa359–610 Hsp70359–610 is responsible for stimulating Th1-polarizing cytokine (IL-12 and TNF-α, C-C chemokine release, and acts as an adjuvant (Wang et al. 2002). Immunization of nonhuman primates with Hsp70359–610 induced the production of RANTES and IL-12, and acted as an adjuvant when loaded with CC5-peptide (Wang et al. 2002) suggesting a possible alternative vaccine strategy for HIV infection (Lehner and Anton 2002). Hsp60 suppresses Th1-mediated hepatitis by down regulating T-bet, NF-κB, and NFATp and upregulating GATA-3, thereby resulting in the abrogation of TNF-α and IFN-γ secretion and a concomitant increase in secretion of IL-10 *via* a TLR2-dependent mechanism (Zanin-Zhorov et al. 2005a, b). IL-1β induces the expression as well as secretion of Hsp60 in extracellular milieu, which then binds with TLR4 of microglia to exert a chaperokine effect (Swaroop 2017). Further Hsp60 increases the phosphorylation of ERK, JNK, and p38 MAPKs in microglia during inflammation but specific inhibition of p38 only occurs in decreased inflammation by Hsp60. Inflammation plays an important role in the pathophysiology of sickle cell disease (SCD) (Adewoye and McMahon 2005). A significant increase in circulating serum Hsp70 levels in SCD during vaso-occlusive crisis (VOC) suggesting that circulating serum Hsp70 might be a marker for VOC in SCD (Adewoye and McMahon 2005).

1.1.6 Chaperokine Activity and Aging

A progressive decline in serum Hsp70 levels at older ages in 60 individuals ranging in age from 20 to 96 years (Calderwood et al. 2009; Jin et al. 2004). Serum Hsp70 levels are lower at older ages, our studies showed that there is a decline in Hsp70 function over time. Our working hypothesis is summarized in Fig. 1.3. There are several proposed explanations as to why serum Hsp70 levels are low at very old age (Fig. 1.2). It may be that cells in long lived individuals are exposed to similar stress but do not mount the same response from heat shock factor 1 (Hsf1), the transcription factor that regulates Hsp70 synthesis. Data from both rat hepatocytes (Wu et al. 1993) and human T-lympocyte (Effros et al. 1994) cultures that demonstrate a decrease in Hsp70 synthesis with age are consistent with this notion. However, a study of *Caenorhabditis elegans* makes this notion less plausible since an overexpression of the hsf1 gene in this organism leads to a 40% greater life-span compared to wild type animals (Hsu et al. 2003).

Another possible explanation for the low serum Hsp70 level in very old individuals and their offspring is that their cells synthesize similar amounts of intracellular Hsp70 but the cell walls are less permeable resulting in less leakage into the extracellular space. Consistent with this, there is a direct relationship between intact cell membranes and the release of Hsp70 (Bausero et al. 2005). Lastly, it may be that survivors to very old age have always had low serum Hsp70, which could be attributed to a lack of exposure to cellular stress or it could be because of decreased exposure to inflammation (Terry et al. 2003).

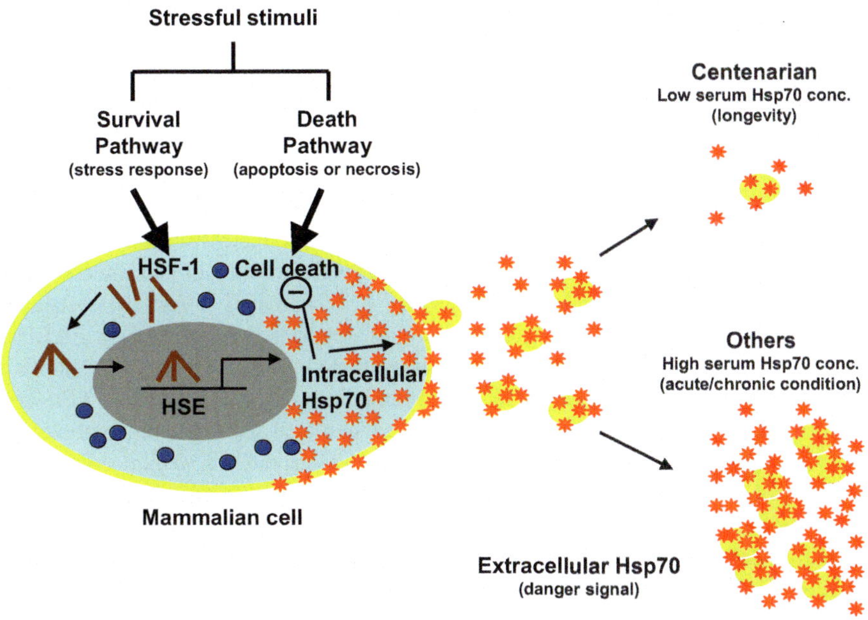

Fig. 1.3 Schematic representation of the working hypotheses. Under normal physiological conditions low levels of heat shock proteins are expressed. However, in response to various kinds of stressful stimuli, pathological stimuli or physiological stimuli, two pathways can be activated. During the stress response heat shock factor-1 (HSF-1; rods) found in the cytoplasm of all cells trimerizes and migrates from the cytoplasm to the nucleus. Once inside the nucleus the trimerized HSF-1 binds to a specific region of the DNA known as the heat shock element (HSE). This binding initiates the synthesis of many heat shock proteins including Hsp70 (stars). During the stress response, Hsp70 is released into the extracellular milieu (extracellular Hsp70). This implies that it will stimulate a potent inflammatory response including the release of pro-inflammatory cytokines, chemokines, nitric oxide production and maturation of dendritic cells, a process known as the chaperokine activity of Hsp70. High concentrations of serum Hsp70 have been reported in individuals with acute and/or chronic conditions including arthritis, acute myocardial infarction, arthrosclerosis, cancer and acute infection

1.1.7 HSP-Peptide Complexes (HSPPC) as Immunotherapy for Cancer

Srivastava et al. first reported that HSPPC-gp96 from tumor cells contain the HSP associated with cancer specific antigenic peptides (Srivastava et al. 1986). Animals were immunized with HSP-peptide complexes (HSPPC) purified from autologous cancer tissue elicited strong tumor specific immune responses and resulted in retarded progression of the primary cancer. Efficacy of autologous cancer derived HSPPC-gp96 in immunotherapy of cancers (such as 3LL lung carcinoma, B16 melanoma, colon carcinoma and spindle cell carcinoma) (Tamura et al. 1997). In addition to gp96, a number of HSP, including Hsp70, Hsp90, Hsp110 and calreticulin, later were identified to display similar antigen specific antitumor activity when purified from tumors.

Prior vaccination with HSP110 or grp170 purified from methylcholanthrene-induced fibrosarcoma caused complete regression of the tumor in mice (Wang et al. 2001). Meanwhile, HSP110 or grp170 purified from Colon 26 tumors led to a significant growth inhibition of this tumor and also extended the lifespan of Colon 26 tumor bearing mice after tumor transplantation. HSP70 peptides were able to trigger a CTL response in vivo in HLA-A*0201-transgenic HHD mice and in vitro in HLA-A*0201+ healthy donors. p391- and p393-specific human and murine CTL recognized human tumor cells overexpressing HSP70 in a HLA-A*0201-restricted manner (Faure et al. 2004). In 2001, Sato et al. found that immunization BALB/c mice with HSP70 and GP96 purified from syngeneic leukemia cell line A20 elicited a specific response of potent CD8(+) T lymphocytes cytotoxic against A20 and prolonged mice survival after A20 inoculation (Sato et al. 2001). Later, Jimbo et al. also reported that immunization with A20 leukemia derived Hsp70 induced the production of anti-A20-antibodies against leukemia cell specific peptides and play a crucial role in the eradication of leukemia cells in mice. Autologous, HLAA0201 dendritic cells were pulsed with gp96 derived from HLA-A0201 human myeloma cell line U266 or primary myeloma cells. Specific CTL cell lines were obtained after repeatedly stimulating T cells with these pg96-pulsed dendritic cells.

HSPPC-gp96 vaccination elicited a significant increase in CD8+ T-cell response against colon cancer (Mazzaferro et al. 2003). Phase I/II clinical studies using tumor derived HSPPC-gp96 as vaccines are ongoing for melanoma, colorectal carcinoma and kidney cancer is safe and elicited potent antigenspecific CTL responses against the autologous tumors (Gordon and Clark 2004). The well known but occasional occurrence of spontaneous or nephrectomy induced regression of metastasis in patients with RCC and the existence of tumor reactive and tumor infiltrating CTL imply that RCC is an immunogenic tumor (Marcus et al. 1993). Investigations into the role of heat shock proteins (HSP) in immune response have progressed well into a third decade, and use of HSPPC-gp96 for the treatment of renal cell carcinoma (RCC) in the adjuvant setting already showed good efficacy. In the first study, this HSPPC-gp96 vaccine was easy to administer, appears well tolerated, shows antitumor activity and significantly improves the recurrence free survival rated in patients with metastatic RCC. Reconstituted HSP-peptide complexes and HSP-peptide fusion proteins, are developed to overcome the limitations of tumor derived HSPPC. Manjili et al. took advantage of more binding efficiency of large HSP to reconstitute Hsp110 and intracellular domain (ICD) of human HER-2/neu complexes, which is a fourfold more efficient in binding to protein substrates compared to Hsp70 had elicited effective IFN-γ-producing T cells against spontaneous mammary tumors in FVB-neu transgenic mice (Manjili et al. 2002a, b, c, 2003). Suzue et al. first developed recombinant mycobacterial Hsp70 and a large fragment of ovalbumin fusion protein. An HSP fusion protein called HSPE7 composed of *Mycobacterium bovis* BCG Hsp65 linked to E7 protein of human papillomaviruses type 16 (HPV16) was developed and showed that HSP-E7 can primes potent E7-specific CD8+ T cells with cytolytic and cytokine secretion activities. Even fusion protein containing truncated Hsp70 (N-terminal or C-terminal domain) linked o HPV16-E7 are capable of inducing potent antigen specific CTL activity in animal models.

Insights into Mycobacterium tuberculosis (Mtb) factors modulating host response are required for the discovery of novel vaccine antigen targets as well as a better understanding of dynamic interactions between the bacterial factors and host cells. Functional role of Mtb GrpE, a cofactor of heat shock protein 70 (HSP70), in promoting naïve CD4+/CD8+T cell differentiation toward Th1-type T-cell immunity through interaction with dendritic cells (DCs) and GrpE-induced DC maturation by up regulating the expression of cell surface molecules (CD80, CD86, and MHC class I and II) and production of several pro-inflammatory cytokines (TNF-α, IL-1β, IL-6, and IL-12p70). GrpE-stimulated DCs induced the proliferation of GrpE-specific Th1-type effector/memory CD4+/CD8+CD44 highCD62L low T cells from the spleen of Mtb-infected mice in a TLR4-dependent manner. These results demonstrate that GrpE is a novel immune activator that interacts with DCs, in particular, via TLR4, to generate Th1-biased memory T cells in an antigen specific manner, which is a rational basis for the discovery of new potential targets to develop an effective tuberculosis vaccine (Kim et al. 2018). Researchers have been trying to inhibit Hsp70 expression for therapeutic gain. Small molecule 2-phenylethyenesulfonamide (PES), also referred to as pifithrin-mu, a known as an Hsp70 inhibitor was used to inhibit tumor growth in a mouse xenograft model of lung cancer by A549 cells (Zhou et al. 2017).

1.2 Conclusions

It is clear from numerous clinical trials that modulating heat shock responses are of considerable therapeutic benefit in a wide range of diseases and disorders. The passive release mechanism occurs when cells are infected, trauma or necrosis, resulting in the release of HSP into systemic circulation. We hypothesize that passive release of HSP results in high levels of free HSP and low levels of HSP contained within exosomes (Fig. 1.4). The active release mechanism occurs in response to psychological stress, stressful stimuli, in response to inflammatory mediators or receptor mediated events including ligation of HSP to its receptor(s). The active release mechanism is mediated by stimulation of the stress response which induces monomeric heat shock factor (HSF) which is in the cytosol to trimerize and translocate to the nucleus, where it binds to the heat shock element (HSE) and activates the synthesis of HSP (Fig. 1.1). HSP released into the circulation as free HSP and within exosomes act as a danger signal and cytokine. Studies on the signal transduction pathway activated by extracellular HSPA1A has begun to shed some light on the nature of the HSPA1A-receptor and seems to suggest that there might be more than one receptor for this unique protein. The next frontier is to elucidate the exact role of elevated circulating serum HSPA1A in various physiological situations including acute psychological stress and exercise. Successful completion of these studies

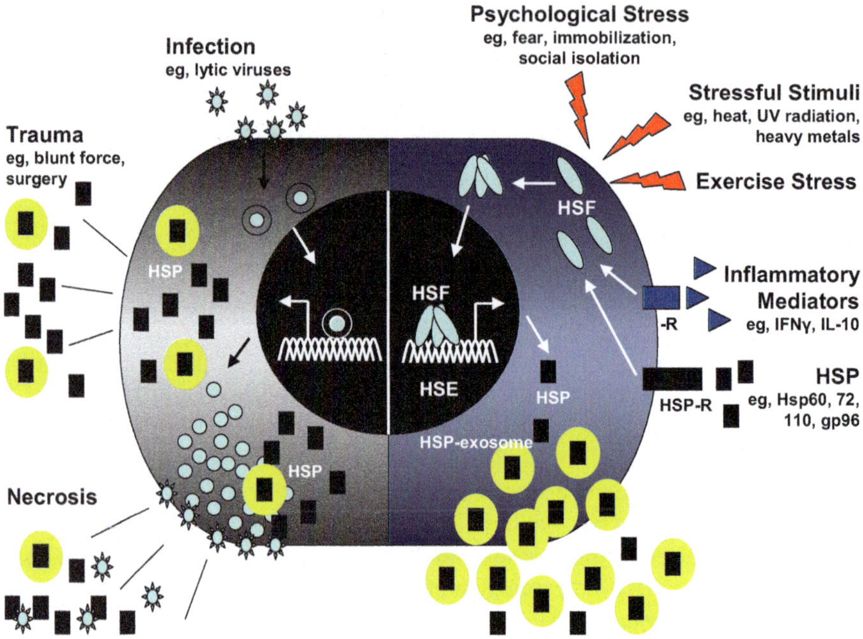

Fig. 1.4 Schematic representation of mechanisms by which HSP is released in circulation. Currently, there are two mechanisms of HSP released; passive (left grey half of the cell) and active (right blue half of the cell). The passive release mechanism occurs when cells are infected or trauma or necrosis, resulting in the release of HSP into systemic circulation. The active release mechanism occurs in response to psychological stress or following stressful stimuli or in response to inflammatory mediators or receptor mediated events including ligation of HSP to its receptor(s). The active release mechanism is mediated by stimulation of the stress response which induces monomeric heat shock factor (HSF) which is in the cytosol to trimerize and translocate to the nucleus, where it binds to the heat shock element (HSE) and activates the synthesis of HSP (filled squares). HSP are released into the circulation as free HSP (filled squares) and within exosomes (yellow circles) to act as a danger signal

promises to greatly further our current understanding of the role of heat shock proteins and help decipher how the host communicates when it perceives danger.

Acknowledgments We thank all students, faculty and staff of the Asea Lab through the years. This work was supported in part the US National Institutes of Health grant RO1CA91889, Dana Faber Cancer Institute, Harvard Medical School, Boston University School of Medicine, Scott & White Hospital and Clinic, the Texas A&M Health Science Center College of Medicine, the Central Texas Veterans Health Administration, an Endowment from the Cain Foundation and the University of Toledo College of Medicine and Life Sciences (to A.A.). The funders had no role in study design, data collection and analysis, decision to publish or preparation of the chapter.

References

Adewoye AH, McMahon L (2005) Chaperones and disease. N Engl J Med 353:2821–2822; author reply 2821-2822

Akide-Ndunge OB, Tambini E, Giribaldi G, McMillan PJ, Muller S, Arese P, Turrini F (2009) Co-ordinated stage-dependent enhancement of *Plasmodium falciparum* antioxidant enzymes and heat shock protein expression in parasites growing in oxidatively stressed or G6PD-deficient red blood cells. Malar J 8:113

Asea A (2003) Chaperokine-induced signal transduction pathways. Exerc Immunol Rev 9:25–33

Asea A, Brown IR (2007) Heat shock proteins and the brain: implications for neurodegenerative diseases and neuroprotection. Springer, Dordrecht

Asea AA, Calderwood SK (2000) HSP70 binds to the cell surface of monocytes and induces cytokine expression. Cell Stress Chaperones 5:374–374

Asea P, Gunning JW (2002) AERC plenary session December 1999 – privatisation and corporate governance – preface. J Afr Econ 11:ii–ii

Asea A, Ara G, Teicher BA, Stevenson MA, Calderwood SK (2000a) Effects of the flavnoid drug quercetin on the response of human prostate tumors to hyperthermia in vivo. Cell Stress Chaperones 5:500–500

Asea A, Ara G, Teicher BA, Calderwood SK (2000b) Cyclooxygenase inhibitors are potent sensitizers of DU-145 prostate tumors to hyperthermia in vivo. Cell Stress Chaperones 5:501–501

Asea A, Kabingu E, Stevenson MA, Calderwood SK (2000c) HSP70 peptide-bearing and peptide-negative preparations act as chaperokines. Cell Stress Chaperones 5:425–431

Asea A, Kabingu E, Stevenson MA, Calderwood SK (2000d) Hsp70 peptide-bearing and peptide-negative preparations function as chaperokines. Cell Stress Chaperones 5:492–492

Asea A, Kabingu E, Stevenson MA, Calderwood SK (2000e) HSP70 peptidembearing and peptide-negative preparations act as chaperokines. Cell Stress Chaperones 5:425–431

Asea A, Kraeft SK, Kurt-Jones EA, Stevenson MA, Chen LB, Finberg RW, Koo GC, Calderwood SK (2000f) HSP70 stimulates cytokine production through a CD14-dependant pathway, demonstrating its dual role as a chaperone and cytokine. Nat Med 6:435–442

Asea A, Kraeft SK, Kurt-Jones EA, Stevenson MA, Chen LB, Finberg RW, Koo GC, Calderwood SK (2000g) Hsp70 stimulates cytokine production via a CD14-dependent pathway: a "chaperokine". Cell Stress Chaperones 5:491–491

Asea A, Rehli M, Kabingu E, Boch JA, Bare O, Auron PE, Stevenson MA, Calderwood SK (2002a) Novel signal transduction pathway utilized by extracellular HSP70 – role of toll-like receptor (TLR) 2 AND TLR4. J Biol Chem 277:15028–15034

Asea A, Rehli M, Kabingu E, Boch JA, Bare O, Auron PE, Stevenson MA, Calderwood SK (2002b) Novel signal transduction pathway utilized by extracellular HSP70: role of toll-like receptor (TLR) 2 and TLR4. J Biol Chem 277:15028–15034

Barreto A, Gonzalez JM, Kabingu E, Asea A, Fiorentino S (2003) Stress-induced release of HSC70 from human tumors. Cell Immunol 222:97–104

Bausero MA, Gastpar R, Multhoff G, Asea A (2005) Alternative mechanism by which IFN-gamma enhances tumor recognition: active release of heat shock protein 72. J Immunol 175:2900–2912

Becker T, Hartl FU, Wieland F (2002) CD40, an extracellular receptor for binding and uptake of Hsp70-peptide complexes. J Cell Biol 158:1277–1285

Bergmeier LA, Babaahmady K, Pido-Lopez J, Heesom KJ, Kelly CG, Lehner T (2010) Cytoskeletal proteins bound to heat-shock protein 70 may elicit resistance to simian immunodeficiency virus infection of CD4(+) T cells. Immunology 129:506–515

Binder RJ, Anderson KM, Basu S, Srivastava PK (2000) Cutting edge: heat shock protein gp96 induces maturation and migration of CD11c+ cells in vivo. J Immunol 165:6029–6035

Binder RJ, Karimeddini D, Srivastava PK (2001a) Adjuvanticity of alpha 2-macroglobulin, an independent ligand for the heat shock protein receptor CD91. J Immunol 166:4968–4972

Binder RJ, Blachere NE, Srivastava PK (2001b) Heat shock protein-chaperoned peptides but not free peptides introduced into the cytosol are presented efficiently by major histocompatibility complex I molecules. J Biol Chem 276:17163–17171

Boudesco C, Cause S, Jego G, Garrido C (2018) Hsp70: a cancer target inside and outside the cell. Methods Mol Biol 1709:371–396

Calderwood SK, Neckers L (2016) Hsp90 in cancer: transcriptional roles in the nucleus. Adv Cancer Res 129:89–106

Calderwood SK, Mambula SS, Gray PJ Jr (2007a) Extracellular heat shock proteins in cell signaling and immunity. Ann N Y Acad Sci 1113:28–39

Calderwood SK, Mambula SS, Gray PJ Jr, Theriault JR (2007b) Extracellular heat shock proteins in cell signaling. FEBS Lett 581:3689–3694

Calderwood SK, Sherman ML, Ciocca DR (2007c) Heat shock proteins in cancer. Springer, Dordrecht

Calderwood SK, Sherman MY, Ciocca D (2007d) Heat shock proteins in cancer. Springer, Dordrecht

Calderwood SK, Sherman YM, Ciocca D (2007e) Heat shock proteins in cancer. Springer, Dordrecht

Calderwood SK, Sherman MY, Ciocca DR (2007f) Heat shock proteins in cancer. Springer, Dordrecht

Calderwood SK, Theriault J, Gray PJ, Gong J (2007g) Cell surface receptors for molecular chaperones. Methods 43:199–206

Calderwood SK, Murshid A, Prince T (2009) The shock of aging: molecular chaperones and the heat shock response in longevity and aging – a mini-review. Gerontology 55:550–558

Calderwood SK, Gong J, Murshid A (2016) Extracellular HSPs: the complicated roles of extracellular HSPs in immunity. Front Immunol 7:159

Campisi J, Fleshner M (2003) Role of extracellular HSP72 in acute stress-induced potentiation of innate immunity in active rats. J Appl Physiol 94:43–52

Campisi J, Hansen MK, O'Connor KA, Biedenkapp JC, Watkins LR, Maier SF, Fleshner M (2003a) Circulating cytokines and endotoxin are not necessary for the activation of the sickness or corticosterone response produced by peripheral *E. coli* challenge. J Appl Physiol 95:1873–1882

Campisi J, Leem TH, Fleshner M (2003b) Stress-induced extracellular Hsp72 is a functionally significant danger signal to the immune system. Cell Stress Chaperones 8:272–286

Campisi J, Leem TH, Greenwood BN, Hansen MK, Moraska A, Higgins K, Smith TP, Fleshner M (2003c) Habitual physical activity facilitates stress-induced HSP72 induction in brain, peripheral, and immune tissues. Am J Phys Regul Integr Comp Phys 284:R520–R530

Clayton A, Turkes A, Navabi H, Mason MD, Tabi Z (2005) Induction of heat shock proteins in B-cell exosomes. J Cell Sci 118:3631–3638

Delneste Y, Magistrelli G, Gauchat J, Haeuw J, Aubry J, Nakamura K, Kawakami-Honda N, Goetsch L, Sawamura T, Bonnefoy J, Jeannin P (2002) Involvement of LOX-1 in dendritic cell-mediated antigen cross-presentation. Immunity 17:353–362

van Eden W (2006) Immunoregulation of autoimmune diseases. Hum Immunol 67:446–453

Effros RB, Zhu X, Walford RL (1994) Stress response of senescent T lymphocytes: reduced hsp70 is independent of the proliferative block. J Gerontol 49:B65–B70

Escola JM, Kleijmeer MJ, Stoorvogel W, Griffith JM, Yoshie O, Geuze HJ (1998) Selective enrichment of tetraspan proteins on the internal vesicles of multivesicular endosomes and on exosomes secreted by human B-lymphocytes. J Biol Chem 273:20121–20127

Farkas B, Hantschel M, Magyarlaki M, Becker B, Scherer K, Landthaler M, Pfister K, Gehrmann M, Gross C, Mackensen A, Multhoff G (2003) Heat shock protein 70 membrane expression and melanoma-associated marker phenotype in primary and metastatic melanoma. Melanoma Res 13:147–152

Faure O, Graff-Dubois S, Bretaudeau L, Derre L, Gross DA, Alves PM, Cornet S, Duffour MT, Chouaib S, Miconnet I, Gregoire M, Jotereau F, Lemonnier FA, Abastado JP, Kosmatopoulos K (2004) Inducible Hsp70 as target of anticancer immunotherapy: identification of HLA-A*0201-restricted epitopes. Int J Cancer 108:863–870

Gardiner G, Guindon L, Perez L, Evans-Molina C, Blum J (2015) Heat shock protein 90, a potential biomarker for type I diabetes, is secreted by human pancreatic beta cells in response to cytokine stress (BA7P.161). J Immunol 194:115.21

Gastpar R, Gehrmann M, Bausero MA, Asea A, Gross C, Schroeder JA, Multhoff G (2005) Heat shock protein 70 surface-positive tumor exosomes stimulate migratory and cytolytic activity of natural killer cells. Cancer Res 65:5238–5247

Gehrmann M, Schmetzer H, Eissner G, Haferlach T, Hiddemann W, Multhoff G (2003) Membrane-bound heat shock protein 70 (Hsp70) in acute myeloid leukemia: a tumor specific recognition structure for the cytolytic activity of autologous NK cells. Haematologica 88:474–476

Gordon NF, Clark BL (2004) The challenges of bringing autologous HSP-based vaccines to commercial reality. Methods 32:63–69

Gross C, Hansch D, Gastpar R, Multhoff G (2003a) Interaction of heat shock protein 70 peptide with NK cells involves the NK receptor CD94. Biol Chem 384:267–279

Gross C, Koelch W, DeMaio A, Arispe N, Multhoff G (2003b) Cell surface-bound heat shock protein 70 (Hsp70) mediates perforin-independent apoptosis by specific binding and uptake of granzyme B. J Biol Chem 278:41173–41181

Gross C, Schmidt-Wolf IG, Nagaraj S, Gastpar R, Ellwart J, Kunz-Schughart LA, Multhoff G (2003c) Heat shock protein 70-reactivity is associated with increased cell surface density of CD94/CD56 on primary natural killer cells. Cell Stress Chaperones 8:348–360

Gross C, Holler E, Stangl S, Dickinson A, Pockley AG, Asea AA, Mallappa N, Multhoff G (2008a) An Hsp70 peptide initiates NK cell killing of leukemic blasts after stem cell transplantation. Leuk Res 32:527–534

Gross M, Ramirez C, Luthringer D, Nepomuceno E, Vollmer R, Burchette J, Freedland SJ (2008b) Expression of androgen and estrogen related proteins in normal weight and obese prostate cancer patients. Prostate 69:520–527

Guzhova I, Kislyakova K, Moskaliova O, Fridlanskaya I, Tytell M, Cheetham M, Margulis B (2001) In vitro studies show that Hsp70 can be released by glia and that exogenous Hsp70 can enhance neuronal stress tolerance. Brain Res 914:66–73

Hightower LE, Guidon PT Jr (1989) Selective release from cultured mammalian cells of heat-shock (stress) proteins that resemble glia-axon transfer proteins. J Cell Physiol 138:257–266

Hori S, Nomura T, Sakaguchi S (2003a) Control of regulatory T cell development by the transcription factor Foxp3. Science 299:1057–1061

Hori S, Takahashi T, Sakaguchi S (2003b) Control of autoimmunity by naturally arising regulatory CD4+ T cells. Adv Immunol 81:331–371

Hsu DH, Paz P, Villaflor G, Rivas A, Mehta-Damani A, Angevin E, Zitvogel L, Le Pecq JB (2003) Exosomes as a tumor vaccine: enhancing potency through direct loading of antigenic peptides. J Immunother 26:440–450

Hunter-Lavin C, Hudson PR, Mukherjee S, Davies GK, Williams CP, Harvey JN, Child DF, Williams JH (2004) Folate supplementation reduces serum hsp70 levels in patients with type 2 diabetes. Cell Stress Chaperones 9:344–349

Jin X, Wang R, Xiao C, Cheng L, Wang F, Yang L, Feng T, Chen M, Chen S, Fu X, Deng J, Wang R, Tang F, Wei Q, Tanguay RM, Wu T (2004) Serum and lymphocyte levels of heat shock protein 70 in aging: a study in the normal Chinese population. Cell Stress Chaperones 9:69–75

Kawabata Y, Udono H, Honma K, Ueda M, Mukae H, Kadota J, Kohno S, Yui K (2002) Merozoite surface protein 1-specific immune response is protective against exoerythrocytic forms of *Plasmodium yoelii*. Infect Immun 70:6075–6082

Kempaiah P, Dokladny K, Karim Z, Raballah E, Achieng AO, Ong'echa JM, Moseley PL, Lambert CG, Perkins DJ (2017) Reduced HSP70 and glutamine in pediatric severe malaria anemia: role of hemozoin in suppressing HSP70 and NF-κb activation. Am J Trop Med Hyg 97:517–518

Kim WS, Jung ID, Kim JS, Kim HM, Kwon KW, Park YM, Shin SJ (2018) Mycobacterium tuberculosis GrpE, a heat-shock stress responsive chaperone, promotes Th1-biased T cell immune response via TLR4-mediated activation of dendritic cells. Front Cell Infect Microbiol 8:95

Kleinjung T, Arndt O, Feldmann HJ, Bockmuhl U, Gehrmann M, Zilch T, Pfister K, Schonberger J, Marienhagen J, Eilles C, Rossbacher L, Multhoff G (2003) Heat shock protein 70 (Hsp70) membrane expression on head-and-neck cancer biopsy-a target for natural killer (NK) cells. Int J Radiat Oncol Biol Phys 57:820–826

Lackie RE, Maciejewski A, Ostapchenko VG, Marques-Lopes J, Choy WY, Duennwald ML, Prado VF, Prado MAM (2017) The Hsp70/Hsp90 chaperone machinery in neurodegenerative diseases. Front Neurosci 11:254

Lehner T, Anton PA (2002) Mucosal immunity and vaccination against HIV. AIDS 16(Suppl 4):S125–S132

Manjili MH, Henderson R, Wang XY, Chen X, Li Y, Repasky E, Kazim L, Subjeck JR (2002a) Development of a recombinant HSP110-HER-2/neu vaccine using the chaperoning properties of HSP110. Cancer Res 62:1737–1742

Manjili MH, Wang XY, Park J, Facciponte JG, Repasky EA, Subjeck JR (2002b) Immunotherapy of cancer using heat shock proteins. Front Biosci 7:d43–d52

Manjili MH, Wang XY, Park J, Macdonald IJ, Li Y, Van Schie RC, Subjeck JR (2002c) Cancer immunotherapy: stress proteins and hyperthermia. Int J Hyperth 18:506–520

Manjili MH, Wang XY, Chen X, Martin T, Repasky EA, Henderson R, Subjeck JR (2003) HSP110-HER2/neu chaperone complex vaccine induces protective immunity against spontaneous mammary tumors in HER-2/neu transgenic mice. J Immunol 171:4054–4061

Marcus R, Culver DH, Bell DM, Srivastava PU, Mendelson MH, Zalenski RJ, Farber B, Fligner D, Hassett J, Quinn TC et al (1993) Risk of human immunodeficiency virus infection among emergency department workers. Am J Med 94:363–370

Mazzaferro V, Coppa J, Carrabba MG, Rivoltini L, Schiavo M, Regalia E, Mariani L, Camerini T, Marchiano A, Andreola S, Camerini R, Corsi M, Lewis JJ, Srivastava PK, Parmiani G (2003) Vaccination with autologous tumor-derived heat-shock protein gp96 after liver resection for metastatic colorectal cancer. Clin Cancer Res 9:3235–3245

Moser C, Schmidbauer C, Gurtler U, Gross C, Gehrmann M, Thonigs G, Pfister K, Multhoff G (2002) Inhibition of tumor growth in mice with severe combined immunodeficiency is mediated by heat shock protein 70 (Hsp70)-peptide-activated, CD94 positive natural killer cells. Cell Stress Chaperones 7:365–373

Multhoff G (2006) Heat shock proteins in immunity. Handb Exp Pharmacol 172:279–304

Multhoff G (2007) Heat shock protein 70 (Hsp70): membrane location, export and immunological relevance. Methods 43:229–237

Multhoff G, Pfister K, Gehrmann M, Hantschel M, Gross C, Hafner M, Hiddemann W (2001) A 14-mer Hsp70 peptide stimulates natural killer (NK) cell activity. Cell Stress Chaperones 6:337–344

Murshid A, Gong J, Stevenson MA, Calderwood SK (2011a) Heat shock proteins and cancer vaccines: developments in the past decade and chaperoning in the decade to come. Expert Rev Vaccines 10:1553–1568

Murshid A, Theriault J, Gong J, Calderwood SK (2011b) Investigating receptors for extracellular heat shock proteins. Methods Mol Biol 787:289–302

Neckers L, Blagg B, Haystead T, Trepel JB, Whitesell L, Picard D (2018) Methods to validate Hsp90 inhibitor specificity, to identify off-target effects, and to rethink approaches for further clinical development. Cell Stress Chaperones 23:467–482

van Noort JM, Bugiani M, Amor S (2017) Heat shock proteins: old and novel roles in neurodegenerative diseases in the central nervous system. CNS Neurol Disord Drug Targets 16:244–256

Panjwani NN, Popova L, Srivastava PK (2002) Heat shock proteins gp96 and hsp70 activate the release of nitric oxide by APCs. J Immunol 168:2997–3003

Prado M (2017) New mechanisms regulating neuronal resilience in neurodegeneration. J Neurochem 142:24

Sato K, Torimoto Y, Tamura Y, Shindo M, Shinzaki H, Hirai K, Kohgo Y (2001) Immunotherapy using heat-shock protein preparations of leukemia cells after syngeneic bone marrow transplantation in mice. Blood 98:1852–1857

Schneider EM, Niess AM, Lorenz I, Northoff H, Fehrenbach E (2002) Inducible hsp70 expression analysis after heat and physical exercise: transcriptional, protein expression, and subcellular localization. Ann N Y Acad Sci 973:8–12

Sevin M, Girodon F, Garrido C, de Thonel A (2015) HSP90 and HSP70: implication in inflammation processes and therapeutic approaches for myeloproliferative neoplasms. Mediat Inflamm 2015:970242

Srivastava P (2002a) Interaction of heat shock proteins with peptides and antigen presenting cells: chaperoning of the innate and adaptive immune responses. Annu Rev Immunol 20:395–425

Srivastava P (2002b) Roles of heat-shock proteins in innate and adaptive immunity. Nat Rev Immunol 2:185–194

Srivastava P (2004) Heat shock proteins and immune response: methods to madness. Methods 32:1–2

Srivastava PK, DeLeo AB, Old LJ (1986) Tumor rejection antigens of chemically induced sarcomas of inbred mice. Proc Natl Acad Sci U S A 83:3407–3411

Stangl S, Gross C, Pockley AG, Asea AA, Multhoff G (2008) Influence of Hsp70 and HLA-E on the killing of leukemic blasts by cytokine/Hsp70 peptide-activated human natural killer (NK) cells. Cell Stress Chaperones 13:221–230

Stangl S, Gehrmann M, Dressel R, Alves F, Dullin C, Themelis G, Ntziachristos V, Staeblein E, Walch A, Winkelmann I, Multhoff G (2011a) In vivo imaging of CT26 mouse tumours by using cmHsp70.1 monoclonal antibody. J Cell Mol Med 15:874–887

Stangl S, Gehrmann M, Riegger J, Kuhs K, Riederer I, Sievert W, Hube K, Mocikat R, Dressel R, Kremmer E, Pockley AG, Friedrich L, Vigh L, Skerra A, Multhoff G (2011b) Targeting membrane heat-shock protein 70 (Hsp70) on tumors by cmHsp70.1 antibody. Proc Natl Acad Sci U S A 108:733–738

Stangl S, Themelis G, Friedrich L, Ntziachristos V, Sarantopoulos A, Molls M, Skerra A, Multhoff G (2011c) Detection of irradiation-induced, membrane heat shock protein 70 (Hsp70) in mouse tumors using Hsp70 Fab fragment. Radiother Oncol 99:313–316

Swaroop S (2017) HSP60 plays a regulatory role in IL-1β-induced microglial inflammation via TLR4-p38 MAPK axis. J Neurochem 142:104

Tamura Y, Peng P, Liu K, Daou M, Srivastava PK (1997) Immunotherapy of tumors with autologous tumor-derived heat shock protein preparations. Science 278:117–120

Terry DF, Wilcox M, McCormick MA, Lawler E, Perls TT (2003) Cardiovascular advantages among the offspring of centenarians. J Gerontol A Biol Sci Med Sci 58:M425–M431

Theriault JR, Mambula SS, Sawamura T, Stevenson MA, Calderwood SK (2005) Extracellular HSP70 binding to surface receptors present on antigen presenting cells and endothelial/epithelial cells. FEBS Lett 579:1951–1960

Theriault JR, Adachi H, Calderwood SK (2006) Role of scavenger receptors in the binding and internalization of heat shock protein 70. J Immunol 177:8604–8611

Walsh RC, Koukoulas I, Garnham A, Moseley PL, Hargreaves M, Febbraio MA (2001) Exercise increases serum Hsp72 in humans. Cell Stress Chaperones 6:386–393

Wang XY, Kazim L, Repasky EA, Subjeck JR (2001) Characterization of heat shock protein 110 and glucose-regulated protein 170 as cancer vaccines and the effect of fever-range hyperthermia on vaccine activity. J Immunol 166:490–497

Wang Y, Kelly CG, Singh M, McGowan EG, Carrara AS, Bergmeier LA, Lehner T (2002) Stimulation of Th1-polarizing cytokines, C-C chemokines, maturation of dendritic cells, and adjuvant function by the peptide binding fragment of heat shock protein 70. J Immunol 169:2422–2429

Whittall T, Peters B, Rahman D, Kingsley CI, Vaughan R, Lehner T (2011) Immunogenic and tolerogenic signatures in human immunodeficiency virus (HIV)-infected controllers compared with progressors and a conversion strategy of virus control. Clin Exp Immunol 166:208–217

Wu B, Gu MJ, Heydari AR, Richardson A (1993) The effect of age on the synthesis of two heat shock proteins in the hsp70 family. J Gerontol 48:B50–B56

Zanin-Zhorov A, Bruck R, Tal G, Oren S, Aeed H, Hershkoviz R, Cohen IR, Lider O (2005a) Heat shock protein 60 inhibits Th1-mediated hepatitis model via innate regulation of Th1/Th2 transcription factors and cytokines. J Immunol 174:3227–3236

Zanin-Zhorov A, Tal G, Shivtiel S, Cohen M, Lapidot T, Nussbaum G, Margalit R, Cohen IR, Lider O (2005b) Heat shock protein 60 activates cytokine-associated negative regulator suppressor of cytokine signaling 3 in T cells: effects on signaling, chemotaxis, and inflammation. J Immunol 175:276–285

Zhou Y, Ma J, Zhang J, He L, Gong J, Long C (2017) Pifithrin-mu is efficacious against non-small cell lung cancer via inhibition of heat shock protein 70. Oncol Rep 37:313–322

Zitvogel L, Fernandez N, Lozier A, Wolfers J, Regnault A, Raposo G, Amigorena S (1999) Dendritic cells or their exosomes are effective biotherapies of cancer. Eur J Cancer 35(Suppl 3):S36–S38

Chapter 2
Extracellular HSP70, Neuroinflammation and Protection Against Viral Virulence

Michael Oglesbee, Mi Young Kim, Yaoling Shu, and Sonia Longhi

Abstract The major inducible 70 kDa heat shock protein (hsp70) is induced by and supports intracellular replication of viruses belonging to diverse families. Paradoxically, this virus-hsp70 interaction is *protective* in mouse models of viral neurovirulence, enhancing T cell mediated immune clearance in an interferon β (IFN-β)-dependent manner. Protection reflects early release of hsp70 from viable infected neurons and induction of strong innate immune responses in uninfected brain macrophages, including the induction of IFN-β through Toll-like receptor 4. Potency of the response is inherent in the fact that hsp70 is released at a time when pathogen-associated molecular patterns (PAMPs) are in low abundance, and that the innate response is driven by uninfected cells, free from viral interference. Release of hsp70 from viable cells is primarily exosomal, and infection enhances total exosome release and hsp70 content on the surface of exosomes. Exosome content of hsp70 reflects levels of hsp70 in the infected cell. Findings have broad virological relevance and support a protective role for fever, a potent stimulus for hsp70 induction. While protective in the context of microbial infection, recent findings support potential untoward effects of inappropriate extracellular hsp70 release in non-infectious neuroinflammatory conditions.

Keywords 70 kDa heat shock protein · Exosome · Hsp70 · Interferon beta · Neurovirulence · Virus clearance

M. Oglesbee (✉)
Department of Veterinary Biosciences, The Ohio State University, Columbus, OH, USA
e-mail: oglesbee.1@osu.edu

M. Y. Kim
Pharmaceutical Safety Evaluation Division, Ministry of Food and Drug Safety, Cheongju, Republic of Korea

Y. Shu
Department of Neuroscience, Center for Brain and Spinal Cord Repair, The Ohio State University, Columbus, OH, USA

S. Longhi
CNRS, Architecture et Fonction des Macromolécules Biologiques (AFMB), UMR, Aix-Marseille Univ, Marseille, France

A. A. A. Asea, P. Kaur (eds.), *Chaperokine Activity of Heat Shock Proteins*, Heat Shock Proteins 16, https://doi.org/10.1007/978-3-030-02254-9_2

Abbreviations

ANOVA	Analysis of variance
CHUK	Conserved helix-loop-helix ubiquitous ligase
CNS	Central nervous system
CSF	Cerebrospinal fluid
DAMP	Damage-associated molecular pattern
DMEM	Dulbecco's Modified Eagle medium
Ed CAM/RB	Rodent brain adapted Ed-MeV
Ed-MeV	Edmonston measles virus
ELISA	Enzyme-linked immunosorbent assay
FCS	Fetal calf serum
H-2	Mouse major histocompatibility complex
HSF	Heat shock factor
hsp70	70 kDa heat shock protein
IFN	Interferon
IFNAR	Type 1 interferon receptor
ILV	Intralumenal vesicle
IRF3	Interferon regulatory factor 3
L	Viral polymerase protein
LDH	Lactate dehydrogenase
LPS	Lipopolysaccharide
MEF	Mouse embryo fibroblasts
MeV	Measles virus
MHC	Major histocompatibility complex
MVB	Multivesicular body
N	Nucleocapsid protein
N2a-HSP	Mouse neuroblastoma cells that constitutively express hsp70
N2a-V	Vector transfected control mouse neuroblastoma cells
NBD	Nucleotide binding domain
NDV	Newcastle disease virus
NSE	Neuron specific enolase
N_{TAIL}	Carboxyl terminus of the N protein
PAMP	Pathogen-associated molecular pattern
PBD	Peptide binding domain (also known as the substrate binding domain, SBD)
RSV	Respiratory syncytial virus
RT-PCR	Reverse transcription polymerase chain reaction
SBD	Substrate binding domain
STAT-1	Signal transducer and activator of transcription 1
TLR	Toll-like receptor
TRAM	Toll-like receptor adapter molecule 2
TRIF	TIR-domain-containing adapter-inducing interferon-β
VSV	Vesicular stomatitis virus

2.1 Introduction

There is an abundance of literature showing that the major stress inducible 70 kDa heat shock protein (hsp70) supports intracellular replication of viruses belonging to multiple families (Reviewed by Mayer 2005; Nagy et al. 2011). These include RNA viruses in the family *Paramyxoviridae* (Oglesbee et al. 1996; Vasconcelos et al. 1998a; Zhang et al. 2002a; Brown et al. 2005), *Rhabdoviridae* (Lahaye et al. 2012; Kim et al. 2013b), *Flaviviridae* (Das et al. 2009; Ye et al. 2013; Taguwa et al. 2015), and *Picornaviridae* (Macejak and Sarnow 1992), and DNA viruses in the family *Herpesviridae* (Tanguy and Boehmer 2002; Nozawa et al. 2004), *Adenoviridae* (Madara et al. 2005), and *Papillomaviridae* (Song et al. 2010). This stands in contrast to the paucity of *in vivo* studies addressing the significance of hsp70 in viral pathogenesis and immunity. Those studies fall into two general categories. The first group focused on the role of hsp70 as an adjuvant, enhancing the immune response to vaccination. For example, DNA vaccines for Japanese encephalitis virus and herpes simplex virus elicit enhanced immune responses when co-expressed with hsp70, where efficacy is measured as increased protection against subsequent lethal viral challenge (Chen et al. 2000; Pack et al. 2005). The second focuses on viral virulence, which is the ability of a virus to spread throughout tissues and cause disease. The latter has obvious biological and medical relevance, but also poses a challenge to study. A tractable model is one in which cells supporting infection are distinct from cells mediating innate and adaptive immune responses to infection, and manipulation of hsp70 levels is restricted to virus infected cells. In such a system, we can distinguish between effects of virus-hsp70 interaction within the infected cells and secondary innate and adaptive immune responses of uninfected immune cells that are free from viral interference in the form of either general virus-induced cytopathic effects or specific disruption of innate immune signaling pathways. In this model, hsp70 levels expressed within immune cells are not altered and thus not a variable to be considered in evaluating immune cell function. An additional characteristic of a tractable model is that the viral response to hsp70 within infected cells is well-characterized. Such a model exists in virus infection of the mouse central nervous system (CNS) and will be the focus of this discussion.

Using measles virus (MeV) infection of neurons in the mouse brain, we showed that hsp70 promotes antiviral immune responses leading to viral clearance, despite the potential for hsp70 to enhance viral gene expression (Kim et al. 2013a; Carsillo et al. 2009). These findings support a pivotal role for extracellular hsp70. In the current chapter, we will delineate the seminal findings that led to the elucidation of a model for this hsp70 protective effect: the role of hsp70 in supporting intracellular viral replication in neuronal cells; enhanced immune-mediated viral clearance from brain when neurons express hsp70; the pivotal role of IFN-β in mediating hsp70-dependent viral clearance; induction of IFN-β in uninfected brain macrophages (microglia) through hsp70 that is released from viable infected neurons; evidence that hsp70 release from viable cells is exosomal. The model will be developed in such a way that we can appreciate the broad virological and host species relevance, placing this in context of a more general role for hsp70 in neuroinflammatory responses. Before doing this, we must first develop a deeper understanding of the specific infection model that was employed and relevance to naturally occurring disease.

2.1.1 Non-cytolytic Neuronal Infection by Measles Virus

Neurovirulence is a consequence of non-cytolytic virus neuronal infection for multiple virus families including *Paramyxoviridae* (of which MeV is a member), *Picornaviridae, Retroviridae, Arenaviridae, Coronaviridae* and *Togaviridae* (Patterson et al. 2002a). In this virus-cell relationship, viability of the infected cell is largely maintained and virus spread is cell associated, mediated by neuraxonal transport and direct cell-to-cell transmission in the absence of significant cell-free infectious viral progeny. Protection against viral spread is immune-mediated and dependent upon the production of type I interferon, which is primarily IFN-β in the brain (Detje et al. 2009). A major role of the IFN-β is to activate virus specific T cells as they encounter viral antigen-MHC I complexes (Pien et al. 2002). Activated T cells release IFN-γ that in turn mediates non-cytolytic viral clearance from neurons (Pien et al. 2002; Patterson et al. 2002b). A challenge to this clearance mechanism is that neurons are deficient in both MHC and IFN-β expression (Patterson et al. 2002b; Dhib-Jalbut et al. 1999). Macrophages are the primary antigen presenting cells in brain and the predominant source of IFN-β following brain infection (Kallfass et al. 2012). Needed is a signal between infected neurons and uninfected brain macrophages that is capable of IFN-β induction and stimulation of antigen presentation. Release of pathogen associated molecular patterns (PAMPs) is limited by the cell-associated and non-cytolytic nature of infection. This raises the possibility of a mediator that is of cellular origin – a damage associated molecular pattern (DAMP) that can be released from viable infected cells early during the course of infection. Heat shock proteins are one such candidate. The ability of hsp70 to cross-present antigen in support of strong antiviral T cell responses (Noessner et al. 2002; Tischer et al. 2012) and the capacity of murine microglia to cross-present exogenous antigen (Beauvillain et al. 2008) would help explain how uninfected activated microglia could present viral antigen to naïve T cells and any T cells that have been primed in the periphery.

Measles virus (MeV) infection of humans exemplifies the effectiveness of antiviral immunity in suppressing viral neurovirulence (reviewed by Oglesbee and Niewiesk 2011). In the pathogenesis of measles, infection is initiated in lymphoid tissues associated with the respiratory tract, resulting in a primary viremia. Infection of lymphoid tissues throughout the body ensues, amplifying the viremia and spreading the infection to multiple organ systems. Viral shedding follows infection of mucosal epithelium and while this is most effective in the respiratory tract, infection of the gastrointestinal and urogenital tracts may contribute. Viral transmission between infected and uninfected susceptible individuals is highly efficient, with measles being one of the most highly contagious infections that is known. Viral clearance commences with the onset of virus-specific T cell responses that coincide with the onset of fever. Mortality is generally attributed to secondary bacterial infection of compromised mucosal barriers.

A potential by-product of viremia is MeV entry into the central nervous system (CNS). Although virus-induced neurological disease is rare, the incidence of viral RNA in brains of MeV infected individuals indicates that viral invasion of the CNS occurs with much greater frequency than the level of virus-induced neurological disease would suggest (Brooks et al. 1998; Katayama et al. 1995, 1998; Katz 1995; Bellini et al. 2005). Studies in a closely related viral system (i.e., canine distemper virus infection of dogs and ferrets) indicate multiple pathways of entry. These include trafficking of virus-infected leukocytes across the blood brain barrier, infection of endothelial cells that are components of the blood brain barrier, infection of choroid plexus epithelium with viral shedding into the cerebrospinal fluid, and infection of olfactory neurons in the nasal mucosa with spread to the olfactory bulbs of the brain (Krakowka 1989; Rudd et al. 2006). These findings highlight the importance of the CNS immune response in limiting viral spread and supporting clearance for viruses that are inherently neuroinvasive. For MeV, this is illustrated by the fact that even the Edmonston (Ed-MeV) vaccine strain can exhibit a high degree of neurovirulence in severely immune suppressed individuals, giving rise to a disease known as measles inclusion body encephalitis (Bitnum et al. 1999). We know that both IFN-β and IFN-γ are expressed in brains of patients with MeV encephalitis (Joncas et al. 1976; Hofman et al. 1991). But unknown are the details of how an effective CNS immune response is orchestrated, this being the first step in understanding how those mechanisms might fail. This question is by no means restricted to measles. Mumps virus is highly neuroinvasive yet poorly neurovirulent (Rozina et al. 1984), and the number of agents that exhibit this biological property will no doubt grow as we begin to characterize the human brain virome (Strong et al. 2016).

Mouse models have been used to advance our understanding of mechanisms underlying control of neurovirulence in general, and for MeV in particular. Neonatal C57BL/6 mice support a neuron-specific infection following intracranial inoculation. An infectious dose of 4×10^4 infectious virus particles of Ed-MeV induces 15–35% mortality within 25 days of inoculation (Carsillo et al. 2006a, 2009). The intracranial route of inoculation ensures that the initial viral load is constant, such that infection outcomes are a measure of the degree of spread (i.e., neurovirulence). Priming of the adaptive immune response occurs in deep cervical lymph nodes, reflecting trafficking of cells in the cerebrospinal fluid through a lymphatic system in the *dura mater* (Louveau et al. 2015). Survival is dependent upon the efficiency of viral clearance that is mediated by IFN-γ released from both virus-specific CD4 and CD8 T cells (Patterson et al. 2002b; Finke et al. 1995). Activation of these T cells is supported by IFN-β from microglia and not virus infected neurons (Patterson et al. 2002b; Dhib-Jalbut et al. 1999; Kalfass et al. 2012). Mortality reflects the magnitude of brain viral burden, particularly at 14 days post infection when adaptive immune responses determine the outcome of infection (Carsillo et al. 2009). The intermediate level of mortality is such that the model is amenable to identifying variables that either enhance or diminish viral clearance. We can further modify the strength of adaptive antiviral immune responses by manipulating the MHC (H-2) haplotype; H-2^d mice are resistant to infection compared to mice carrying the H-2^b or H-2^k allele (Neumeister and Niewiesk 1998; Niewiesk et al. 1993; Weidinger et al. 2000). Differences reflect the affinity of MHC I complexes for viral peptides and thus strength of subsequent T cell responses (Niewiesk et al. 1993).

2.1.2 HSP70 Expression Levels and the Study of Neurovirulence

Differences in hsp70 expression between mice and humans presents an opportunity to study the specific contribution of hsp70 to viral neurovirulence. In humans, hsp70 is readily induced by protein denaturing stresses that include febrile temperatures (Morrison-Bogorad et al. 1995), but it is also expressed constitutively (Pardue et al. 2007). Mice and rats lack significant constitutive expression of hsp70, although the protein is induced by the same stimuli observed in humans (Pardue et al. 2007) and the genes encoding hsp70 in humans (HSPA1A) and mice (Hspa1a) are highly homologous (Daugaard et al. 2007). Species-specific differences in constitutive expression were demonstrated by our previous quantitation of hsp70 levels in brain total proteins by ELISA (Oglesbee and Niewiesk 2011). Samples from humans were obtained commercially and were derived from patients without neurological disease. The average level of hsp70 in the cerebrum, hippocampus and diencephalon was 130 ng/mg total proteins, reflecting hsp70 expression in all CNS cell types. In contrast, hsp70 in C57BL/6 mice was undetectable by ELISA. This finding allows us to determine the significance of basal levels of hsp70 on the outcome of infection. Mice lack a febrile response to intracranial MeV infection, allowing us to use whole body hyperthermia to demonstrate the influence of elevated hsp70 levels on virus-infected neurons, comparing infection outcomes in hyperthermia treated and non-treated animals (Carsillo et al. 2004). Interpretations are necessarily limited by the fact that hyperthermia induces hsp70 in both neurons and glia and other heat shock protein family members are induced as well. A more refined approach is achieved through creation of transgenic mice, where the neuron specific enolase (NSE) promoter is used to restrict hsp70 expression to neurons. Such mice have been generated, linking the human HSPA1A gene to the rat NSE promoter (Carsillo et al. 2006a, 2009; Kim et al. 2013a, b). An epitope tag (from the G glycoprotein of vesicular stomatitis virus) was introduced to distinguish between transgenic human and endogenous mouse hsp70. Immunohistochemistry confirmed selective neuronal expression of the construct and elevated levels of hsp70 were confirmed by ELISA of brain total proteins (Oglesbee and Niewiesk 2011). The concentration of hsp70 in brains of NSE-hsp70 transgenic mice was 1 ng/mg brain total proteins. This contrasts to 14 ng/mg brain total proteins measured in transgenic mice where global expression of hsp70 from the human HSPA1A gene is driven by the β actin promoter (Oglesbee and Niewiesk 2011). The difference in hsp70 levels between NSE-hsp70 and β actin-hsp70 mice is consistent with the fact that neurons represent approximately 10% of the total brain cell population. Parallel *in vitro* systems have been created in mouse neuroblastoma (N2a) cells. The N2a cell line is not only a well-established model for MeV neuronal infection, but also neuronal exosome formation and release, faithfully reproducing exosome release observed in primary cultures of mouse cortical neurons (Yuyama et al. 2012). The parent N2a cell line does not express readily detectable levels of basal hsp70, and stably transfected cell lines were created that constitutively express hsp70 through the β-actin promoter (Vasconcelos et al. 1998a; Carsillo et al. 2006a; Kim et al. 2013a, b). Combined, the hsp70 transgenic mice and stably transfected cell lines represent powerful tools to dissect the role of hsp70 in both viral and immune parameters associated with neuronal infection.

2.1.3 HSP70 Supports Virus Gene Expression

Measles virus particles are enveloped. Envelope glycoproteins mediate attachment to and fusion with the plasmalemma that in turn releases virus core particles (i.e., nucleocapsid) into the cytoplasm to initiate viral replication. Nucleocapsids are composed of a single stranded negative sense RNA genome packaged by nucleocapsid (N) proteins. Approximately 2500 N protein monomers combine with the 15.9 kb genome to form an extended helical nucleocapsid that is the template for the virus RNA-dependent RNA polymerase (L). The L protein is tethered to N by an additional viral structural protein, P. The viral polymerase mediates both transcription and genome replication, with polymerase activity supported by host factors that include hsp70. Hsp70 stimulates both transcription and genome replication through binding reactions with the surface exposed N protein carboxyl terminus (N_{TAIL}) (reviewed in Longhi and Oglesbee 2010; Zhang et al. 2002a, 2005). The hsp70 binding sites on N_{TAIL} overlaps that of the P protein, with both hsp70 and P recognizing a conserved hydrophobic motif known as Box 2. Based upon data from the P protein (i.e., the X domain within the C-terminus), binding to Box 2 induces limited folding of N_{TAIL}, which is predominantly disordered in the absence of a binding partner (Bourhis et al. 2005). The mechanism by which hsp70 enhances transcription is thought to rely on remodeling of the nucleocapsid template to achieve conformations that are conducive to RNA synthesis by the viral polymerase (reviewed by Oglesbee 2007; Oglesbee et al. 1989, 1990; Vasconcelos et al. 1998a). This could include conformations conducive to recruitment of additional cofactors for polymerase activity, or simply polymerase access to the encapsidated RNA template. Indirect mechanisms also include the contribution that hsp70 may play in the encapsidation of the viral RNA genome, the N protein-RNA complex serving as template for the viral RNA-dependent RNA polymerase (Oglesbee 2007). Hsp70-N protein interactions and stimulation of virus gene expression have been shown for the paramyxoviruses canine distemper (Oglesbee et al. 1996) and respiratory syncytial virus (Brown et al. 2005; Munday et al. 2015), and for rabies virus (Lahaye et al. 2012), a neurovirulent rhabdovirus with a similar nucleocapsid architecture.

Importance of hsp70 to the viral replication cycle is reflected by the incorporation of this stress protein in cell-free virions for both paramyxoviruses (Ren et al. 2012) and rhabdoviruses (Lahaye et al. 2012), and the fact that infection induces hsp70 early during the course of the viral replication cycle (Lahaye et al. 2012; Oglesbee et al. 2002). MeV induces hsp70 in mouse N2a cells, with levels of hsp70 rising in advance of that of the viral N protein (Oglesbee et al. 2002). The N protein is the first and most abundantly expressed viral structural protein within infected cells. The role of virus-induced hsp70 on replication was shown by infection of mouse embryo fibroblasts in which there was a targeted disruption of heat shock factor one (HSF-$1^{-/-}$) (Zhang et al. 2002b). HSF-1 is a transcription factor for hsp70 induction that is active when present as a phosphorylated trimer. The non-phosphorylated monomeric form is maintained through binding to hsp70/hsc70. The ability of N_{TAIL} to bind hsp70 suggests that nucleocapsid/N protein might out-compete HSF-1 for hsp70 binding, releasing HSF-1 for phosphorylation activation. Cells were infected with a rodent-brain adapted strain of MeV (Ed CAM/RB). hsp70 induction was observed in the HSF-$1^{+/+}$ (wild type) but not the HSF-$1^{-/-}$ cells (Fig. 2.1). Viral copy number was reduced by 2.3-fold in the HSF-$1^{-/-}$ cells at 48 h post-infection relative to HSF-$1^{+/+}$, and the N antigen exhibited a dispersed pattern. In contrast, N antigen formed large inclusions in wild type cells, typical of highly productive infections, and the N protein was extensively colocalized with hsp70 (Fig. 2.1). *In vivo* relevance of hsp70 induction by virus infection was shown for canine distemper virus in the brain of a naturally infected dog, with elevated expression of hsp70 in virus-infected cells and extensive co-localization of hsp70 and N proteins (Oglesbee and Krakowka 1993).

Whereas virus-induced hsp70 supports basal virus replication, further elevations in hsp70 levels enhance virus gene expression in ways that alter the infection phenotype. This was shown using either heat shock or sodium arsenite induction of hsp70 in mouse N2a cells persistently infected with canine distemper virus (Oglesbee et al. 1993). Persistent infection is characterized by the absence of virus-mediated cytopathic effects and minimal to absent production of cell-free infectious viral progeny. The infection phenotype reflects restrictions in the level of virus gene expression, particularly that of viral membrane glycoproteins and matrix proteins necessary for viral particle assembly and responsible for much of the cytopathic effects. In this system, induction of hsp70 enhanced virus gene expression, increasing both cytopathic effects and progeny release, converting a stable non-productive persistent infection to one that is both productive and lytic. Similar findings were observed for MeV (Vasconcelos et al. 1998b). Here, hsp70 levels were induced prior to infection, resulting in a stress-conditioned phenotype where cells are tolerant to subsequent noxious stimuli. MeV gene expression was enhanced in the stress-conditioned cells relative to non-shocked controls, the primary effect here being on infectious viral progeny release. That hsp70 alone was capable of supporting these changes was shown in N2a cells stably transfected to overexpress hsp70 (Vasconcelos et al. 1998a). Infection of N2a cells that overexpress hsp70 resulted in increased viral gene expression and infectious progeny release compared to infected vector transfected controls.

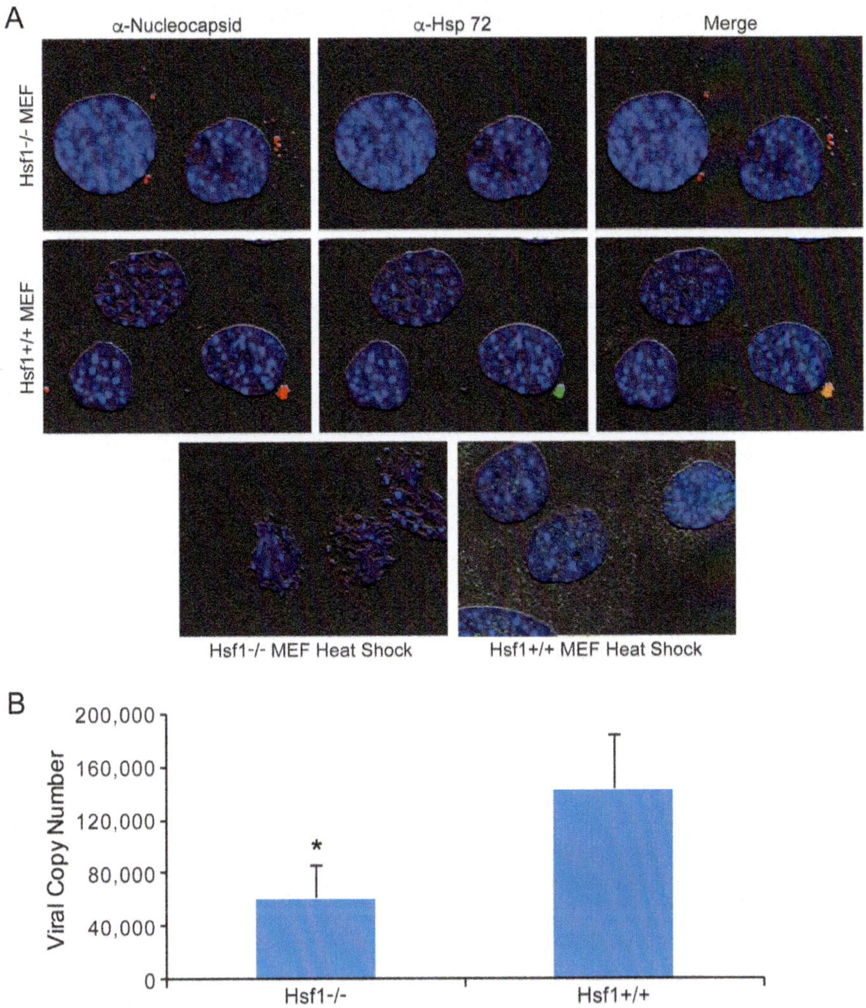

Fig. 2.1 MeV induction is dependent upon heat shock factor 1 (Hsf-1) in mouse embryo fibroblasts (MEF). Induction is associated with extensive co-localization of hsp70 and viral N protein in large inclusions, and a higher burden of viral genomic RNA within infected cells. (**a**) Immunohistochemical staining of N protein (red) and hsp70 (green) at 48 h post-infection of Hsf-$1^{+/+}$ (wild type) and Hsf-$1^{-/-}$ MEF with a rat-brain adapted strain of Ed-MeV (Hoechst stained nuclei are blue). Cells were infected at a multiplicity of infection of 0.1. Uninfected MEF cells at 24 h post heat shock (43 °C, 1.5 h) served as controls for hsp70 induction (bottom two panels). (**b**) Viral N gene levels were quantified by real time RT-PCR at 48 h post infection in Hsf-$1^{+/+}$ and Hsf-$1^{-/-}$ MEF. Differences were statistically significant (t-Test, $p < 0.05$)

Virulence reflects a sufficient level of viral gene expression to support spread, but those levels must be moderated in order to avoid robust induction of host innate and adaptive immune responses. Conditions that increase virus gene expression thus have the potential to enhance viral clearance. Evidence that elevated hsp70 levels may be detrimental to viral virulence is the fact that MeV isolates can be divided into two general categories based upon the ability of N_{TAIL} Box 2 to bind hsp70. Vaccine strains have a Box 2 sequence that binds hsp70 with high affinity, and elevations in cellular hsp70 levels support increases in viral transcription for these viruses (Zhang et al. 2002a, 2005; Carsillo et al. 2006a). In contrast, the vast majority of virulent (wild type) isolates contain a Box 2 sequence in which hsp70 binding is disrupted by a single amino acid substitution (i.e., an asparagine to aspartic acid at amino acid 522). Using reverse genetics, we generated both minireplicons (i.e., subgenomic viral core particles expressing a reporter gene) and recombinant infectious Ed-MeV containing either an asparagine (N) or an aspartic acid (D) at position 522 of the N protein. Elevations in hsp70 stimulate transcription in the 522N but not the 522D virus/replicon (Zhang et al. 2002a; Carsillo et al. 2006b). Dual infections by the 522N and 522D variant viruses show that the 522D virus exhibits reduced fitness relative to 522N both *in vitro* (i.e., N2a cells) or *in vivo* (i.e., airway infection of the cotton rat) (Carsillo et al. 2006b). The prevalence of the 522D Box 2 sequence in natural isolates virulent for humans therefore suggests that this is an adaptive response to a selectional pressure that is not reflected when simply measuring intracellular replication parameters. We believe this selectional pressure to be immunological.

2.1.4 HSP70 Protects Against Measles Virus Neurovirulence in an IFN-β Dependent Manner

Evidence that hsp70 can support immune-mediated clearance of MeV from brain was first shown in stress conditioned neonatal Balb/c mice (Carsillo et al. 2004). Balb/c (H-2^d) mice support persistent neuronal infection by Ed-MeV following intracranial inoculation (Blixenkrone-Moller et al. 1998), where persistence is defined by the presence of virus at 21 days post infection. Infections are not associated with mortality, and the cell-associated nature of the neuronal infection is such that brain viral burden is measured by real time RT-PCR of viral genomic RNA (versus titration of cell free infectious viral progeny). Six hours prior to infection, a 41 °C 30 min whole body hyperthermia was used to induce hsp70, this being confirmed by Western blot analysis of brain total proteins. The mice were thus stress conditioned, and outcome of infection with 4×10^4 plaque-forming units was compared between stress-conditioned and non-conditioned mice. As stated previously, infection does not result in a febrile response that could otherwise alter expression of hsp70 in the latter group. Analysis of animal groups at 2 and 7 days post infection showed that 100% of the animals were infected, and there was no significant difference in viral burden between stress-conditioned versus non-conditioned animals at these times. Evidence of clearance was first detected at 14 days post infection in both stress-conditioned and non-conditioned mice and corresponded to the detection of cell mediated antiviral immune responses (i.e., blastogenic responses of splenocytes to purified MeV antigen). A dramatic difference between groups was evident at 21 days post-infection, where viral clearance was present in 95% of stress-conditioned mice, compared to only 53% of the infected non-conditioned controls. Persistence at this time point was associated with a lymphocytic meningoencephalitis and occasional foci of selective neuronal necrosis and reactive gliosis, changes that were conspicuously absent in mice where the infection was cleared. These results demonstrate the challenge of clearing neuronal infections that are characterized by a low level of virus gene expression, the pathological sequels to virus persistence, and the host protective role of an hsp70-inducing hyperthermic event. The results are consistent with a beneficial role of fever, a phylogenetically conserved response to microbial infection that elevates hsp70 levels in CNS neurons and glia (Tytell et al. 1998), and a defining feature of measles in humans.

The specific contribution of neuronal hsp70 in promoting viral clearance was shown in MeV infection studies using neonatal H-2^d congenic C57BL/6 NSE-hsp70 mice (Carsillo et al. 2009). Mechanistic insights are made possible not only by the fact that elevated hsp70 expression is restricted to neurons, but also the lack of change in levels of other heat shock protein family members that can otherwise be influenced by hyperthermic treatment. The choice of mouse strain for the transgenic construct was based upon consideration of studies where the specific contribution of H-2 haplotype to the outcome of infection would be examined (described later in this chapter), and the desirability of being able to compare results to that of the H-2^d Balb/c mice. C57BL/6 mice carry the H-2^b haplotype and therefore have a relatively weak cell-mediated immune response to MeV infection compared to H-2^d congenic mice. Intracranial inoculation of 4×10^4 plaque forming units of Ed-MeV in H-2^d congenic C57BL/6 mice caused 35% mortality between 14–24 days post infection. Infections were monitored through 28 days post infection. Mortality was correlated to elevated brain viral RNA burdens, T and B cell infiltrates of both brain parenchyma and meninges, neuronal necrosis and reactive gliosis in the hippocampus and granular cell layer of the dentate gyrus. Spread of the infection from the inoculation tract to involve these latter anatomic sites explains the proximate cause of death. Dead neurons were identified based upon characteristic morphology on hematoxylin and eosin stained tissue sections and were negative for cleaved (activated) caspase 3 based upon immunohistochemistry, supporting the interpretation of necrosis. Necrotic neurons would be expected to release significant levels of both PAMPs and DAMPs, and despite the infiltrates of T cells, the host response was inadequate to control viral spread in 35% of the non-transgenic animals. In contrast, there was *zero* mortality in NSE-hsp70 transgenic mice and no cytopathic effect or inflammatory infiltrates in the hippocampus at 28 days post infection. Virus was not completely cleared by 28 days as in the Balb/c mice, but the level of infection was controlled to the point that there was no mortality, morbidity, or active virus-induced cytopathic effects.

The dramatic protection mediated by neuronal hsp70 in H-2^d congenic C57BL/6 mice was similar to the protection observed in stress-conditioned Balb/c mice in that the establishment of infection was not influenced. Rather, it was the antiviral immune responses that controlled the spread of infection. Multiple lines of evidence pointed to a significant innate immune component. First was the fact that hsp70 reduced the reliance on T cells for protection. Protection is heavily reliant on CD4+ T cells in both Balb/c and C57BL/6 mice (Finke and Liebert 1994). Antibody mediated depletion of CD4+ T cells resulted in 95% mortality in MeV infected non-transgenic C57BL/6 mice (Carsillo et al. 2009). This was in contrast to 41% mortality in infected transgenic mice where CD4+ T cells were depleted, and depletion of both CD4+ and CD8+ T cells still resulted in 27% survival. Early hsp70-mediated increases in virus gene expression could be one means by which innate immune responses are enhanced, and this was tested by comparing outcome of infection between mice infected with the 522D Ed-MeV relative to those infected with the parent (522N) virus. Hsp70 enhanced the probability of survival in 522D infected mice, but these results did not achieve statistical significance. While inconclusive, the result does raise the possibility that protection can occur independently of the level of PAMP expression.

Direct evidence of enhanced innate immune activation in infected H-2^d congenic C57BL/6 NSE-hsp70 mice was shown in transcriptome analyses performed at 5 and 10 days post infection with Ed-MeV (Kim et al. 2013a). We know from histological examination that T cell infiltrates are not present at 5 days whereas they are present at 10. As such, 5 days represents a time point in which innate immune responses can be observed, whereas 10 days will represent a combination of innate and adaptive responses. There were no baseline differences in gene expression when transcriptomes of uninfected transgenic and non-transgenic mice were compared. Pathway analyses showed that the major difference between infected transgenic and non-transgenic mice at 5 days post infection was interferon signaling and macrophage activation. Hsp70 enhanced expression of eight interferon inducible genes, seven of which were specific to type I interferon signaling (i.e., USP18, IFIT1L, IFI27, RIG-I, IRF7, IRF9, RNase L). This result was correlated to enhanced expression of IFN-β at 1 day post infection in transgenic mice based upon RT-PCR analysis of total brain RNA. There was no significant induction of IFN-β in infected non-transgenic mice at this time point. Non-transgenic mice did ultimately mount an interferon response of comparable magnitude; it was simply delayed. The essential role of IFN-β to hsp70-mediated protection was shown in survival analysis of infected hsp70 transgenic and non-transgenic mice in which the type I IFN receptor was deleted (IFNAR$^{-/-}$). Disruption of IFN signaling eliminated the survival difference between hsp70 transgenic and non-transgenic mice, with approximately 80% mortality being observed in both groups.

Transcriptome analysis showed the impact of hsp70 on macrophage activation to be apparent at both 5 and 10 days post infection. At 5 days, hsp70 was associated with enhanced expression of STAT-1 and CHUK, the latter being part of the NF-κB signaling pathway needed for TLR-mediated microglial activation. Differences in MHC expression were dramatic and persistent, with hsp70 associated with significantly increased expression of both MHC I (H2-D1, H2-K1) and MHC II (H2-EA) at both 5 and 10 days post infection. Increases in MHC expression were confirmed by immunohistochemistry, and were consistent with previous studies showing hsp70 associated increases in macrophage activation markers by RT-PCR in infected mice (Carsillo et al. 2009). Those markers included CD68, MHC II, and TLR 2 and 4. Induction of TLR 2 and 4 was detected in the transcriptome analysis, but there was no significant difference in transcript levels between infected transgenic and non-transgenic mice. Of interest was the fact that infection was not associated with changes in the level of expression of other TLRs, specifically TLR3-5, 7, 9, 12 and 13.

Collectively, these results show that hsp70 enhances an early type I IFN response to MeV infection of neurons and robust activation of microglia, with IFN-β expression being key to survival. Any stimulatory effect of hsp70 on virus gene expression is offset by the enhancement of innate and adaptive immune responses, and evidence suggests that hsp70-mediated virus gene expression is not required for the immune stimulatory effect. Neurons do possess receptors for type I IFN, but the temporal changes in brain viral burden suggest that the primary effect of IFN-β is not through the direct induction of an antiviral state within neurons but rather the support of adaptive immune responses – T cell mediated clearance through the release of IFN-γ (Patterson et al. 2002a; Pien et al. 2002). Activated microglia also produce IFN-γ (Kawanokuchi et al. 2006), which may explain the T cell independent protection that is mediated by hsp70. We have observed a threefold difference in IFN-γ expression by RT-PCR in infected transgenic mice relative to infected non-transgenic mice at 5 days post infection, and a twofold difference at 10 days, although differences did not achieve statistical significance in the small group sizes that were examined (Carsillo et al. 2009).

2.1.5 HSP70 Protects Against Vesicular Stomatitis Virus Neurovirulence in an IFN-β Dependent Manner

Before exploring mechanisms by which hsp70 in neurons enhances activation of brain macrophages and IFN-β production, it is important to consider the broader virological relevance of the phenomenon. Is the phenomenon restricted to MeV, which is a poorly cytopathic neuronal infectious agent whose spread relies upon direct cell-to-cell transfer, or can it also be demonstrated in a highly cytopathic infection by a member of a different virus family? Is the phenomenon restricted to neonates, or is it independent of age? Is the phenomenon a unique response to an intracranial inoculation, or does it also apply to a natural route of neuroinvasion? To address these questions we used vesicular stomatitis virus (VSV) infection of weanling male mice, 4–6 weeks of age (Kim et al. 2013b). VSV is a member of the *Rhabdoviridae* family, a group of negative-strand RNA viruses that includes rabies. Intranasal administration of 1×10^6 plaque forming units of VSV to weanling H-2^d congenic C57BL/6 mice induces 91% mortality between 7 and 9 days post infection. Unlike MeV, the neuronal infection is highly cytopathic, cell death is via apoptosis and there is significant production of cell-free infectious viral progeny. Viral spread is mediated by these cell free progeny, with titration of infectious virus particles being used to document changes in brain viral burden. Like MeV, host protection is IFN-β dependent and clearance is T cell-mediated (Detje et al. 2009). Others have suggested that the unique mouse susceptibility to VSV brain infection (relative to other animal species) may in fact reflect deficient IFN-β production (Letchworth et al. 1999).

Infection of H-2^d congenic C57BL/6 NSE-hsp70 mice showed that hsp70 is host-protective following intranasal VSV infection, reflected in a reduction in mortality from 91% to 50% (Kim et al. 2013b). As with MeV infection, protection was correlated to an early and robust induction of IFN-β in brains of infected transgenic mice that was not detected in infected non-transgenic mice by RT-PCR, 1 day post infection. TLR4 transcript levels were similarly increased in infected transgenic mice at 1 day. Brain viral burdens were equivalent between infected transgenic and non-transgenic mice at 3 and 6 days post infection, indicating that hsp70 (and the induced IFN-β) is not influencing the establishment of infection. Brain viral burden stabilized in transgenic mice at 7 days, in contrast to a significant continued rise in non-transgenic mice. For animals surviving to 8 days post-infection, a significantly increased rate of viral clearance was observed in the hsp70 transgenic mice.

Using the VSV system, we also showed that it is hsp70 in the virus-infected neurons that is responsible for the host-protective effect, with no contribution of hsp70 from uninfected neurons. This was achieved by constructing a recombinant VSV in which hsp70 is expressed from the viral genome. Non-transgenic mice were infected with the VSV-hsp70 virus or a control virus expressing a norovirus capsid protein (i.e., a genomic insert of comparable size). Outcome of infection was monitored as before. Again, hsp70 expression mediated host protection that was correlated to an early and robust IFN-β response. Mortality in mice infected with 1×10^6 plaque forming units of control virus was 40%, compared to no mortality with the hsp70 expressing virus. Survival differences reflected clearance and not differences in the establishment of infection. In two additional infection studies, we showed that a 100-fold increase in the hsp70-virus challenge dose was required to achieve the 40% mortality observed in mice infected with the control virus. *In vitro* analyses of virus gene expression and replication parameters illustrated another important point. When hsp70 is overexpressed by the host cell, the result is to increase VSV gene expression and infectious viral progeny release, similar to our observations with MeV. But when hsp70 is expressed from the viral genome, no such effect is observed. The difference may reflect the fact that when hsp70 is expressed from the viral genome, levels are always in proportion to that of upstream viral gene products, including the N protein. This is in contrast to cells that constitutively express hsp70, where there will be an excess of hsp70 in relation to viral protein levels early in the course of infection. Profiles of virus gene expression between the hsp70 expressing virus were identical to that of the control virus, proof that the hsp70-protective effect shown *in vivo* is not dependent on hsp70 stimulation of virus gene expression.

From this we conclude that it is hsp70 expressed in the virus infected neurons that induces a protective type I IFN response by activated brain macrophages. The phenomenon has potentially broad virological relevance, including at minimum a representative each of the paramyxovirus and rhabdovirus families, and is not an age-dependent phenomenon. The effect is mediated at the level of virus clearance, and not the establishment of infection. Induction of the protective IFN response is not reliant on hsp70 mediated increases in virus gene expression, suggesting that it is not a response to increased expression and host detection of PAMPs. This raises the possibility that it is hsp70 itself that directly mediates the protective effect, acting as a DAMP.

2.1.6 Extracellular Hsp70 Induces IFN-β in Brain Macrophages

Hsp70 activates microglia/macrophages predominantly through TLR4 (Kakimura et al. 2002), resulting in the release of proinflammatory cytokines such as interleukin 1. TLR4 signaling also triggers IFN-β expression in macrophages (Toshchakov et al. 2002), which raises the possibility that if virus-infected neurons release hsp70, then IFN-β can be induced in uninfected microglia. Mouse neuroblastoma cells (N2a) model behavior of primary cortical neurons, and have been used to show that virus infection releases hsp70 in proportion to the intracellular hsp70 concentration. It should be noted that these cell lines are actually superior to primary cultures when modeling differences in hsp70 expression patterns, in that primary neuronal and glial cultures exhibit a stress-conditioned phenotype (Marini et al. 1990). As previously addressed, important differences exist between MeV and VSV in terms of the nature of the virus-cell relationship.

MeV infection of N2a cells is non-cytolytic. Infection at a multiplicity of infection of one (i.e., one infectious viral particle per target cell) results in no cell death up to 60 h post infection, the time at which a peak in viral antigen expression and infectious viral progeny release is observed (Carsillo et al. 2006b). Release of hsp70 into the culture media by infected and uninfected cells was measured by ELISA, and the release correlated to levels of intracellular expression. Intracellular expression levels were defined by either Western blot analysis of cell total proteins or ELISA of cell lysates. Cells lacking basal hsp70 expression (i.e., vector transfected control N2a, i.e. N2a-V) show that virus infection induces low levels of endogenous hsp70 (Oglesbee et al. 2002), resulting in a very modest rise in hsp70 in the culture media by 10 h post infection (Kim et al. 2013a). Changes of culture media at defined intervals showed that this release was restricted to this early post-infection interval, a time corresponding to primary viral transcription – transcription of the incoming viral genome, with the nucleocapsid protein being the first and most abundantly expressed viral protein. Secondary transcription (i.e., transcription mediated by newly formed transcriptases on newly replicated genomic templates) begins at 12 h post infection, after which viral protein levels begin to rise dramatically within cells. Kinetics of viral replication are similar in N2a cells that constitutively express hsp70 (N2a-HSP), although the magnitude of secondary transcription is significantly greater, with resultant increases in virus gene expression and infectious progeny release. Release of hsp70 is significantly increased in infected N2a-HSP cells, but again only in the 0–10 h post infection interval and again in the absence of any cell death. Levels of released hsp70 achieved a concentration of approximately 80 nM, over half of the concentration released by uninfected N2a-HSP cells lysed with detergent. It is important to note that release of hsp70 from cells require the virus infection stimulus. Increases in intracellular concentration of hsp70 alone were not sufficient.

The mouse microglial cell line BV-2 is used extensively to model inflammatory responses observed in primary mouse microglial cultures (Henn et al. 2009). Culture medium from infected N2a-HSP cells collected during the 0–10 h interval induced IFN-β transcripts in BV-2 cells, whereas corresponding samples from the infected N2a-V cells did not. Antibody against hsp70 neutralized IFN induction by the supernatant from infected N2a-HSP cells, proof that it was hsp70 mediating the effect. The significance of these results is the demonstration that *viable* infected neurons can release large amounts of hsp70 early during the course of infection, with levels of release proportionate to intracellular concentrations, and there being a concentration threshold required to elicit responses in microglial cells. The latter again emphasizes the relevance of the febrile response in modulating the outcome of virus infection in brain.

Results with VSV were surprising, in that cell death is by apoptosis and the conventional belief is that cells undergoing apoptosis do not release hsp70. This view is based upon studies in tumor cells lines and is consistent with the idea that programmed cell death will not (and should not) elicit an inflammatory response (Basu et al. 2000). Prior to our studies, virus-induced release of hsp70 was shown in parvovirus infections where cell death is by pyroptosis and/or lysis, both representing pro-inflammatory forms of cell death where hsp70 release might be anticipated (Moehler et al. 2003). Secretory release has been shown in Japanese encephalitis virus infected BHK cells, although the nature of hsp70 release and the phenotype of infection were not characterized (Wu et al. 2011). Similar to MeV, early release of hsp70 was observed in viable N2a cells infected with VSV, with the level of release proportionate to intracellular hsp70 concentration (Kim et al. 2013b). The onset of apoptosis was 16–24 h post infection based upon the detection of activated caspase 3 and release of LDH into the culture supernatant, and this was associated with tremendous increases in extracellular hsp70 levels associated with the infected N2a-HSP cells. Levels of cleaved caspase 3 were actually greater in VSV infected N2a-HSP cells compared to infected N2a-V cells, this being attributed to increased levels of virus gene expression in the former. The result indicates that hsp70 has no inhibitory effect on apoptosis in this setting, contrasted to observations in other (non-viral) settings (Steel et al. 2004). Programmed cell death would not be "silent" in this scenario based upon the magnitude of hsp70 release. The observation is consistent with the recent finding that VSV-induced apoptosis can progress to secondary necrosis/pyroptosis (Rogers et al. 2017) – inflammatory forms of cell death.

Studies using purified recombinant human hsp70 confirmed the dosage-dependent induction of IFN-β transcripts in brain macrophages through TLR4 (Kim et al. 2013a). The hsp70 was highly purified and used originally to study structure-function relationships between hsp70 and the MeV N protein (Couturier et al. 2010). When applied to culture supernatants of BV-2 cells, NR-9460 cells (a second mouse microglial cell line), and primary cultures of mouse bone marrow macrophages, strong induction of IFN-β transcripts was observed following 6 h of incubation. Induction was not observed when the hsp70 was heat denatured, and induction was unaffected by the addition of polymyxin B, proof that results were not influenced by LPS contaminants. Dose responsiveness was shown, with as little as 9 nM hsp70 mediating a statistically significant induction of IFN-β transcripts. A role for TLR4 in mediating the hsp70 signaling was shown by antibody-mediated inhibition, and more recently this has been confirmed with knockout microglial cell lines (Fig. 2.2). The NR-9560 line is derived from wild type mice and served as a positive control. NR-9900 is a TLR3 knockout, NR-9901 is a TLR4 knockout, and NR-9903 is a TRAM knockout cell line (Glaser et al. 2011). TRAM is an adapter molecular that mediates TLR4 mediated activation of IRF-3 in a MyD88-independent manner, and this in turn drives IFN-β transcription (Qin et al. 2005). Hsp70 (80 nM) induced IFN-β in wild type (NR-9560) cells and in TLR3 knockout microglial cells, but not in TLR4 or TRAM knockout cells. Similar results were achieved with 100 ng/ml LPS. TLR3 detects intracellular double stranded RNA that is an intermediate of RNA virus replication, activating IRF-3 through TRIF, independent of the TRAM adapter molecule. As such, results with the TLR-3$^{-/-}$ cell line confirm integrity of TLR signaling pathways that mediate IFN-β induction through IRF-3. Results with Newcastle disease virus (NDV) provide further support for the integrity of IRF-3 signaling. Infection of all cell lines with NDV at a multiplicity of infection of 1.0 induced IFN-β transcription, where induction occurs in a TLR3 and TLR-4 independent manner.

Studies of hsp70-MeV N protein interaction also examined the role of specific hsp70 functional domains in the induction of IFN-β transcription. Hsp70 consists of an N-terminal nucleotide binding domain (NBD amino acids 1-384), followed by a substrate binding domain (SBD amino acids 385-543) and a C-terminal lid (amino acids 544-641) (Fig. 2.3). Constructs used were the full-length hsp70, hsp70 1-543 (i.e., hsp70 lacking the C-terminal lid), and hsp70 384-641 (i.e., hsp70 lacking the N-terminal NBD) (Couturier et al. 2010). All proteins were used at a concentration of 80 nM in culture supernatants of BV-2 cells, together with 30 μg/l polymyxin B. IFN-β transcript levels were analyzed following a 6-h incubation. Results show that the N-terminal 541 amino acids of the hsp70 molecule are necessary and sufficient for IFN-β induction, while the C-terminal lid is dispensable. Heat denaturation of constructs eliminated the ability to induce IFN-β, further proof that the result does not reflect contamination of the preparations by LPS. The latter must be emphasized because of the challenges that other groups have experienced in obtaining LPS-free recombinant protein preparations (Luong et al. 2012). Lack of IFN induction by the hsp70 384-641 construct should further allay any concerns regarding the potential for sample contamination. The significance of these findings relate to possible mechanisms of hsp70 release from virus-infected neurons.

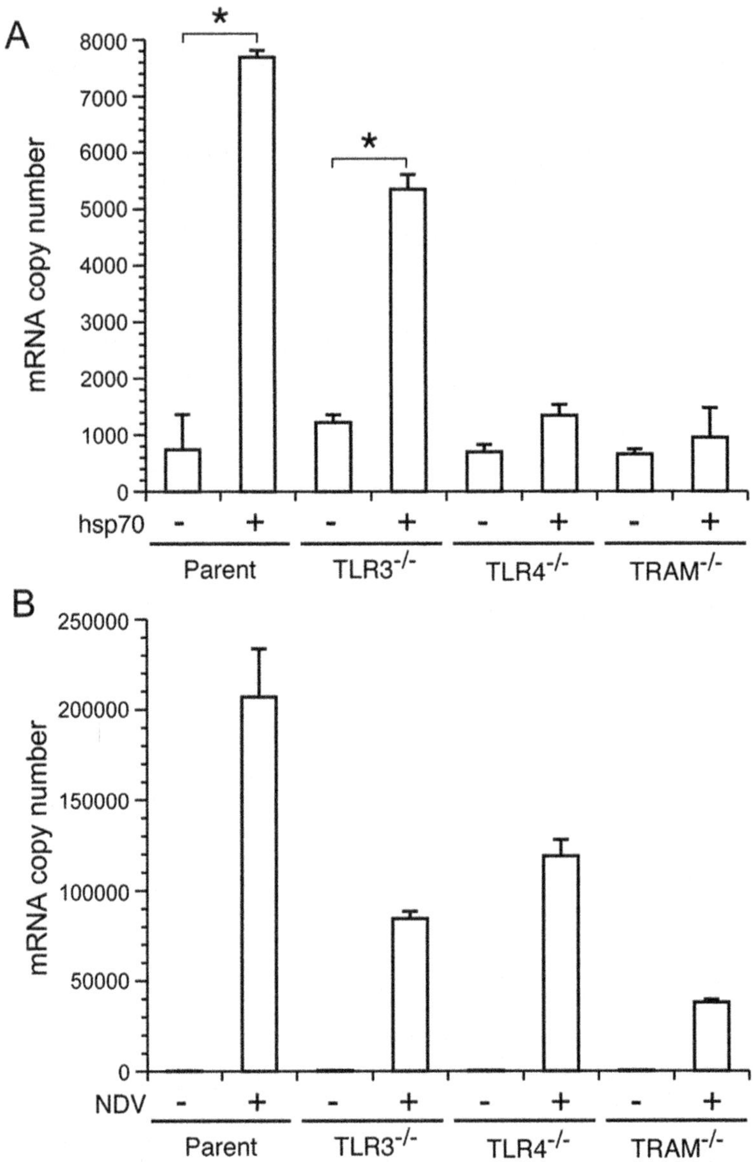

Fig. 2.2 Hsp70 induces IFN-β transcription in mouse microglial cells through TLR4 via a TRAM-dependent pathway (i.e., MyD88 independent). Cells lines were derived from wild type (NR-9460, parent), TLR3 knockout (NR-9900), TLR4 knockout (NR-9901) or TRAM knockout mice (NR-9903). (**a**) Cells were incubated with 80 nM hsp70 for 6 h in the presence of 30 μg/ml polymyxin B, and IFN-β transcript levels quantified by real time RT-PCR. (**b**) Cells were infected with Newcastle disease virus at an MOI of 1.0, to confirm the integrity of IRF3-dependent IFN-β induction pathways

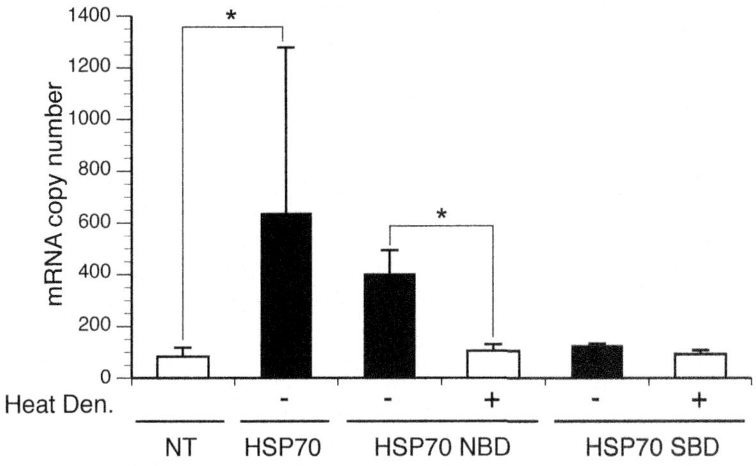

Fig. 2.3 The hsp70 N-terminal 543 amino acids is sufficient for IFN-β induction in a mouse microglial cell line (BV-2). This portion of the of the 641 amino acid protein is inclusive of the hsp70 nucleotide binding domain (NBD) and peptide binding domain (PBD). The C-terminal 98 amino acids represent the lid of the peptide-binding domain. Protein constructs were added to culture media at a concentration of 80 nM, together with 30 μg/ml polymyxin B, and incubated for 6 h prior to extraction of total RNA for quantification of IFN-β transcripts by real time RT-PCR. Heat denatured hsp70 constructs served as negative controls to further address the absence of LPS contamination. Background signal observed in non-treated cells and cells treated with denatured protein were identical. Significant induction over background was observed only with full-length hsp70 and hsp70 1–543. The asterisk denotes a statistically significant ($p < 0.05$) difference (t-Test)

2.1.7 *Exosomal Release of Hsp70 from Infected Cells*

The preponderance of non-cytolytic neuron infections highlights the importance of understanding how a viable neuron can release hsp70 in response to that infection. Hsp70 release from viable cells may reflect virus-induced exosomal secretion or microvesicular shedding (De Maio 2011; Lancaster and Febbraio 2005). Focus by our group was placed on the exosomal mechanism due to the relevance of exosomal release in neuronal cells, the prevalence of hsp70 in exosomes, and the ability of microbial infection to influence hsp70 content of exosomes. Exosomes are 40–100 nm endosome-derived vesicles released from various cell types under both normal and disease conditions. They originate from early endosomes, where inward budding forms intralumenal vesicules (ILVs) of multivesicular bodies (MVBs). There are two distinct classes of MVBs including degradative MVBs which lead to lysosomal degradation, and exocytic MVBs which fuse with the plasma membrane and release ILVs from the cell in the form of exosomes (Meckes and Raab-Traub 2011). The molecular composition of exosomes is largely dependent on cell type, with exosomes from various origins containing both conserved proteins and cell type specific proteins. Most exosomes contain cytoplasmic proteins including proteins of endocytosis pathways and signal transduction proteins, with hsp70 being most consistently identified as an exosomal constituent (Simpson et al. 2009). The latter is consistent with hsp70 localization to lipid rafts within cells, including those associated with early endosomes (Chen et al. 2005).

Neurons release significant levels of exosomes under normal physiological conditions, where exosomes are used to communicate with other neurons and glia (Fruhbeis et al. 2012). An emerging body of literature also suggests that exosomal secretion of hsp70 is part of a basic inflammatory response. Levels of exosome release and hsp70 content of exosomes are increased in macrophages following infection with Mycobacterium species (Anand et al. 2010). The potential of microbial infection to modify neuronal exosome release is currently unknown, as is the potential of neuronal exosomes to stimulate innate immune and/or inflammatory responses.

MeV infection of N2a cells is particularly well-suited to study virus induced exosome release of hsp70, with N2a cells faithfully reproducing exosome release observed in primary cultures of mouse cortical neurons (Yuyama et al. 2012). Our current efforts have focused on the isolation of exosomes from infected and uninfected N2a cells, both in the presence (N2a-HSP) and absence (N2a-V) of constitutive hsp70 expression. The resultant exosomes were characterized both in terms of hsp70 content and the ability to induce IFN-β in the mouse BV-2 microglial cell line.

Exosomes were purified from the culture supernatants of MeV infected and uninfected N2a-HSP and N2a-V cells at 20 h post infection using a commercially available polymer-based recovery system (ExoQuick-TC™, Systems Biosciences). This interval follows the early release of hsp70 from viable infected cells at 0–10 h post infection (Kim et al. 2013a) and precedes the onset of cell-free infectious viral progeny release (Carsillo et al. 2006b). MeV infection increased exosome protein yield in N2-HSP cells by 17%, compared to a 4% increase in N2a-V cells. Analysis of total proteins by 10% SDS-PAGE and Coomassie blue staining showed that the protein profiles observed in exosomes from both infected and uninfected cells are different from those of cell lysates, but are similar to each other. To confirm that isolated samples had characteristics of exosomes, cell lysates and exosomes from infected N2a-HSP were probed for the presence of exosome markers (i.e., hsp70, Alix, and Flotillin-1) and the exclusion of the luminal endoplasmic reticulum marker Grp94 that would represent cell organelle contamination as a result of cell lysis (Stoorvogel et al. 2002; Thery et al. 2002). Exosomes contained hsp70, Alix and Flotillin-1 but lacked Grp94, whereas all proteins were detected in cell lysates (Fig. 2.4a). MeV N protein was not detected in exosomes from infected N2a-HSP and N2a-V cells (data not shown).

The size of the hsp70 immunoreactive protein in exosomes was approximately 64 kDa, in contrast to the 70 kDa band in cell lysates (Fig. 2.4a). The 64 kDa exosome protein and the 70 kDa cytosolic protein bands from MeV infected N2a-HSP cells were excised from the the Coomassie blue stained gels and subjected to mass spectral analysis. A match between the 64 kDa exosomal protein and human hsp70 was obtained. Four peptides matched macaque hsp70-1 like isoform 2 in the Mascot mammalian protein database covering 9% of the sequence, yielding an ion score of 253. The results identify the 64 kDa band as primate hsp70 with a high degree of

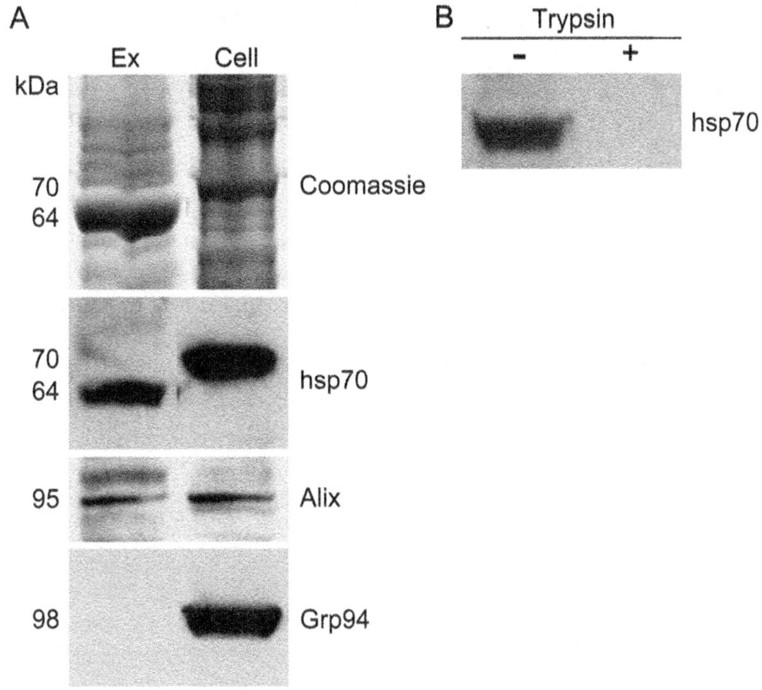

Fig. 2.4 Exosome and cell lysate total proteins from infected N2a-HSP cells. (**a**) Samples were analyzed by 10% SDS-PAGE and Coomassie blue staining (top panel) and Western blots analysis (bottom three panels) with antibody against exosome markers (i.e., hsp70, Alix, and Flotillin-1) and the endoplasmic reticulin marker Grp94. hsp70, Alix, and Flotillin-1 but not Grp94 were detected in the exosomes. The size of hsp70 detected in the exosomes was 64 kDa, in contrast to the 70 kDa form detected in cell lysates. (**b**) Exosomes from infected N2a-HSP cells, representing 10 μg total protein, were digested with trypsin for 5 min at 37 °C (+). Exosomes incubated in the absence of trypsin (−) served as treatment negative controls. Western blot analysis showed that exosomal hsp70 was accessible to trypsin, consistent with a surface (solvent exposed) location

confidence, given that only two peptide matches and an ion score >20 is required for a reliable match. Sequence coverage by the 64 kDa hsp70-derived peptides came within 25 amino acids of the amino terminus of the reference hsp70 in the Mascot database. A 6 kDa difference between the exosomal and cytosolic hsp70 would represent 55 amino acids, assuming an average of 110 Daltons per amino acid. If we assume that a single truncation accounts for the difference, the truncation would have to be present on the carboxyl terminus of the parent 70 kDa hsp70. Bioinformatic analysis of the parent hsp70 was performed using PeptideCutter software and the MeDor metaserver for the prediction of intrinsic disorder (Lieutaud et al. 2008). The parent protein is 641 amino acids in length, and 55 amino acids from the carboxyl terminus (i.e., position 586) fall within an intrinsically disordered solvent exposed domain containing numerous protease recognition sites. Given the modular domain structure of hsp70, cleavage at position 586 would preserve the hsp70 peptide binding domain, yet remove approximately 56% of the lid domain that stabilizes hsp70 peptide substrate binding and would completely remove the extreme C-terminal EEVD sequence that mediates hsp70 oligomer formation (Couturier et al. 2010). Collectively, the results suggest that the 64 kDa exosomal hsp70 is a C-terminal truncation of the parent hsp70, reflecting loss of a multimerization domain and a portion of the lid.

Densities of total and 64 kDa hsp70 bands in the Coomassie blue stained gels were quantified in order to compare the level of exosomal hsp70 between infected and uninfected cells. Infection increased the relative 64 kDa band intensity in both N2a-HSP and N2a-V cells, with infected N2a-HSP cells having the greatest exosomal hsp70 content (i.e., approximately 15% of exosome total proteins compared to 6% of exosome total proteins for infected N2a-V cells). Infection increased the exosome hsp70 content by 43% in N2a-HSP cells relative to uninfected controls, and 13% in N2a-V cells. In order to demonstrate that hsp70 is located on the surface of exosomes, where it can engage TLRs on the cell surface (Anand 2010), exosomes from N2a-HSP infected cells were incubated with trypsin for 5 min at 37 °C or in buffer lacking trypsin as a treatment negative control. Trypsin treatment eliminated the 64 kDa hsp70 species isolated from infected N2a-HSP cells based upon Western blots analysis of exosome total protein (Fig. 2.4b).

Exosomes from both infected and uninfected N2a-HSP and N2a-V cells were added to the mouse microglial cell line BV-2 for 24 h, and total BV-2 cell RNA was analyzed by real-time RT-PCR for the presence of IFN-β transcripts. Significant induction of IFN-β transcripts in BV-2 cells was only observed following treatment with exosomes derived from infected N2a-HSP cells, using BV-2 cells incubated with exosome-free media as a negative control (Fig. 2.5).

Results confirm that MeV infection of N2a cells is associated with exosomal release of hsp70, and that exosomes derived from infected N2a cells that constitutively express hsp70 are capable of IFN-β induction in uninfected microglial cells. Constitutive expression of hsp70 increases the proportion of hsp70 in exosome total proteins, and MeV infection results in an additional 43% increase in exosome hsp70 levels. Infection also increases the yield of exosome total proteins, this being used by other groups to reflect level of exosome release (Anand 2010). The protease sensitivity of the exosomal hsp70 supports localization to the exosome surface, enabling it to directly engage cell surface TLRs, and we know that the N terminus of hsp70 is sufficient for IFN-β induction. The ability of MeV to increase exosome release and the hsp70 content of exosomes thus parallels observations in macrophages following infection with mycobacteria (Anand et al. 2010).

The hsp70-containing exosomes are recovered from uninfected N2a-HSP cells, and also from infected and uninfected N2a-V cells. Neurons can support high levels of constitutive exosomal release, explaining the recovery from uninfected N2a cells. Hsp70 contained in exosomes from N2a-V cells would represent endogenous hsp70, which is highly homologous to human hsp70. But only exosomes from MeV infected N2a-HSP cells were able to induce IFN-β transcripts in uninfected microglial cells. This observation may reflect the concentration of hsp70 on the surface of exosomes that is required to activate TLR-4 signaling. We know that purified hsp70 can induce IFN-β through TLR4, and that there is an hsp70 dose response to the IFN

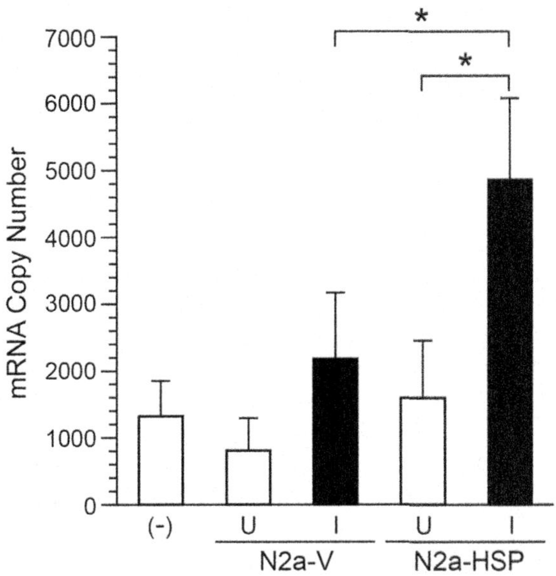

Fig. 2.5 Exosomes released from infected N2a-HSP cells induce IFN-β transcription in a mouse microglial cell line (BV-2). Exosomes from uninfected (U) and infected (I) N2a-V and N2a-HSP, representing 20 μg total proteins, were incubated with BV-2 cells for 24 h, after which BV-2 IFN-β transcripts were quantified using real-time RT-PCR analysis of total cell RNA. DMEM containing 5% FCS was added to BV-2 cells as negative control (−). Significant induction of IFN-β transcripts was observed in BV-2 cells stimulated with exosomes from infected N2a-HSP, relative to cells treated with exosomes from uninfected N2a-HSP and infected N2a-V cells ($p < 0.05$, ANOVA)

induction (Kim et al. 2013a). But there may also be qualitative differences in the exosomes between groups that explain the differential ability to induce IFN-β. Infected N2a cells may direct trafficking of both viral gene products (PAMPs) and other endogenous ligands (DAMPs) into exosomes, and these additional molecules may directly support TLR signaling or other pathways of IFN induction (Simpson et al. 2009). We do not observe intact MeV N proteins in exosomes, this being the earliest and most abundant viral protein that is expressed in infected cells, although we still have to examine the possible content of viral RNAs or viral protein subunits. Exosomes from cells infected with hepatitis C virus and Epstein-Barr virus contain viral RNAs that are capable of inducing innate immune responses in uninfected cells (Dreux et al. 2012; Pegtel et al. 2010). We must also consider the contribution of other DAMPs, present in exosomes, which may contribute to TLR signaling. Exosomes contain multiple heat shock proteins under normal conditions, including hsp27, hsp60, hsp70, hsc70, and hsp90, although only hsp70 levels are increased by intracellular bacterial infection (De Maio 2011). The diverse molecules present in exosomes that may contribute to cell surface TLR signaling and the efficient transfer of RNAs from infected to uninfected cells may explain how an even modest increase in the release of hsp70 containing exosomes can have such a dramatic effect on IFN induction in uninfected brain macrophages.

The potential significance of the hsp70 C-terminal truncation is not clear at present, although our deep understanding of the interaction between hsp70 and the MeV N protein offers insights that allow us to construct a model for virus induction of hsp70 trafficking to exosomes. Protease sensitivity of the hsp70 C-terminus is supported by others, where activated caspase-3 was shown to cleave the C-terminus of a variant hsp70, giving rise to a 62 kDa product (Ran et al. 2004). An approximately 60 kDa exosomal hsp70 is also identified in product information for the ExoQuick™ exosome purification system (System Biosciences). However, little has been done in the literature to understand the nature and incidence of hsp70 post-translational modifications (Freeman et al. 1995; Liu and Hendrickson 2007), let alone the possibility of alternate transcriptional or translational gene products. The exclusive presence of 64 kDa hsp70 on the exosomes from our work suggests that the truncation is vital to trafficking of hsp70 from endosomes to multivesicular bodies and ultimately exosomes. Studies in neurons show that hsp70 binding to α-synuclein and aggregation of α-synuclein is a stimulus for extracellular hsp70 release, although the hsp70 that was released was not subject to biochemical characterization (Danzer et al. 2011). Hsp70 binds the MeV N protein, and the N-protein self-assembles to form nucleocapsid helices (Oglesbee 2007), which might provide a stimulus for extracellular hsp70 release similar to that observed in cells expressing α-synuclein. Hsp70 also co-localizes with MeV N on endosomal membranes, placing hsp70 directly in the exosomal pathway. We propose that release of the hsp70-bound substrate may be necessary for exosomal trafficking and extracellular hsp70 release. We know that removal of the C-terminal "lid" of the hsp70 peptide binding domain facilitates release of bound N (Couturier et al. 2010), and it may be that both 70 and 64 kDa hsp70 in the cytoplasmic pool is capable of N binding and endosomal membrane localization, but only the 64 kDa hsp70 is capable of exosomal trafficking. The short interval of hsp70 release from MeV infected cells may reflect subsequent hsp70 sequestration as intracellular N protein concentrations rise during the course of infection.

2.1.8 Biological Implications of Hsp70 as a Virus-Induced Chaperokine

The concept of hsp70 as a cytokine (i.e., chaperokine) is not new, but recognition of its role in virus infections of the CNS provides support for that role being essential to host protection. Virus-induced exosome release of hsp70 from neurons provides the danger signal to trigger innate immune responses by uninfected microglia, this being vital for host defense against viruses engaged in non-cytolytic infection. We have barely begun to investigate the number of viruses for which this relationship is true, but the number is certain to grow. As mentioned above, we know that Japanese encephalitis virus induces hsp70 release from infected neuronal cultures, although the mechanism of release has not been investigated (Wu et al. 2011). The efficiency of the pathway is inherent to the fact that many viruses must induce hsp70 in support of virus replication, and physiologic responses to infection frequently includes fever, a potent inducer of hsp70. The early release of a DAMP means that we can have efficient stimulation of innate immunity, in contrast to simple reliance on the recognition of PAMPs. Relying on the latter means that innate immune responses are essentially playing catch-up to the spreading infection. Moreover, the mechanism allows for induction of robust innate immune responses in uninfected cells, where there is no possibility of virus-mediated inhibition of innate immune signaling. The latter is a fundamental feature of replication for multiple virus families. The only challenge is the need to facilitate antigen cross-presentation, and here too hsp70 may play a role. In the case of MeV, the N protein alone drives immune-mediated viral clearance and the epitopes mediating this response lay within the hsp70 binding domain of N (Weidinger et al. 2000). This positions hsp70 to not only induce IFN-β and activate microglia, but also to cross-present viral antigen that originates from virus infected neurons. This relationship is summarized in Fig. 2.6.

As relates to cytolytic neuronal infections, there is no challenge to understanding how cell death may result in the release of hsp70 to drive a strong innate immune response. Apoptosis is frequently the mechanism incriminated in virus-induced cell death and many view apoptosis to be a process that does not trigger inflammatory or immune responses. However, it should be remembered that apoptosis is also a defense mechanism against virus infections, a means of destroying viral factories. For VSV, a particularly robust activation of caspase 3 may facilitate the progression to secondary necrosis/pyroptosis (Rogers et al. 2017).

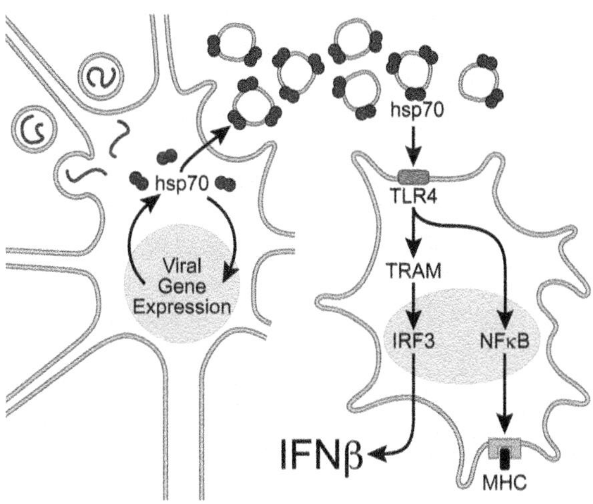

Fig. 2.6 A model for virus-induced exosomal release of hsp70 from viable infected neurons, where extracellular hsp70 can activate uninfected microglia to release IFN-β and become antigen-presenting cells – innate immune responses that are vital to cell-mediated immune clearance of virus from brain. Virus induces endogenous hsp70 to support replication, and hsp70 can be further elevated by physiologic responses to infection such as fever, with the level of extracellular hsp70 release reflecting intracellular concentrations

So now the question becomes one of understanding how viruses like rabies escape such an innate immune mechanism. For rabies virus that is well adapted to its host, the neuronal infection is exquisitely non-cytolytic. It also induces hsp70 in support of replication and forms large cytoplasmic inclusions composed in large part by hsp70 and nucleocapsid (Lahaye et al. 2012). This raises the intriguing possibility that rabies virus is sequestering hsp70, preventing the release that is central to the defense pathway. To address this possibility we must develop a better understanding of mechanisms by which viruses successfully stimulate exosomal release of hsp70. From a therapeutic standpoint, we may also consider the potential efficacy of exosomes as means of treating an otherwise fatal viral encephalitis.

There is a potential cost to an hsp70-dependent immune response. First, the model is adaptive provided that the host is immune competent. What happens if host immune competence is impaired? We addressed this in experiments using hsp70 transgenic H-2^{b} C57BL/6 mice (Carsillo et al. 2006). Recall that these mice are not immune deficient *per se*, but they do lack the ability to efficiently present MeV epitopes that drive host protective immune responses because of the H-2^{b} haplotype. In this instance, the stimulatory effect of hsp70 on virus gene expression outweighs the beneficial effect on innate immune stimulation, and the result is an hsp70-dependent increase in virus-induced mortality. In these H-2^{b} mice, hsp70 enhanced MeV brain burden by two orders of magnitude relative to non-transgenic mice, and mortality was fivefold greater. Using the 522D variant of Ed-MeV, we showed no-significant difference in brain viral burdens or mortality between transgenic and non-transgenic mice infected with the 522D virus, or between 522D infected transgenic and 522N (i.e., parent Ed-MeV) infected non-transgenic mice – proof that it was hsp70-mediated stimulation of viral gene expression that was driving the increased mortality in hsp70 transgenic relative to non-transgenic H-2^{b} mice.

A second consideration relates to hsp70 release that can be detrimental to the host. Cellular insults or degenerative processes that mimic a virus infection (e.g., protein denaturing insults that cause aggregation of hsp70 bound proteins) may trigger such release. We know that ischemia-reperfusion injury of the spinal cord in both humans and dogs causes the release of high levels of hsp70 into the cerebrospinal fluid (CSF) and these high levels persist for days (Awad et al. 2008; Hecker et al. 2008). In this capacity, hsp70 may contribute to the neuroinflammatory response that is primarily responsible for the functional deficits. Here, we are reminded that hsp70 not only induces type I IFN, but also proinflammatory cytokines such as interleukin 1 (Vabulas et al. 2002). The source of hsp70 does not appear to be degenerative neurons, but rather viable ependymal cells in the ischemic penumbra (Awad et al. 2008). As such, the release appears to be secretory. In a second scenario, hsp70 is elevated in CSF of dogs during the clinical phase of an idiopathic immune mediated disorder known as steroid responsive meningitis and arteritis (Moore et al. 2012). Dogs go into clinical remission following steroid treatment, and this is accompanied by a return of hsp70 CSF levels to baseline. Studies also support potential of the hsp70-TLR4 pathway to contribute to seizure activity (von Rüden et al. 2018), an intriguing possibility given the linkage between histories of viral encephalitis and a proconvulsant state. In the end, the benefits of the hsp70-TLR4 axis may outweigh the costs considering the challenges of life in a microbial world.

This discussion has focused on the hsp70-TLR4 axis in the CNS because we have established a tractable model. Components of the model have probable relevance to other organ systems. For example, we know that human respiratory syncytial virus (RSV) induces hsp70 (Brasier et al. 2004), that hsp70 supports RSV replication (Brown et al. 2005), and in RSV-infected infants there is evidence of increased extracellular hsp70 in bronchoalveolar lavage fluids and evidence of TLR4 signaling in leukocytes recovered from the respiratory tract (Wheeler et al. 2009). The degree to which this and other virus-hsp70 interactions contribute to protection (or virulence) remains a fertile area of future investigation.

2.2 Conclusions

Our understanding of how hsp70 influences virus infection of neurons has evolved from intracellular chaperonin functions in support of virus replication to an appreciation of the profound impact of extracellular hsp70 as a chaperokine capable of driving immune-mediated viral clearance from the CNS. We have established the potential broad virological relevance of this hsp70-dependent protective mechanism, the potential untoward effects of inappropriate hsp70 release in in sterile neuroinflammatory disorders, and how mechanistic insights may lead to development of novel therapeutic approaches for virus infection of the CNS. This potential warrants inclusion of hsp70 in CSF cytokine profiles that are part of the diagnosis and characterization of neuroinflammatory processes in general. As we have discussed, significant elevations of CSF hsp70 are documented in sterile inflammatory processes for both humans and dogs (i.e., species that constitutively express hsp70). Similar analyses have yet to be performed in virus-infected patients in order to address the degree to which regional release in the parenchyma might impact the low basal levels otherwise observed in the CSF. Regarding treatment of viral encephalitis, manipulation of the hsp70-immune axis requires consideration of immune status. For the immune competent host, increasing intracellular concentrations of hsp70 results in increased hsp70 release by virus-infected cells. Even for a virus that uses intracellular hsp70 in support of replication, the result is to promote clearance. Fever is the physiological approach, raising questions as to the potential unintentional consequences of therapeutically suppressing fever. In contrast, failure of an *immune compromised* patient to contain increases in viral replication that are mediated by intracellular hsp70 could have life-threatening consequences. To avoid this possibility, extracellular hsp70 or its subunits could be administered via the CSF (i.e., intrathecal), possibly displayed on microvesicles. The approach should be further investigated, particularly considering the paucity of direct antiviral treatment options for the majority of neurovirulent viruses. Perhaps the most significant consideration raised in this discussion is the use of mice when modeling viral neurovirulence in humans, or any inflammatory process in the CNS for which extracellular hsp70 may be an important component. As we have discussed, mice are unique in lacking significant constitutive expression of hsp70 and this in turn impacts the magnitude of release. Considering the hsp70 dose-dependent responsiveness of TRL4 signaling pathways, activity of the hsp70-immune axis may thus be missed or significantly under-estimated. The consideration is not restricted to models of human disease, but models of disease of multiple animal species for which constitutive expression is observed.

Acknowledgements This work was supported in part by funds from the National Institute of Neurological Disorders and Stroke (R01NS31693). We thank Dr. Mamuka Kvaratskhelia (The Ohio State University) for assistance in the proteome analysis of exosomes and Dr. Prosper Boyaka (The Ohio State University) for suggestions in preparing the manuscript.

References

Anand PK (2010) Exosomal membrane molecules are potent immune response modulators. Commun Integr Biol 3:405–408

Anand PK, Anand E, Bleck CK, Anes E, Griffiths G (2010) Exosomal Hsp70 induces a pro-inflammatory response to foreign particles including mycobacteria. PLoS One 5:e10136

Awad H, Suntres Z, Heijmans J, Smeak D, Bergdall-Costell V, Cristofi FL, Magro C, Oglesbee M (2008) Intracellular and extracellular expression of the major inducible 70 kDa heat shock protein in experimental ischemia-reperfusion injury of the spinal cord. Exp Neurol 212:275–284

Basu S, Binder R, Suto R, Anderson K, Srivastava P (2000) Necrotic but not apoptotic cell death releases heat shock proteins, which deliver a partial maturation signal to dendritic cells and activate the NF-kappa B pathway. Int Immunol 12(11):1539–1546

Beauvillain C, Donnou S, Jarry U, Scotet M, Gascan H, Delneste Y, Guermonprez P, Jeannin P, Couez D (2008) Neonatal and adult microglia cross-present exogenous antigens. Glia 56:69–77

Bellini W, Rota J, Lowe L, Katz R, Dyken P, Zaki S, Shieh W, Rota P (2005) Subacute sclerosing panencephalitis: more cases of this fatal disease are prevented by measles immunization than was previously recognized. J Infect Dis 192(10):1686–1693

Bitnum A, Shannon P, Durward A, Rota P, Bellini W, Graham C, Wang E, Ford-Jones E, Cox P, Becker L, Fearon M, Petric M, Tellier R (1999) Measles inclusion body encephalitis caused by the vaccine strain of measles virus. Clin Infect Dis 29(4):855–861

Blixenkrone-Moller M, Bernard A, Bencsik A, Sixt N, Diamond L, Logan J, Wild T (1998) Role of CD46 in measles virus infection in CD46 transgenic mice. Virology 249:238–248

Bourhis JM, Receveur-Bréchot V, Oglesbee M, Zhang X, Buccellato M, Darbon H, Canard B, Finet S, Longhi S (2005) The intrinsically disordered C-terminal domain of the measles virus nucleoprotein interacts with the C-terminal domain of the phosphoprotein via two distinct sites and remains predominantly unfolded. Protein Sci 14:1975–1992

Brasier A, Spratt H, Wu Z, Boldogh I, Zhang Y, Garofalo R, Casola A, Pashmi J, Haag A, Luxon B, Kurosky A (2004) Nuclear heat shock response and novel nulear domain 10 reorganization in respiratory syncytial virus-infected A549 cells identified by high-resolution two-dimensional gel electrophoresis. J Virol 21:11461–11476

Brooks G, Butel J, Morse S (1998) Paramyxovirus and rubella virus. In: Butler J, Ransom J, Ryan E (eds) Adelberg's microbiology. Appleton and Lange, Stanford, pp 507–527

Brown G, Rixon H, Steel J, McDonald T, Pitt A, Graham S, Sugrue R (2005) Evidence for an association between heat shock protein 70 and the respiratory syncytial virus polymerase complex within lipid-raft membranes during virus infection. Virology 338:69–80

Carsillo T, Carsillo M, Niewiesk S, Vasconcelos D, Oglesbee M (2004) Hyperthermic preconditioning promotes measles virus clearance from brain in a mouse model of persistent infection. Brain Res 1004:73–82

Carsillo T, Zhang X, Vasconcelos D, Niewiesk S, Oglesbee M (2006) A single codon in the nucleocapsid protein C terminus contributes to in vitro and in vivo fitness of Edmonston measles virus. J Virol 80(6):2904–2912

Carsillo T, Traylor Z, Choi C, Niewiesk S, Oglesbee M (2006a) Hsp72, a host determinant of measles virus neurovirulence. J Virol 80:11031–11039

Carsillo T, Zhang X, Vasconcelos D, Niewiesk S, Oglesbee M (2006b) A single codon in the nucleocapsid protein C terminus contributes to in vitro and in vivo fitness of Edmonston measles virus. J Virol 80:2904–2912

Carsillo T, Carsillo M, Traylor Z, Rajala-Schultz P, Popovich P, Niewiesk S, Oglesbee M (2009) Major histocompatibility phenotype determines hsp70-dependent protection against measles virus neurovirulence. J Virol 83(11):5544–5555

Chen W, Lin Y, Liao C, Hsieh S (2000) Modulatory effects of the human heat shock protein 70 on DNA vaccination. J Biomed Sci 7:412–419

Chen S, Bawa D, Besshoh S, Gurd JW, Brown IR (2005) Association of heat shock proteins and neuronal membrane components with lipid rafts from the rat brain. J Neurosci Res 81:522–529

Couturier M, Buccellato M, Costanzo S, Bourhis J, Shu Y, Nicaise M, Desmadril M, Flaudrops C, Longhi S, Oglesbee M (2010) High affinity binding between Hsp70 and the C-terminal domain of the measles virus nucleoprotein requires an Hsp40 co-chaperone. J Mol Recognit 23:301–315

Danzer KM, Ruf WP, Putcha P, Joyner D, Hashimoto T, Glabe C, Hyman BT, McLean PJ (2011) Heat-shock protein 70 modulates toxic extracellular alpha-synuclein oligomers and rescues trans-synaptic toxicity. FASEB J 25:326–336

Das S, Laxminarayana S, Chandra N, Ravi V, Desai A (2009) Heat shock protein 70 on Neuro2a cells is a putative receptor for Japanese encephalitis virus. Virology 385(1):47–57

Daugaard M, Rohde M, Jäättelä M (2007) The heat shock protein 70 family: highly homologous proteins with overlapping and distinct functions. FEBS Lett 581:3702–3710

De Maio A (2011) Extracellular heat shock proteins, cellular export vesicles, and the Stress Observation System: a form of communication during injury, infection, and cell damage. It is never known how far a controversial finding will go! Dedicated to Ferruccio Ritossa. Cell Stress Chaperones 16:235–249

Detje C, Meyer T, Schmidt H, Kruez D, Rose J, Bechmann I, Prinz M, Kalinke U (2009) Local type 1 IFN receptor signaling protects against virus spread within the central nervous system. J Immunol 182:2297–2304

Dreux M, Garaigorta U, Boyd B, Decembre E, Chung J, Whitten-Bauer C, Wieland S, Chisari FV (2012) Short-range exosomal transfer of viral RNA from infected cells to plasmacytoid dendritic cells triggers innate immunity. Cell Host Microbe 12:558–570

Duib-Jalbut S, Xia J, Rangaviggula H, Fang Y, Lee T (1999) Failure of measles virus to activate nuclear factor-κB in neuronal cells: implications on the immune response to viral infections in the central nervous system. J Immunol 162:4024–4029

Finke D, Liebert U (1994) CD4(+) T cells are essential in overcoming experimental measles encephalitis. Immunology 83:184–189

Finke D, Brinckmann U, ter Meulen V, Liebert U (1995) Gamma interferon is a major mediator of antiviral defense in experimental measles virus-induced encephalitis. J Virol 69:5469–5474

Freeman BC, Myers MP, Schumacher R, Morimoto RI (1995) Identification of a regulatory motif in Hsp70 that affects ATPase activity, substrate binding and interaction with HDJ-1. EMBO J 14:2281–2292

Frühbeis C, Fröhlich D, Krämer-Albers EM (2012) Emerging roles of exosomes in neuron-glia communication. Front Physiol 3:119

Glaser K, Hagos B, Molestina R (2011) Effects of Toxoplasma gondii genotype and absence of host MAL/Myd88 on the temporal regulation of gene expression in infected microglial cells. Exp Parasitol 129:409–413

Hecker J, Sundram H, Zou S, Praestgaard A, Bavaria J, Ramchandren S, McGarvey M (2008) Heat shock proteins HSP70 and HSP27 in the cerebrospinal fluid of patients undergoing thoracic aneurysm repair correlate with the probability of postoperative paralysis. Cell Stress Chaperones 13:435–446

Henn A, Lund S, Hedtjarn M, Schrattenholz A, Porzgen P, Leist M (2009) The suitability of BV2 cells as alternative model system for primary microglia cultures or for animal experiments examining brain inflammation. ALTEX 26:83–94

Hofman F, Hinton D, Baemayr J, Weil M, Merrill J (1991) Lymphokines and immunoregulatory molecules in subacute sclerosing panencephalitis. Clin Immunol Immunopathol 58(3):331–342

Joncas J, Robillard L, Boudreault A, Leyritz M, McLaughlin B (1976) Letter: interferon in serum and cebrospinal fluid in subacute sclerosing panencephalitis. Can Med Assoc 115(4):309

Kakimura J, Kitamura Y, Takata K, Umeki M, Suzuki S, Shibagaki K, Taniguchi T, Nomura Y, Gebicke-Haerter P, Smith M, Perry G, Shimohama S (2002) Microglial activation and amyloid-beta clearance induced by exogenous heat-shock proteins. FASEB J 16:601–603

Kallfass C, Ackerman A, Lienenklaus S, Weiss S, Heimrich B, Staeheli P (2012) Visualizing production of Beta interferon by astrocytes and microglia in brain of la crosse virus-infected mice. J Virol 86:11223–11230

Katayama Y, Hotta H, Nishimura A (1995) Detection of measles virus nucleoprotein mRNA in autopsied brain tissues. J Gen Virol 76(Pt 12):3201–3204

Katayama Y, Kosho K, Nichimura A, Tatsuno Y, Homma M, Hotta H (1998) Detection of measles virus mRNA from autopsied human tissues. J Clin Microbiol 36(1):299–301

Katz M (1995) Clinical spectrum of measles. Curr Top Microbiol Immunol 191:1–12

Kawanokuchi J, Mizuno T, Takeuchi H, Kato H, Wang J, Mitsuma N, Suzumura A (2006) Production of interferon gamma by microglia. Mult Scler 12(5):558–564

Kim MY, Shu Y, Carsillo T, Zhang J, Yu L, Peterson C, Longhi S, Girod S, Niewiesk S, Oglesbee M (2013a) Hsp70 and a novel axis of type 1 interferon-dependent antiviral immunity in the measles virus-infected brain. J Virol 87(2):998–1009

Kim MY, Ma Y, Zhang Y, Li J, Shu Y, Oglesbee M (2013b) hsp70-dependent antiviral immunity against cytopathic neuronal infection by vesicular stomatitis virus. J Virol 87(19):10668–10678

Krakowka S (1989) Canine distemper virus infectivity of various blood fractions for central nervous system vasculature. J Neuroimmunol 21:75–80

Lahaye X, Vidy A, Fouquet B, Blondel D (2012) Hsp70 protein positively regulates rabies virus infection. J Virol 86(9):4743–4751

Lancaster GI, Febbraio MA (2005) Exosome-dependent trafficking of HSP70: a novel secretory pathway for cellular stress proteins. J Biol Chem 280:23349–23355

Letchworth G, Rodriguez L, Del cbarrera J (1999) Vesicular stomatitis. Vet J 157:239–260

Lieutaud P, Canard B, Longhi S (2008) MeDor: a metaserver for predicting protein disorder. BMC Genomics 9(Suppl 2):S25

Liu Q, Hendrickson WA (2007) Insights into Hsp70 chaperone activity from a crystal structure of the yeast Hsp110 Sse1. Cell 131:106–120

Longhi S, Oglesbee M (2010) Structural disorder within the measles virus nucleoprotein and phosphoprotein. Protein Pept Lett 17(8):961–978

Louveau A, Smirnov I, Keyes T, Eccles J, Rouhani S, Peske D, Derecki N, Castle D, Mandell J, Lee K, Harris T, Kipnis J (2015) Structural and functional features of central nervous system lymphatic vessels. Nature 523(7560):337–341

Luong M, Zhang Y, Chamberlain T, Zhou T, Wright J, Dower K, Hall JP (2012) Stimulation of TLR4 by recombinant HSP70 requires structural integrity of the HSP70 protein itself. J Inflamm 9:11

Macejak D, Sarnow P (1992) Association of heat shock protein 70 with enterovirus capsid precursor P1 in infected human cells. J Virol 66:1520–1527

Madara J, Krewet J, Shah M (2005) Heat shock protein 72 expression allows permissive replication of oncolytic adenovirus dl1520 (ONYX-015) in rat glioblastoma cells. Mol Cancer 4:12

Marini A, Kozuka M, Lipsky R, Nowak T (1990) 70-kilodalton heat shock protein induction in cerebellar astrocytes and cerebellar granule cells in vitro: comparison with immunocytochemical localization after hyperthermia in vivo. J Neurochem 54:1509–1516

Mayer M (2005) Recruitment of Hsp70 chaperones: a crucial part of viral survival strategies. Rev Physiol Biochem Pharmacol 153:1–46

Meckes DG, Raab-Traub N (2011) Microvesicles and viral infection. J Virol 85:12844–12854

Moehler M, Zeidler M, Schede J, Rommelaere J, Galle P, Cornelis J, Heike M (2003) Oncolytic parvovirus H1 induces release of heat shock protein HSP72 in susceptible human tumor cells but may not affect primary immune cells. Cancer Gene Ther 10:477–480

Moore S, Kim MY, Maiolini A, Tipold A, Oglesbee M (2012) Extracellular hsp70 release in canine steroid responsive meningitis-arteritis. Vet Immunol Immunopathol 145(1–2):129–133

Morrison-Bogorad M, Zimmerman A, Pardue S (1995) Heat shock 70 messenger RNA levels in human brain: correlation with agonal fever. J Neurochem 64:235–246

Munday D, Wu W, Smith N, Fix J, Noton S, Galloux M, Touzelet O, Armstrong S, Dawson J, Aljabr W, Easton A, Rameix-Welti MA, de Oliveira A, Simabuco F, Ventura A, Hughes D, Barr J, Fearns R, Digard P, Eléouët JF, Hiscox J (2015) Interactome analysis of the human respiratory syncytial virus RNA polymerase complex identifies protein chaperones as important cofactors that promote L-protein stability and RNA synthesis. J Virol 89(2):917–930

Nagy P, Wang R, Pogany J, Hafren A, Makinen K (2011) Emerging picture of host chaperone and cyclophilin roles in RNA virus replication. Virology 411:374–382

Neumeister C, Niewiesk S (1998) Recognition of measles virus-infected cells by CD8+ T cells dependent upon the H-2 molecule. J Gen Virol 79:2583–2591

Niewiesk S, Brinckmann U, Bankamp B, Sirak S, Liebert U, ter Meulen V (1993) Susceptibility to measles virus-induced encephalitis in mice correlates with impaired antigen presentation to cytotoxic lymphocytes. J Virol 67:75–81

Noessner E, Gastpar R, Milani V, Brandl A, Hutzler P, Kuppner M, Roos M, Kremmer E, Asea A, Calderwood S, Issels R (2002) Tumor derived heat shock protein complexes are cross-presented by human dendritic cells. J Immunol 169(10):5424–5432

Nozawa N, Yamauchi Y, Ohtsuka K, Kawaguchi Y, Nishiyama Y (2004) Formation of aggresome-like structures in herpes simplex virus type 2-infected cells and a potential role in virus assembly. Exp Cell Res 299(2):486–497

Oglesbee M (2007) Nucleocapsid protein interactions with the major inducible 70 kDa heat shock protein. In: Longhi S (ed) Measles virus nucleoprotein. Nova Science Publishers, Hauppauge, pp 53–98

Oglesbee M, Krakowka S (1993) The cellular stress response induces selective intranuclear trafficking and accumulation of morbillivirus major core protein. Lab Investig 68(1):109–117

Oglesbee M, Niewiesk S (2011) Measles virus neurovirulence and host immunity. Future Virol 6(1):85–99

Oglesbee M, Tatalick L, Rice J, Krakowka S (1989) Isolation and characterization of canine distemper virus nucleocapsid variants. J Gen Virol 70:2409–2419

Oglesbee M, Ringler S, Krakowka S (1990) Interaction of canine distemper virus nucleocapsid variants with 70 kDa heat shock proteins. J Gen Virol 71:1585–1590

Oglesbee M, Kenney H, Kenney T, Krakowka S (1993) Enhanced production of morbillivirus gene-specific RNAs following induction of the cellular stress response in stable persistent infection. Virology 192:556–567

Oglesbee M, Liu Z, Kenney H, Brooks C (1996) The highly inducible member of the 70 kDa family of heat shock proteins increases canine distemper virus polymerase activity. J Gen Virol 77:2125–2135

Oglesbee M, Pratt M, Carsillo T (2002) Role for heat shock proteins in the immune response to measles virus infection. Viral Immunol 15:399–416

Pack C, Kumaraguru U, Suvas S, Rouse B (2005) Heat-shock protein 70 acts as an effective adjuvant in neonatal mice and confers protection against challenge with herpes simplex virus. Vaccine 23:3526–3534

Pardue S, Wang S, Miller M, Morrison-Bogorad M (2007) Elevated levels of inducible heat shock 70 proteins in human brain. Neurobiol Aging 28(2):314–324

Patterson CE, Daley JK, Rall GF (2002a) Neuronal survival strategies in the face of RNA viral infection. J Infect Dis 186(Suppl 2):S215–S219

Patterson CE, Lawrence DM, Echols LA, Rall G (2002b) Immune mediated protection from measles virus-induced central nervous system disease is non-cytolytic and γ interferon dependent. J Virol 76:4497–4506

Pegtel DM, Cosmopoulos K, Thorley-Lawson DA, van Eijndhoven MA, Hopmans ES, Lindenberg JL, de Gruijl TD, Wurdinger T, Middeldorp JM (2010) Functional delivery of viral miRNAs via exosomes. Proc Natl Acad Sci U S A 107:6328–6333

Pien G, Nguyen K, Malmgaard L, Satoskar A, Biron C (2002) A unique mechanism for innate cytokine promotion of T cell responses to viral infections. J Immunol 169:5827–5837

Qin H, Wilson C, Lee S, Zhao X, Benveniste E (2005) LPS induces CD40 gene expression through the activation of NF-kappaB and STAT-1alpha in macrophages and microglia. Blood 106:3114–3122

Ran R, Zhou G, Lu A, Zhang L, Tang Y, Rigby AC, Sharp FR (2004) Hsp70 mutant proteins modulate additional apoptotic pathways and improve cell survival. Cell Stress Chaperones 9:229–242

Ren X, Xue C, Kong Q, Zhang C, Bi Y, Cao Y (2012) Proteomic analysis of purified Newcastle disease virus particles. Proteome Sci 10:32

Rogers C, Fernandes-Alnemri T, Mayes L, Alnemri D, Cingolani G, Alnemri E (2017) Cleavage of DFNA5 by caspase-3 during apoptosis mediates progression to secondary necrotic/apoptotic cell death. Nat Commun 8:14128

Rozina E, Kaptsova T, Sharova O, Nikolaeva M, Nesterova T (1984) Study of mumps virus invasiveness in monkeys. Acta Virol 28(2):107–113

Rudd P, Cattaneo R, von Messling V (2006) Canine distemper virus uses both the anterograde and hematogenous pathway for neuroinvasion. J Virol 80:9361–9370

Simpson RJ, Lim JW, Moritz RL, Mathivanan S (2009) Exosomes: proteomic insights and diagnostic potential. Expert Rev Proteomics 6:267–283

Song H, Moseley P, Lowe S, Ozbun M (2010) Inducible heat shock protein 70 enhances HPV31 viral genome replication and virion production during the differentiation-dependent life cycle in human keratinocytes. Virus Res 147:113–122

Steel R, Doherty J, Buzzard K, Clemons N, Hawkins C, Anderson R (2004) Hsp72 inhibits apoptosis upstream of the mitochondria and not through interactions with Apaf-1. J Biol Chem 279(49):51490–51499

Stoorvogel W, Kleijmeer MJ, Geuze HJ, Raposo G (2002) The biogenesis and functions of exosomes. Traffic 3:321–330

Strong M, Blanchard E, Lin Z, Morris C, Baddoo M, Taylor C, Ware M, Flemington E (2016) A comprehensive next generation sequencing-based virome assessment in brain tissue suggests no major virus-tumor association. Acta Neuropathol Commun 4(1):71

Taguwa S, Maringer K, Li X, Bernal-Rubio D, Rauch J, Gestwicki J, Andino R, Fernandez-Sesma A, Frydman J (2015) Defining hsp70 subnetworks in Dengue virus replication reveals key vulnerability in flavivirus infection. Cell 163(5):1108–1123

Tanguy Le G, Boehmer P (2002) Activation of the herpes simplex virus type-1 origin binding protein (UL9) by heat shock proteins. J Biol Chem 277:5660–5666

Thery C, Zitvogel L, Amigorena S (2002) Exosomes: composition, biogenesis and function. Nat Rev Immunol 2:569–579

Tischer S, Basila M, Maecker-Kolhoff B, Immenschuh S, Oelke M, Blasczyk R, Eiz-Vesper B (2012) Heat shock protein 70/peptide complexes: potent mediators for the generation of antiviral T cells particularly with regard to low precursor frequencies. J Transl Med 9:175

Toshchakov V, Jones B, Perera P, Thomas K, Cody M, Zhang S, Williams B, Major J, Hamilton T, Fenton M, Vogel S (2002) TLR4, but not TLR2, mediates IFN-beta-induced STAT1alpha/beta-dependent gene expression in macrophages. Nat Immunol 3:392–398

Tytell M, Brown W, Moody D, Challa V (1998) Immunohistochemical assessment of constitutive and inducible heat-shock protein 70 and ubiquitin in human cerebellum and caudate nucleus. Mol Chem Neuropathol 35:97–117

Vabulas R, Ahmad-Nejad P, Ghose S, Kirschning C, Issels R, Wagner H (2002) HSP70 as endogenous stimulus of the Toll/interleukin-1 receptor signal pathway. J Biol Chem 277:15107–15112

Vasconcelos D, Cai X, Oglesbee M (1998a) Constitutive overexpression of the major inducible 70 kDa heat shock protein mediates large plaque formation by measles virus. J Gen Virol 79:2239–2908

Vasconcelos D, Norrby E, Oglesbee M (1998b) The cellular stress response increases measles virus-induced cytopathic effect. J Gen Virol 79:1769–1773

von Rüden E-L, Wolf F, Keck M, Gualtieri F, Nowakowska M, Oglesbee M, Potschka H (2018) Genetic modulation of HSPA1A accelerates kindling progression and exerts pro-convulsant effects. Neuroscience 386:108–120

Weidinger G, Czub S, Neumeister C, Harriott P, ter Meulen V, Niewiesk S (2000) Role of CD4(+) and CD8(+) T cells in the prevention of measles virus-induced encephalitis in mice. J Gen Virol 81:2707–2713

Wheeler D, Chase M, Senft A, Poynter S, Wong H, Page K (2009) Extracellular Hsp70, an endogenous DAMP, is released by virally infected airway epithelial cells and activates neutrophils via Toll-like receptor (TLR)-4. Respir Res 10:31

Wu YP, Chang CM, Hung CY, Tsai MC, Schuyler S, Wang R (2011) Japanese encephalitis virus co-opts the ER-stress response protein GRP78 for viral infectivity. Virol J 8:128

Ye J, Chen Z, Zhang B, Miao H, Zohaib A, Xu Q, Chen H, Cao S (2013) Heat shock protein 70 is associated with replicase complex of Japanese encephalitis virus and positively regulates viral genome replication. PLoS One 8:e75188

Yuyama K, Sun H, Mitsutake S, Igarashi Y (2012) Sphingolipid-modulated exosome secretion promotes clearance of amyloid beta by microglia. J Biol Chem 287:10977–10989

Zhang X, Glendening C, Linke H, Parks C, Brooks C, Udem S, Oglesbee M (2002a) Identification and characterization of a regulatory domain on the carboxyl terminus of the measles virus nucleocapsid protein. J Virol 76(17):8737–8746

Zhang Y, Huang L, Zhang J, Moskophidis D, Mivechi N (2002b) Targeted disruption of *hsf1* leads to lack of thermotolerance and defines tissue-specific regulation for stress-inducible Hsp molecular chaperones. J Cell Biochem 86(2):376–393

Zhang X, Bourhis J, Longhi S, Carsillo T, Buccellato M, Morin B, Canard B, Oglesbee M (2005) Hsp72 recognizes a P binding motif in the measles virus N protein C-terminus. Virology 337:162–174

Chapter 3
Biology of Extracellular HSP60

Brice Nativel, Cynthia Planesse, Philippe Gasque, Christine Robert Da Silva, Olivier Meihac, and Wildriss Viranaïcken

Abstract The exposure of cells or organisms to high temperature leads to the release of alert molecules such as Heat Shock Protein: the HSP. This protein family has been initially described in Drosophila. The cellular response to a heat shock involving HSP is conserved across species, from bacteria to humans and including plants. Other stresses, such as ischemia, heavy metal poisoning, nutrient deprivation, irradiations, infections, oxidative stress and inflammation, can also induce the HSP expression. HSP form a large family of proteins which are classified according to their molecular weight: HSP100, HSP90, HSP70, HSP60, HSP40, HSP from 20 to 30 kDa and HSP10. HSP60 has different functions depending on its localization. Intracellular HSP60 can be found in the cytosol, mitochondria and the chloroplast. Therein, it has a chaperone activity by assisting the proteins folding. On the cell surface or in the extracellular medium, HSP60 acts as a danger signaling molecule. Thus, stressed or damaged cells can stimulate the immune system. Indeed, this extracellular HSP60 are involved in several inflammatory pathologies.

Keywords Apoptosis · Chaperokine · Extracellular · HSP60 · Immune activation · Inflammation

B. Nativel · C. Planesse · C. R. Da Silva · O. Meihac
Université de La Réunion, INSERM, UMR 1188 Diabète athérothombose Thérapies Réunion Océan Indien (DéTROI), Saint-Denis de La Réunion, France

P. Gasque
Université de La Réunion, CNRS UMR9192, INSERM U1187, IRD UMR249, Unité Mixte Processus Infectieux en Milieu Insulaire Tropical (PIMIT), Plateforme Technologique CYROI, Sainte-Clotilde, La Reunion, France

Laboratoire de biologie, Secteur Laboratoire d'immunologie clinique et expérimentale de la zone de l'océan indien (LICE-OI) CHU La Réunion site Félix Guyon, Saint-Denis, La Reunion, France

W. Viranaïcken (✉)
Université de La Réunion, CNRS UMR9192, INSERM U1187, IRD UMR249, Unité Mixte Processus Infectieux en Milieu Insulaire Tropical (PIMIT), Plateforme Technologique CYROI, Sainte-Clotilde, La Reunion, France
e-mail: wildriss.viranaicken@univ-reunion.fr

A. A. A. Asea, P. Kaur (eds.), *Chaperokine Activity of Heat Shock Proteins*, Heat Shock Proteins 16, https://doi.org/10.1007/978-3-030-02254-9_3

Abbreviations

APC	Antigen presenting cells
CCT	"chaperon" containing the T-complex of polypeptide 1
Cpn60	Chaperonin 60 kDa
DAMP	Damage associated molecular pattern
DC	Dendritic cells
HSP	Heat shock protein
LPS	Lipopolysaccharide
LTγδ	Gamma delta T lymphocytes
mtHSP60	Mitochondrial form
PAMP	Pathogen associated molecular pattern
RBP	Rubisco binding protein
sHSP60	Soluble HSP60
Th	Helper T lymphocytes
Treg	Regulatory T lymphocytes
TriC	TCP-1 Complex ring

3.1 Introduction

HSP60 has been discovered in the mitochondria of mammals' cells for the first time in 1989. The researchers highlighted the implication of HSP60 in protein's folding inside the mitochondria (Cheng et al. 1989). HSP60 is also called chaperone in 60 kDa (Cpn60), or GroEL in bacteria (Ellis and van der Vies 1991). In the first set of experiments, GroEL was used (Viitanen et al. 1992). They showed that the presence of yet another HSP, called GroES, was necessary for GroEL function (Goloubinoff et al. 1989, 1997). GroES is considered like a co-chaperone protein. Their structures are very close (Hunt et al. 1996). Later, the co-chaperone was discovered in human, called HSP10 pt Cpn10. The genes sequences encoding for GroEL/GroES and HSP60/HSP10 are very similar. This similarity was also notable in their structures and functions in the mitochondria (Nielsen and Cowan 1998; Nielsen et al. 1999). HSP60 generally refers to the mitochondrial form (mtHSP60), which is the most abundant form in eukaryotic cells. The cytosolic form is called "chaperone" containing the T-complex of polypeptide 1 (CCT). There are many CCT homologues in different species. The CCT proteins constitute a hetero-oligomer called TCP-1 complex ring (TRiC). TRiC is involved in the synthesis of actin and tubulin (Llorca et al. 2000; Yaffe et al. 1992) and the folding of many other proteins (Dunn et al. 2001; Yam et al. 2008). None of the CCTs is known to be induced by thermal shocks. Drosophila has four CCT homologues, while Human has nine different forms. During the evolution of an organism, TRiC chaperone properties evolve. This facilitates the folding of proteins more and more complex in eukaryotic cells. This phenomenon can be explained by the gene duplication and gene derivation during the evolution.

Thus, there are different types of HSP60 subdivided in two groups of chaperone proteins. The first group includes GroEL and mtHSP60. A vegetal HSP60 known as RBP (for Rubisco binding protein) and localized in chloroplast is part this first group (Apuya et al. 2001; Balczun et al. 2006). A second group includes the cytosolic HSP60 (CCTs) of eukaryotic cells. The HSP60 family is mainly described as chaperone proteins, assisting the proteins folding. It also contributes to the unfolding and refolding of the misfielded proteins. Member of the HSP60 family are also involved in many other intracellular processes, such as the DNA replication, mitochondrial DNA transmission (Kaufman et al. 2003) and intracellular trafficking of proteins (Czarnecka et al. 2006; Deocaris et al. 2006) and hormones (Sigal et al. 2001). HSP60 is essential for protein importation into mitochondria, which also requires the participation of co-chaperones HSP10 and HSP70 (Gupta et al. 2008). It is interesting to note that HSP60 KO mice are not viable and do not survive after 7 days of gestation (Christensen et al. 2010). Heterozygous mice are viable but show mitochondrial function deficits leading to brain disorders (Magnoni et al. 2013). These animal models show the importance of HSP60 in the organism. On the cells surface or in the extracellular medium (Merendino et al. 2010), HSP60 acts as a danger signaling. Thus, stressed or damaged cells can stimulate the immune system through HSP60 (Pockley and Multhoff 2008; Pockley et al. 2008). After a brief presentation of cellular function associated with cellular HSP60, this chapter will focus on the "chaperokine" activity of extracellular HSP60.

3.2 Cellular HSP60 Activity

3.2.1 Structure, Modifications and Chaperone Activity

HSP60 is a protein encoded by the HSPD1 gene, which is largely conserved in numerous living organisms such as humans, monkeys, fishes, insects, plants, and bacteria. The same is true for HSP10 encoded by the HSPE1 gene. These genes are localized side by side in the genome (Hansen et al. 2003). They are transcribed and the mRNA are then translocated into the cytosol. The resulting HSP60 protein has 573 amino acids. It has a signal sequence of 26 amino acids in the N-terminal position for mitochondrial localization, allowing its transport into the mitochondria (Singh et al. 1990). Once synthesized, it stays in a transit state from the cytoplasm into the mitochondria and already has its chaperone activity. Once in mitochondria, the signal sequence is cleaved and serial post-translational modifications can occur. These modifications such as phosphorylation (Kotlo et al. 2014), ubiquitination (Tang et al. 2013), sumoylation (Leach et al. 2011), acetylation (Choudhary et al. 2009), malonyation (Peng et al. 2011), glycosylation (Hayoun et al. 2012), will ultimately modulate the function of HSP60.

The functional organization of HSP60 has been initially described in bacteria. This chaperone protein is constituted of 14 identical subunits of GroEL divided into two rings of 7 superimposed subunits. The co-chaperone is also composed of seven

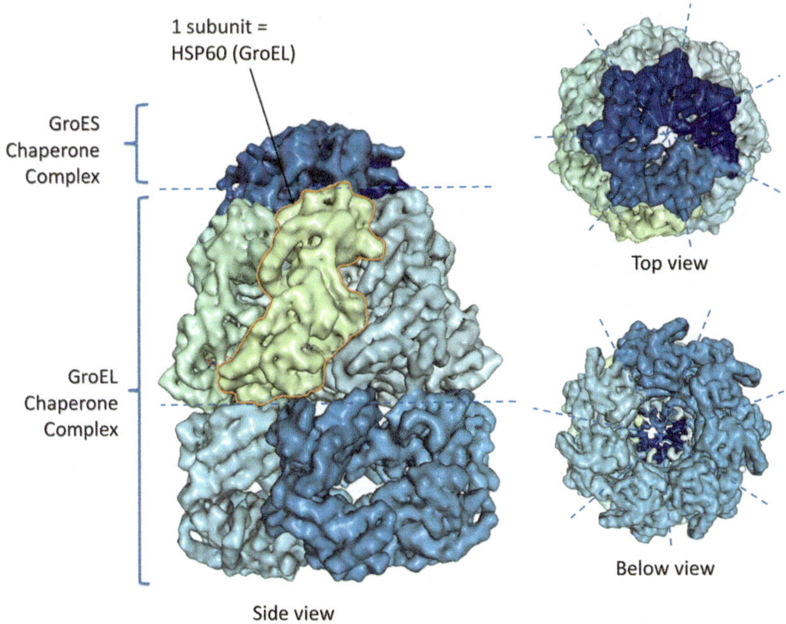

Fig. 3.1 3D structure of chaperone complex HSP60/HSP10 (GroEL/GroES). RCSB PDB3D view 1AON

identical GroES subunits (Fig. 3.1). GroEL contains many highly conserved regions and residues, mainly involved in the binding of substrate (Cheng et al. 1989; Fenton et al. 1994), ATP/ADP (Roseman et al. 1996; Sigler et al. 1998), Mg^{2+}/K^{+} (Fenton et al. 1994) and GroES (Murai et al. 1996). In the presence of ATP and Mg^{2+}, GroEL and GroES quickly constitute a chaperone complex GroEL/ATP/GroES. This complex can also be used to recover an additional co-chaperone in particular conditions of concentration of Mg^{2+}/K^{+} and ATP/ADP ratio (Llorca et al. 1996). The whole forms a barrel (GroEL) with a lid (GroES) at either both or at only one end of the complex (Fig. 3.1). In humans, the HSP60 chaperone exists in the form where seven identical subunits form a ring (Jindal et al. 1989).

The chaperone activity of the HSP60 family (mtHSP60, GroEL, and TRiC) is mainly to participate to the folding of newly formed proteins. The HSP60s contain a chamber in which proteins are folded in their tertiary structure. This process is ATP dependent. The synthesis of the HSP is induced following a high temperature exposure. Some proteins are sensitive to high temperatures, these proteins fail to fold and can form aggregates. HSP60, in addition to its ability to assist the protein folding, also has the ability to unfold them. It has the ability to recognize misfielded proteins and even protein aggregates. Thus, HSP60 unfold these proteins to allow the correct refolding (Lin et al. 2008; Priya et al. 2013). This mechanism protects the body against an abnormal and deleterious accumulation of non-functional proteins and protein aggregates, which are potentially toxic. It therefore promotes increased resistance to high temperatures (Sigler et al. 1998). This activity requires energy from ATP hydrolysis and the presence of their co-chaperone GroES (Mendoza et al. 1991). In the cytosol of eukaryotic cells, the structure of TRiC chaperone complex is slightly different. In fact, it has eight CCT different subunits (Cong et al. 2010). The first interpretations revealed that TRiC acts without co-chaperone (Frydman et al. 1992). Recently, it has been shown that HSP60 can be found in the cytosol, independently of CCTs.

3.2.2 HSP60 in Apoptosis

HSP60 plays a role in apoptosis by interacting with pro-caspase-3. Depending on its origin, HSP60 may have a pro- or anti-apoptotic action (Fig. 3.2) (Chandra et al. 2007). If it has mitochondrial origin and is released into the cytosol following a stress or loss of transmembrane potential, HSP60 activates pro-caspase-3 and acts as a pro-apoptotic factor. HSP60 forms a pre-apoptotic complex with pro-caspase-3 and HSP10. The complex dissociation is followed by caspase-3 activation and apoptosis In contrast, if the cytosolic form of HSP60 is directly neo-produced under a stress stimulation, it inhibits pro-caspase-3 and has an anti-apoptotic effect (Chandra et al. 2007). Co-chaperone HSP10 is also involved in this inhibition of apoptosis (Lin et al. 2001; Samali et al. 1999).

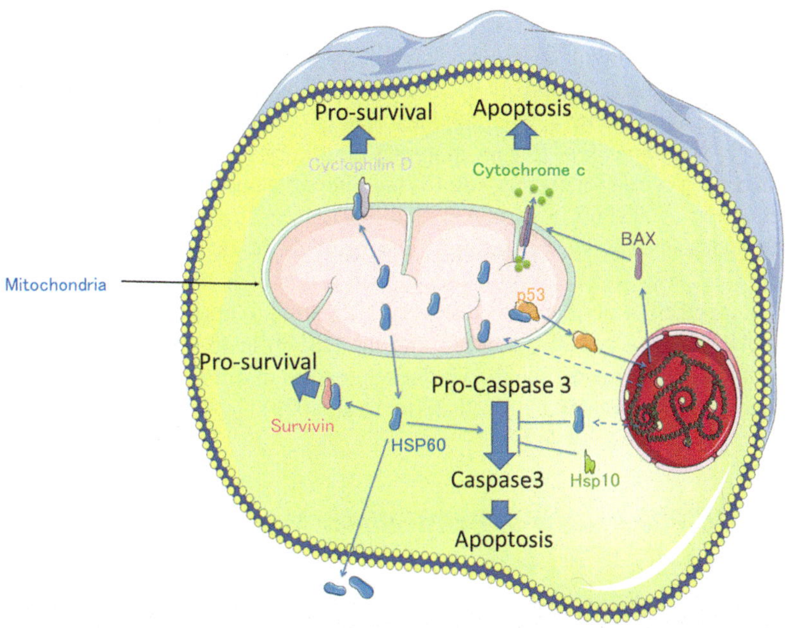

Fig. 3.2 HSP60 function in apoptosis

Other proteins are also involved in the apoptosis pathways. The mitochondrial form of HSP60 interacts with p53, which is another protein involved in apoptosis (Benjamin et al. 2008; Green 2006). HSP60 bound to p53 acts as an inhibitor of apoptosis (Ghosh et al. 2008). p53 is a transcription factor, it activates the expression of the Bax gene. The Bax protein participates in the process of permeabilization of the mitochondrial membrane. This results to the burst of the mitochondria and to the release of cytochrome c and HSP60. The release of cytochrome c from the mitochondria to the cytosol is a central event in the intrinsic apoptotic process. Thus, the separation of HSP60 and p53 leads to the double action of induction of apoptosis: via action p53 and Bax and via the action of HSP60 and pro-caspase-3. In addition, HSP60 also directly interacts and inhibits Bax functions (Gupta and Knowlton 2002). A post-translational modification mechanism of O-linked N-acetylglucosamine type would be involved in this interaction. A high level of O-linked N-acetylglucosamine HSP60 would impede its binding to Bax and leads to the activation of apoptosis (Kim et al. 2006).

Survivin (pro-survival) is another protein involved in apoptosis and identified as interacting with HSP60. This protein participates in the regulation of the cell cycle and apoptosis. HSP60 participates to the stabilization of this protein (Ghosh et al. 2008). A decrease in the expression of HSP60 results in a decrease in survivin level and which will cause the cell apoptosis. A third protein, cyclophilin D is also involved in the regulation of apoptosis. This protein composes the pores of mitochondria membrane (Ghosh et al. 2010). Only described in cancer, a decrease in overexpression of HSP60 results in destabilization of mitochondria membrane permeability by cyclophilin D. This leads to cell apoptosis. Because of its ambivalent involvement in apoptosis, the role of HSP60 is also described like ambiguous in cancers. HSP60 is involved in some cancers by playing either a pro-apoptotic role or a pro-survival role (Cappello et al. 2008; Chaiwatanasirikul and Sala 2011; Di Felice et al. 2005; Ghosh et al. 2008; Pace et al. 2013). In a large number of cancers, HSP60 is over expressed. It is its apoptosis inhibitory function which enhances the pro-survival process of cancer cells. This phenomenon involves particularly the direct action of HSP60 in surviving stabilization, p53 repression, Bax inhibition, mitochondria membrane permeability via cyclophilin D, and HSP60 interaction with pro-caspase-3. However, the mechanism action of HSP60 on pro-caspase-3 is still not precisely described. It is also very interesting to note that there is a correlation between the grade of the tumor and the HSP60 level (Cappello et al. 2005). Exogenous HSP60 are also able to induce apoptosis in different cell type including cardiomyocyte and endothelial cells (Kim et al. 2009).

3.2.3 HSP60 Release in Extracellular Medium

HSP60 can be found on the surface of cells and also in the extracellular medium. In both cases, HSP60 acts mainly as a stimulator of the immune system, of the innate and adaptive immune responses. It is interesting to note that there is only one gene encoding for HSP60, only one mRNA and only one translated protein. However, HSP60 can be found in cytosol, mitochondria and outside the cell. It is supposed that localization of HSP60 molecule is only determined during the post-translational stages of the protein. There are various secretion pathways of soluble HSP60 (sHSP60) (Fig. 3.3). It can be an active secretion, when it is released by a cellular mechanism or a passive secretion, when it is released during apoptosis or necrosis. The activity of HSP60 will be different depending on its origin and its secretion pathway. When HSP60 is derived from damaged cells (self), it is called a Damage associated molecular pattern (DAMP) or alarmins and known to activate the immune system through patter-recognition receptors (PRRs).

One of the mechanisms involved in HSP60 active secretion is the classical pathway of proteins secretion passing by the endoplasmic reticulum and the Golgi apparatus (Arya et al. 2015; Hayoun et al. 2012). Thus, the protein is neosynthetised and directly secreted. Though this mechanism, HSP60 undergoes a post-translational N-glycan modification (Hayoun et al. 2012). However, the initial signal which determines this secretion pathway remains to be described. A second active secretion mechanism has been described; it involved exosomes and lipid channels (Chen et al. 2006; Gupta and Knowlton 2007). HSP60 is anchored to exosomes. Thus, the presence of HSP60 on the cell surface may be a necessary condition for release via

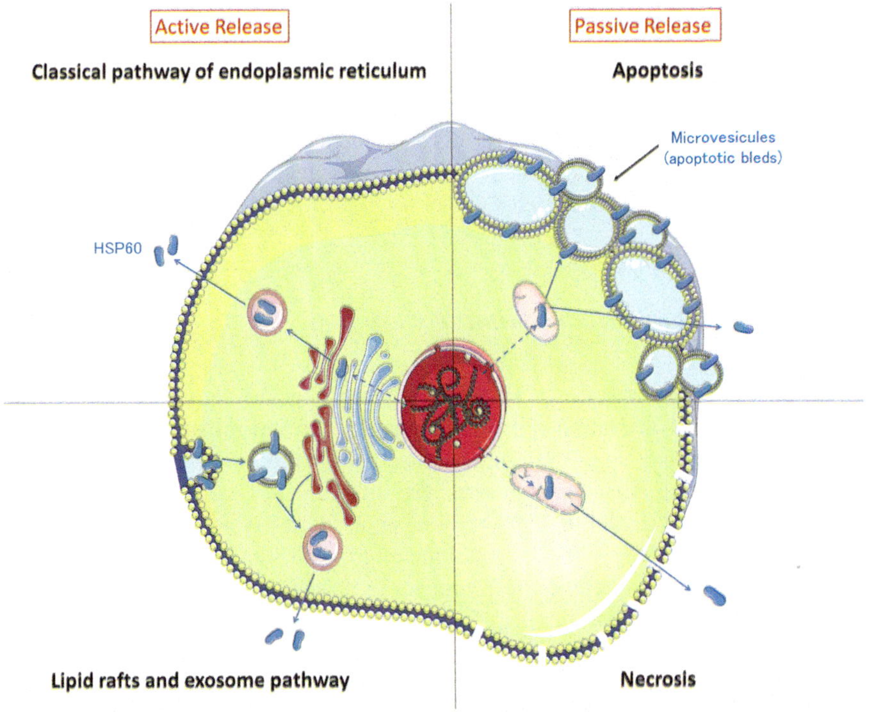

Fig. 3.3 The release pathways of HSP60

the active secretion pathway (Campanella et al. 2012). The HSP60 do not contain a transmembrane domain, so HSP60 should be loaded on the cell membrane through anchored proteins. In addition to these active secretion mechanisms, HSP60 can be released passively during necrosis (Lehnardt et al. 2008). The extracellular release of HSP60 is more important during necrosis than apoptosis (Lehnardt et al. 2008). HSP have a pro-inflammatory effect when released during necrosis but not during apoptosis (Basu et al. 2000). However, the first experiments showed that there is no different effect on neurons when HSP60 is released by necrosis or apoptosis (Lehnardt et al. 2008). Additional studies must be done to define the inflammatory effect of HSP60 released by apoptosis and necrosis.

During apoptosis HSP60 is release in extracellular media trough it loading on apoptotic blebs. This presence of HSP60 enhances the clearance of apoptotic bodies or bodies by phagocytic cells like monocyte (Goh et al. 2011). HSP60 released may end up in the bloodstream and may be measured in plasma of patients (Lewthwaite et al. 2002; Pockley et al. 1999; Shamaei-Tousi et al. 2007a; Xu 2002). The secretion profile of HSP60 is very different depending to individuals (Lewthwaite et al. 2002; Xu et al. 2000; Zhang et al. 2008), and cell types (Bassan et al. 1998). Many factors can influence the HSP60 secretion, such as age for example. Indeed, it has been shown that older is a person, lower is his HSP60 level in the bloodstream (Rea et al. 2001; Terry et al. 2004). An increase of HSP60 level in serum is associated with numerous pathologies, particularly those associated to tissue necrosis. This is the case for example during infarcts which lead to tissue ischemia and then necrosis (Schett et al. 1999). This will lead to a passive release of HSP60 due to necrosis. Besides the release of HSP60 from host cells (i.e. endogenous HSP60), circulating HSP60 can be releasing from pathogen or commensal flora mainly by bacteria. This is an exogenous source of circulating HSP60 in the bloodstream and has a particular importance for HSP 60 dependent pathologies (see below the role of soluble HSP60 in chronic inflammatory diseases).

3.2.4 HSP60 in Innate Immune Response

The main effect of extracellular HSP60 is its ability to activate the innate immune system. It results in activation of pro-inflammatory and anti-inflammatory pathways in different cell types (Fig. 3.4). These cell activation processes are mainly characterized by the induction of numerous pro-inflammatory factors (TNF-α, IL-6, CCL2, CCL5, MIP-2, NO, IL-1β ...) (Flohé et al. 2003; Habich and Burkart 2007; Pfister et al. 2005) (Fig. 3.5). This cytokine effect of HSP60 compared to its chaperone function does not require ATP, cofactor or protein complexes. HSP60 has the capacity to activate a large number of immune cells. For example, HSP60 activates dendritic cells (DC) (Flohé et al. 2003; Habich and Burkart 2007; Pfister et al. 2005) by increasing expression of MHC II and co-stimulatory molecules (CD40, CD54 and CD86) (Flohé et al. 2003; Habich and Burkart 2007). HSP60 can also activate the granulocytes promoting their degranulation and their antimicrobial activity.

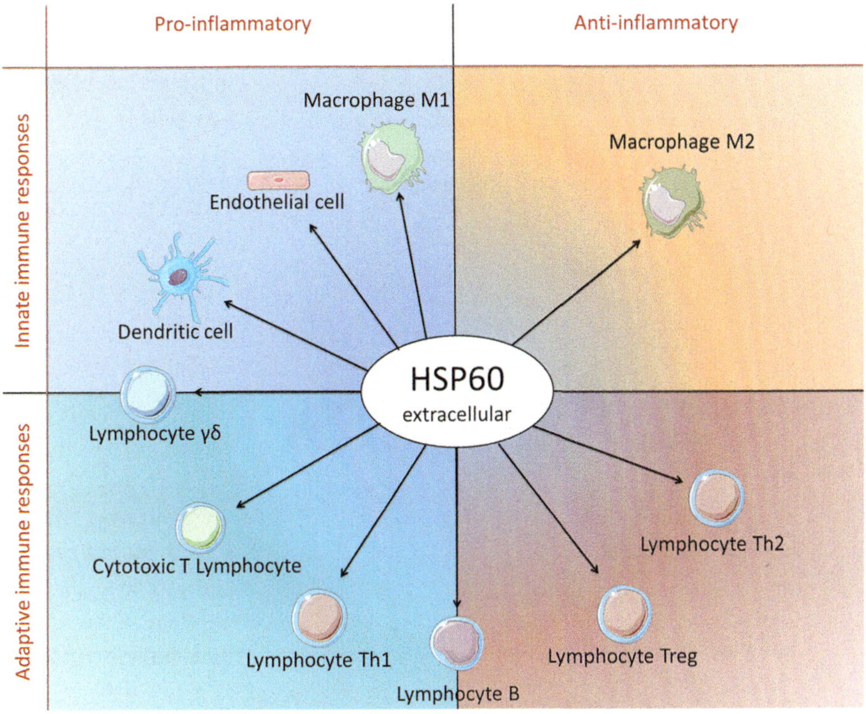

Fig. 3.4 HSP60 induces different immune response

It activates monocytes/macrophages by inducing a pro-inflammatory signaling cascade and an increase in phagocytosis (Goh et al. 2011; Kol et al. 1999; Pfister et al. 2005).

In the extracellular medium, sHSP60 is known to interact with several PRRs such as CD14, Toll like receptors (TLR2 and TLR4) as well as CD40 (Binder et al. 2004; Ohashi et al. 2000; Vabulas et al. 2001) leading to the activation of intracellular signaling pathways such as JNK-1/2, MAPK, ERK-1/2, NF-κB (Habich and Burkart 2007; Märker et al. 2014; Vabulas et al. 2001) and MyD88 (Rosenberger et al. 2015). These signaling cascades will result in the production of pro-inflammatory cytokines such as IL-6, IL-12, IL-15, IL-1β, CCL2 and TNF-α. In addition, HSP60 can act in a positive retroactive loop, inducing its own production. It has been shown that expression of sHSP60 is increased by pro-inflammatory cytokines such as IL-1β, IL-6, and TNF-α (Bajramović et al. 2000). The expression of HSP60, like all proteins in the HSP family, is under the control of the transcription factor NF-κB (Cappello et al. 2011; Wang et al. 2010). The role of HSP60 in cellular regeneration was very recently been described (Pei et al. 2016). HSP60 released to a precise site would act as chemokine by attracting immune cells. It would participate to the dampening of inflammation. Thus, sHSP60 has an anti-inflammatory function on macrophages with an M2 type (anti-inflammatory) polarization. These

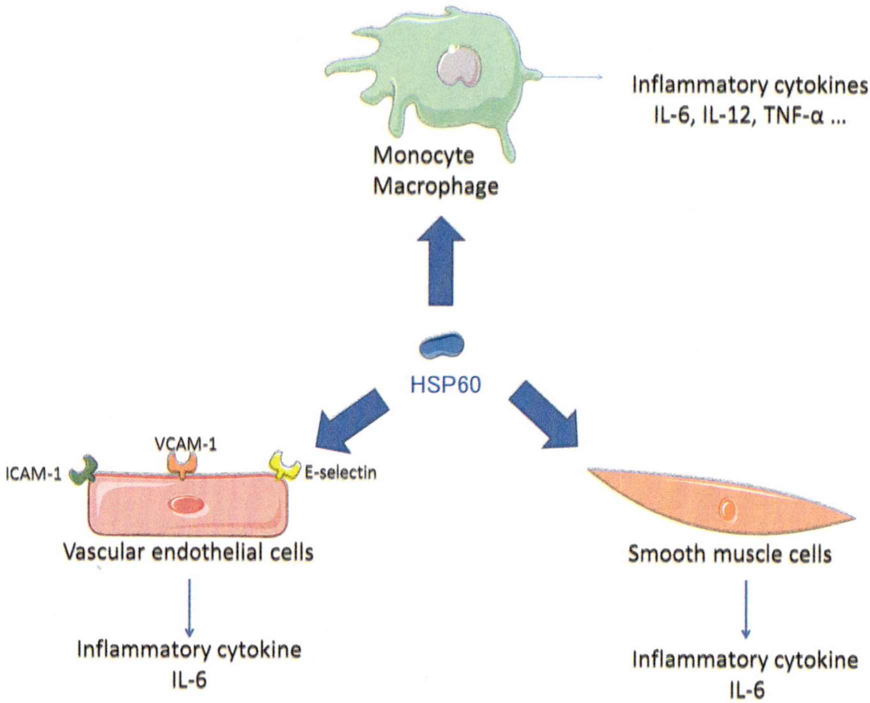

Fig. 3.5 Inflammatory cytokine induced by HSP60 in different cell types

M2 macrophages are characterized by higher phagocytic capacity compare to M1. HSP60 can enhance this capacity and therefore contribute to the anti-inflammatory phenotype and tissue repair (Fig. 3.4).

3.2.5 *Action of HSP60 in Crosstalk Between Innate and Adaptive Immunity*

Antigen presenting cells (APC) which is compose mainly of monocytes, macrophages and dendritic cells present antigen to lymphocytes. These cells are capable to induce a signaling cascade of anti-inflammatory or pro-inflammatory type (Fig. 3.4):

- HSP60 plays a role in the activation and proliferation of regulatory T lymphocytes (Treg). These cells are involved in inhibition of the immune system and are often present during remission following cellular stress. In the same time, an inhibition of cytotoxic T lymphocytes is described (Aalberse et al. 2011; de Kleer et al. 2010).

- HSP60 acts on gamma delta T lymphocytes (LTγδ) regulation. These unconventional lymphocytes straddle innate immunity and adaptive immunity. They are the first activated immune cells against a viral or bacterial infection and induce a pro-inflammatory response. HSP60 could induce apoptosis of LTγδ (Atre et al. 2006).
- HSP60 interacts with helper T lymphocytes (Th). These immune cells help and regulate other lymphocyte functions. It has been shown that during an immune response increasing the expression and active secretion of HSP60, naive T cells are converted into helper T lymphocytes (Ohue et al. 2011). In addition, HSP60 induces a cytokines secretion profile of IL-4 and IL-10 by DC and Treg. Thus, the helper T lymphocytes are switched to a Th2 phenotype (van Roon et al. 1997; Wu 1995), with an anti-inflammatory activity. It has been demonstrated that this HSP60 action occurs via the TLR2 receptor.
- HSP60 is also capable of activating B lymphocytes. This leads to the proliferation of B lymphocytes, the induction of co-stimulatory molecules (CD40, MHC II) and the production of Th2-type cytokine (IL-10). In addition, a decrease of B lymphocytes apoptosis was observed after stimulation by HSP60 (Cohen-Sfady et al. 2009), consistent with the increase in proliferation. Thus, HSP60 positions itself between pro- and anti-inflammatory effects (Fig. 3.4). But, it is interesting to note that the stimulation of B lymphocytes by GroEL, or non-self HSP60, does not lead to B lymphocytes proliferation (Pfister et al. 2005).

These functions on lymphocyte were not mediated directly by HSP60 but trough the binding of sHSP60 to APC. sHSP60 induce maturation of APC cells by increasing surface expression of co-stimulatory receptor or MHC which contribute to presentation of HSP60 peptide or peptide derived from factor associated to sHSP60 to lymphocyte and activation and maturation of lymphocyte (Fig. 3.6). The binding of sHSP60 to APC induce production and secretion of cytokines by APC and this pool of molecules regulate also lymphocyte activation, maturation and proliferation (Fig. 3.6). Finally, sHSP60 mediated the crosstalk between APC, an actor of innate immune response to the lymphocytes, the main actors of adaptive immunity (Fig. 3.6).

3.2.6 Receptor for sHSP60

In central nervous system HSP60 can bind to microglial cells through TREM2 (Triggering Receptor expressed in myeloid); The HSP60 binding to TREM2 increase phagocytic capacity of microglial cells (Stefano et al. 2009). The member of scavenger receptor, LOX1 (lectin-like oxidized low-lipoprotein receptor 1), on microglial cells bind to soluble HSP60 release from dying neurons and increase production of ROS, NO and IL-1β. In this last case HSP60 link neurotoxicity and neuro-inflammation (Zhang et al. 2012). C-type lectin receptor, CD91 and Dectin-1, have been describe to be potent receptor for sHSP60, using blocking antibodies

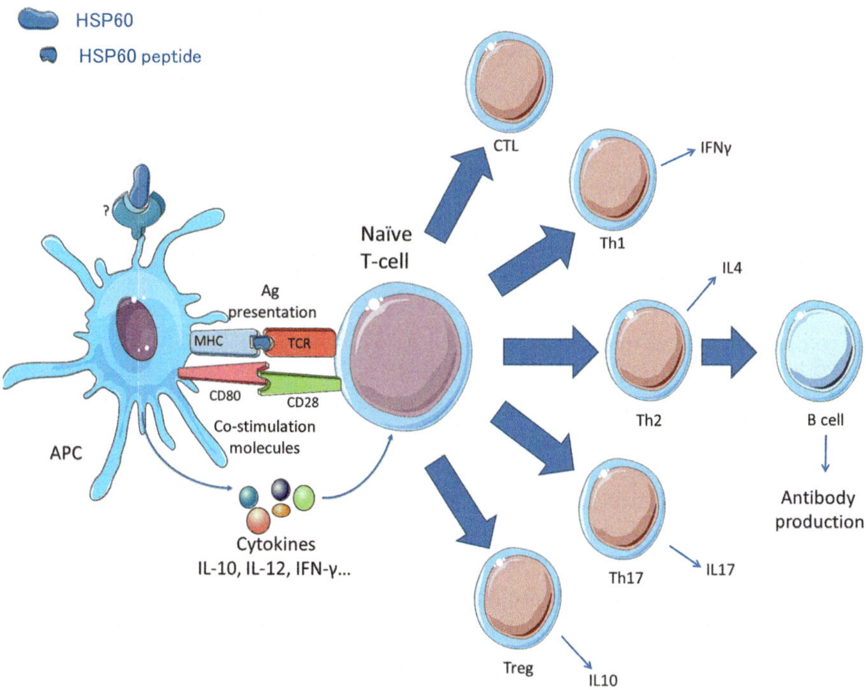

Fig. 3.6 Action of HSP60 on APC maturation and T-cell activation

against CD91, saturation with ligand for CD91 or overexpression of Dectin-1, these interaction appear to be non-specific or weak (Habich and Burkart 2007; Habich et al. 2002; Xie et al. 2010). Extracellular Hsp60 has been reported to bind to several cell surface receptors. The TLR2, TLR4, CD14 and CD40, as describe before can bind to HSP60 but all of them bind to LPS. The pro-inflammatory effect of HSP60 appeared to be related to the effect of LPS or others factors associated to HSP60 (Planesse et al. 2015). Blocking antibodies against TLR4 and CD14 are not able to prevent binding of HSP60 to macrophages (Habich and Burkart 2007). HSP60 binding to APC trough TLR4, CD14, TLR2, CD40 remains controversial.

3.2.7 HSP60 and LPS

Lipopolysaccharide (LPS) are found at the cell surface of Gram$^-$ bacteria like *E. Coli*. LPS are part of pathogen associated molecular pattern (PAMP). HSP60 interact with LPS trough the region located between amino acids 351 and 370 (Habich et al. 2005). However, HSP60 can induce immune stimulation by different mechanisms in a dependent or independent manner with LPS. The cytokine effect of HSP60 is similar to the effect of LPS and bacterial lipoproteins. It is important to note that bacterial and human recombinant HSP60 proteins are produced in

Escherichia coli. Hence, the final HSP60 preparation can be contaminated with bacterial products such as LPS or peptidoglycans. It is interesting to stress that the HSP60 and the LPS receptors are the same: TLR4 (and its co-receptor CD14) and TLR2 in some cases. It was highlighted that the pro-inflammatory properties of HSP60 could be solely due to the effect of LPS (Tsan and Gao 2009). The question whether LPS contaminants or other bacteria-associated molecules could be entirely responsible for the innate immune functions of HSP recombinant proteins remains critical (Pfister et al. 2005). Different techniques were used to validate the absence of LPS in the preparations of HSP60. The first was based on LPS temperature resistance. The LPS is more resistant to high temperatures than HSP60. The researchers also used polymyxin B for its LPS inhibitor activity (Galdiero et al. 1997; Pfister et al. 2005; Retzlaff et al. 1996). Some rare studies used blocking antibodies anti-HSP60 (Lund et al. 2003; Pfister et al. 2005). Following these treatments, the researchers reported that their HSP60 preparations were LPS free. These criteria for the absence of LPS were refuted by the fact that LPS and other contaminants can also be degraded and lose their cytokine activity at high temperatures. So inactivation of HSP60 by the temperature also inactivates the LPS. In addition, the LPS contamination by other LPS-associated molecules may also be involved in this inflammatory response. These contaminants can also be inactivated by the temperature (Gao and Tsan 2003a). Moreover, if other contaminants are present in the preparations, the use of polymyxin B will have no effect or partial effect on the decrease of the pro-inflammatory property imputed to HSP60 preparation (Gao and Tsan 2003a; Pfister et al. 2005; Retzlaff et al. 1996).

Another aspect reinforcing the concerns that LPS may be responsible for the HSP 60 activities is the concentrations used by the researchers in their studies. LPS is very reactive even at very low doses, few hundred pg.mL^{-1} are sufficient to induce an inflammatory response by monocytes and macrophages. Thus, low LPS contamination could be responsible for the alleged HSP60-mediated pro-inflammatory activity. This situation has been verified for HSP70 (Gao and Tsan 2003b). It is interesting to note that HSP60 proteins, without any LPS and contaminants, keep their chaperone activity and ATPase activity (Gao and Tsan 2003a). Therefore, the absence of the cytokine activity is not due to a non-functional protein or a purification problem. In contrast to monocytes/macrophages, activation of T lymphocytes is possible in the presence of HSP60 contaminated or not by endotoxins (Osterloh et al. 2004). Moreover, the activation of T lymphocytes by HSP60 does not pass through the TLR4 receptor and MyD88 (Osterloh et al. 2008). HSP60 activation of T lymphocytes produces IFN-α, while this cytokine is not produced when HSP60 is contaminated with LPS (Fig. 3.7).

The action of LPS to boost the HSP60 cytokine function must be better emphasized in the context of immune system activation. A bacterial infection will result in a release of HSP60. This HSP60 is capable to interact with the bacteria LPS. This phenomenon would in essence facilitate the microbes' recognition by lowering the detection threshold. In addition, HSP60 has a synergistic effect on LPS. Neither highly purified HSP60 preparations nor LPS alone can induce a pro-inflammatory effect comparable to contaminated HSP60 preparations (Tsan and Gao 2004, 2009).

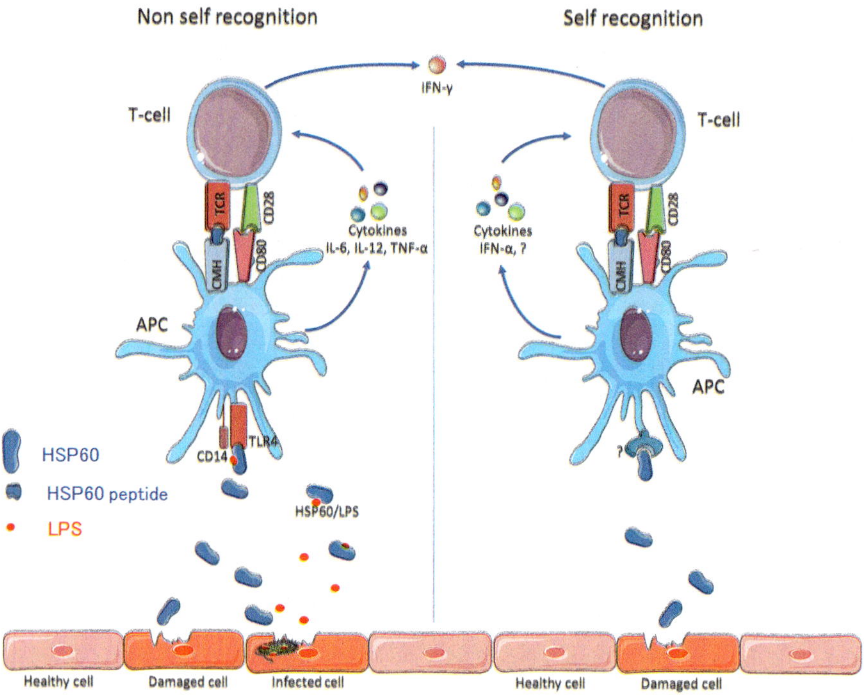

Fig. 3.7 Self and non self HSP60 recognition by APC

Even when LPS is used at concentrations corresponding to those found in contaminated preparations of HSP60. HSP10 would be able to inhibit the inflammatory response of LPS, through interaction with HSP60 (Pfister et al. 2005).

3.3 Role of sHSP60 in Several Chronic Inflammatory Diseases

HSP60 is involved in many chronic inflammatory diseases such as periodontitis, diabetes, cardiovascular disease and atherosclerosis. HSP60 would act as a stimulator of inflammation.

3.3.1 Atherosclerosis

Atherosclerosis is a chronic inflammatory disease of the blood vessel wall. Innate and adaptive immune systems are involved in the development of this pathology. A great involvement of macrophages, T and B lymphocytes has been described in the development of atherosclerosis. In the early stages of atheroma plaque formation, endothelial cells may be stressed by exogenous or endogenous HSP60 molecules. This stress can thus come from the HSP60 present in the bloodstream. The presence of HSP60 has been described on the surface of stressed endothelial cells (Pfister et al. 2005). Indeed, stressed endothelial cells induce the expression on their surfaces of the HSP60 protein and also other adhesion molecules such as, for example, ICAM and VCAM leading to the binding of immune cell to the vasculature (Pfister et al. 2005). In addition, HSP60 can induces an inflammatory process leading to the production of inflammatory cytokines, and hence aiding the recruitment of immune cells (Pfister et al. 2005). HSP60 would ultimately assist in the formation and activation of foam macrophage cells (Grundtman and Wick 2011).

One of the theories for the development of atherosclerosis is the implantation of an autoimmune type of inflammation. HSP60 was also thought to contribute to this autoimmunity (Grundtman and Wick 2011; Wick et al. 1992). This concept is based on the fact that healthy individuals possess adaptive protective immunity against HSP60 which was initially derived from bacteria or modified HSP60 of the self (Grundtman and Wick 2011; Wick et al. 1992) (Fig. 3.8). The great similarity of structure between human and bacterial HSP60 (GroEL), protective immunity against GroEL generate antibody that react with human HSP60 (Ranford et al.

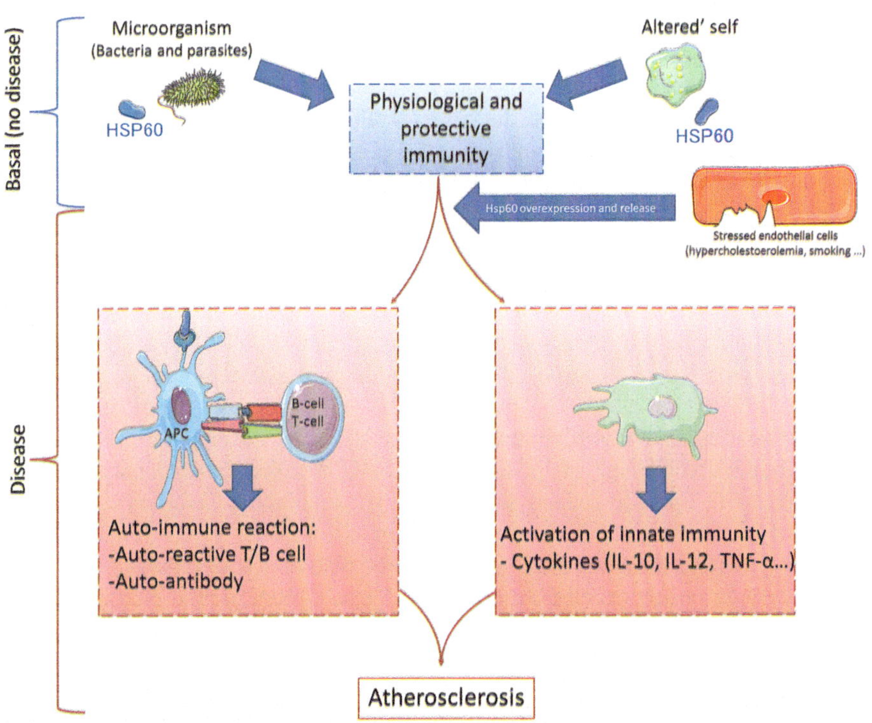

Fig. 3.8 Function of HSP60 in the installation of atherosclerosis disease

2000). Upon endothelial stress like smoking, hypercholesterolemia, endothelial cells may overexpress HSP60 at their cell surface and could also release HSP60. sHSP60 could activate auto-reactive T and B cell, auto-antibody production and also innate immune activation with increased secretion of cytokines. Antibodies against HSP60 will induce increased recognition of cell surface HSP60 and cytokine may provide a pro-inflammatory context thereby promoting atheroma plaque development (Fig. 3.8).

A strong risk factor in the development of atherosclerosis is the involvement of periodontitis. Periodontitis are caused by bacteria present in the gut flora. People with periodontitis have high levels of HSP60 in the blood and this pool of HSP60 can elicit the HSP60 inflammatory model of atherosclerosis (Fig. 3.8) (Shamaei-Tousi et al. 2007b). It is interesting to note that the HSP10 level in these patients is low, and that after treatment of this periodontitis an elevation of HSP10 is observed (Shamaei-Tousi et al. 2007b).

3.3.2 *Diabetes and Obesity*

HSP60 is known to be involved in inflammation related to adipose tissue. HSP60 stimulates adipocytes and induces the production of inflammatory cytokines (Gülden et al. 2008, 2009; Märker et al. 2010, 2014). Of note, in obese individuals, where the adipocytes are more numerous and self-inflamed (Trayhurn and Wood 2005; Vachharajani and Granger 2009), a higher concentration of HSP60 is observed in the bloodstream. Furthermore, there is a correlation between the body mass index and the HSP60 level present in plasma (Märker et al. 2012). Diabetes is also linked to a high level of HSP60 present in plasma for both type 1 diabetes (Devaraj et al. 2009; Juwono and Martinus 2016; Shamaei-Tousi et al. 2006), and to a less extent for type 2 diabetes (Dasu et al. 2010; Juwono and Martinus 2016; Shamaei-Tousi et al. 2006). This is observed in cases of diabetes not associated with obesity, but obesity will increase the rate of sHSP60. HSP60 is also related to insulin resistance (Märker et al. 2012). Hyperglycaemia causes an increase in HSP60 and also apoptosis of β cells (Kim et al. 2006). The action of autoimmune T lymphocytes associated with HSP60 has also been described in the cases of destruction of pancreatic β cells (Birk et al. 1996), which again would lead to the development of diabetes. Here too, the involvement of T lymphocytes must be taken into account according to its Th1 or Th2 profile. Indeed, studies have shown that the modulation of T lymphocytes by HSP60 and leading to a Th2 profile induces protection of β cells (Bockova et al. 1997; Elias et al. 1997). HSP60 also bound to HSP60 receptor on adipocyte and induce secretion of inflammatory mediators (IL-6, CXCL-1, CCL2) (Gülden et al. 2008). These mediators can contribute to adipose tissue inflammation and insulin resistance (Märker et al. 2014; Nativel et al. 2013).

3.3.3 Arthritis-Related Diseases

It has been described that HSP60 is involved in arthritis-related diseases. There are several models of arthritis study in rodents in which the symptoms resemble that of human rheumatoid arthritis. In these models, an adjuvant often associated with constituents of bacteria (Chang et al. 1980; Cromartie et al. 1977; Pearson 1956) is injected. The induction of this disease is due to the action of bacterial HSP60 on the activation of T lymphocytes (van Eden et al. 1988; Hogervorst et al. 1991; Tsan and Gao 2009). Some forms of arthritis such as juvenile arthritis often have remitting stages. Remission is associated with the presence of T lymphocytes reacting against with HSP60 and leading to the production of anti-inflammatory cytokines, Th2 type. In addition, isolated T lymphocytes from patients with rheumatoid arthritis respond to human HSP60 (self) by producing Th2 type anti-inflammatory cytokines, while stimulated with HSP60 from bacteria (non-self), they induce a Th1-type pro-inflammatory response. Similar observations were made on murine models of arthritis (Dominguez et al. 2011; Leavenworth et al. 2013).

3.4 Conclusions

The proteins of the HSP60 family are involved in the cellular response following a stress. These proteins have different functions depending on their location. In the cell, HSP60 is mainly known for its role as chaperone, but it is also involved in apoptosis. It can be found on the cells membrane following apoptosis, promoting the phagocytosis of apoptotic bodies and promoting adhesion of immune cells. In the extracellular medium, HSP60 has an influence on the immune system. Numerous studies have shown the pro-inflammatory capacity of HSP60. Recent findings have shown that this activity also requires the presence of other inflammatory factors such as LPS. However, under normal physiological conditions, endogenous HSP60 may have anti-inflammatory properties. The pro-inflammatory action of HSP60 has been observed in many specialized and non-specialized cell types. T lymphocytes would play a predominant role in the immune action of HSP60. HSP60 is part of a complex network within the immune system by regulating inflammatory events which are important for chronic inflammatory diseases.

Acknowledgements To servier for providing design medical art for figure. Servier Medical Art by Servier is licensed under a Creative Commons Attribution 3.0 Unported Licence.

References

Aalberse JA, Kapitein B, de Roock S, Klein MR, de Jager W, van der Zee R, Hoekstra MO, van Wijk F, Prakken BJ (2011) Cord blood CD4+ T cells respond to self heat shock protein 60 (HSP60). PLoS One 6:e24119

Apuya NR, Yadegari R, Fischer RL, Harada JJ, Zimmerman JL, Goldberg RB (2001) The Arabidopsis embryo mutant schlepperless has a defect in the chaperonin-60alpha gene. Plant Physiol 126:717–730

Arya RK, Singh A, Yadav NK, Cheruvu SH, Hossain Z, Meena S, Maheshwari S, Singh AK, Shahab U, Sharma C et al (2015) Anti-breast tumor activity of Eclipta extract in-vitro and in-vivo: novel evidence of endoplasmic reticulum specific localization of Hsp60 during apoptosis. Sci Rep 5:18457

Atre N, Thomas L, Mistry R, Pathak K, Chiplunkar S (2006) Role of nitric oxide in heat shock protein induced apoptosis of gammadeltaT cells. Int J Cancer 119:1368–1376

Bajramović JJ, Bsibsi M, Geutskens SB, Hassankhan R, Verhulst KC, Stege GJ, de Groot CJ, van Noort JM (2000) Differential expression of stress proteins in human adult astrocytes in response to cytokines. J Neuroimmunol 106:14–22

Balczun C, Bunse A, Schwarz C, Piotrowski M, Kück U (2006) Chloroplast heat shock protein Cpn60 from *Chlamydomonas reinhardtii* exhibits a novel function as a group II intron-specific RNA-binding protein. FEBS Lett 580:4527–4532

Bassan M, Zamostiano R, Giladi E, Davidson A, Wollman Y, Pitman J, Hauser J, Brenneman DE, Gozes I (1998) The identification of secreted heat shock 60 -like protein from rat glial cells and a human neuroblastoma cell line. Neurosci Lett 250:37–40

Basu S, Binder RJ, Suto R, Anderson KM, Srivastava PK (2000) Necrotic but not apoptotic cell death releases heat shock proteins, which deliver a partial maturation signal to dendritic cells and activate the NF-kappa B pathway. Int Immunol 12:1539–1546

Benjamin CL, Ullrich SE, Kripke ML, Ananthaswamy HN (2008) p53 tumor suppressor gene: a critical molecular target for UV induction and prevention of skin cancer. Photochem Photobiol 84:55–62

Binder RJ, Vatner R, Srivastava P (2004) The heat-shock protein receptors: some answers and more questions. Tissue Antigens 64:442–451

Birk OS, Douek DC, Elias D, Takacs K, Dewchand H, Gur SL, Walker MD, van der Zee R, Cohen IR, Altmann DM (1996) A role of Hsp60 in autoimmune diabetes: analysis in a transgenic model. Proc Natl Acad Sci 93:1032–1037

Bockova J, Elias D, Cohen IR (1997) Treatment of NOD diabetes with a novel peptide of the hsp60 molecule induces Th2-type antibodies. J Autoimmun 10:323–329

Campanella C, Bucchieri F, Merendino AM, Fucarino A, Burgio G, Corona DFV, Barbieri G, David S, Farina F, Zummo G et al (2012) The Odyssey of Hsp60 from tumor cells to other destinations includes plasma membrane-associated stages and Golgi and exosomal protein-trafficking modalities. PLoS One:7, e42008

Cappello F, David S, Rappa F, Bucchieri F, Marasà L, Bartolotta TE, Farina F, Zummo G (2005) The expression of HSP60 and HSP10 in large bowel carcinomas with lymph node metastase. BMC Cancer 5:139

Cappello F, Conway de Macario E, Marasà L, Zummo G, Macario AJL (2008) Hsp60 expression, new locations, functions and perspectives for cancer diagnosis and therapy. Cancer Biol Ther 7:801–809

Cappello F, Caramori G, Campanella C, Vicari C, Gnemmi I, Zanini A, Spanevello A, Capelli A, La Rocca G, Anzalone R et al (2011) Convergent sets of data from in vivo and in vitro methods point to an active role of Hsp60 in chronic obstructive pulmonary disease pathogenesis. PLoS One 6:e28200

Chaiwatanasirikul K-A, Sala A (2011) The tumour-suppressive function of CLU is explained by its localisation and interaction with HSP60. Cell Death Dis 2:e219

Chandra D, Choy G, Tang DG (2007) Cytosolic accumulation of HSP60 during apoptosis with or without apparent mitochondrial release: evidence that its pro-apoptotic or pro-survival functions involve differential interactions with caspase-3. J Biol Chem 282:31289–31301
Chang YH, Pearson CM, Abe C (1980) Adjuvant polyarthritis. IV. Induction by a synthetic adjuvant: immunologic, histopathologic, and other studies. Arthritis Rheum 23:62–71
Chen W, Wang J, Shao C, Liu S, Yu Y, Wang Q, Cao X (2006) Efficient induction of antitumor T cell immunity by exosomes derived from heat-shocked lymphoma cells. Eur J Immunol 36:1598–1607
Cheng MY, Hartl F-U, Martin J, Pollock RA, Kalousek F, Neuper W, Hallberg EM, Hallberg RL, Horwich AL (1989) Mitochondrial heat-shock protein hsp60 is essential for assembly of proteins imported into yeast mitochondria. Nature 337:620–625
Choudhary C, Kumar C, Gnad F, Nielsen ML, Rehman M, Walther TC, Olsen JV, Mann M (2009) Lysine acetylation targets protein complexes and co-regulates major cellular functions. Science 325:834–840
Christensen JH, Nielsen MN, Hansen J, Füchtbauer A, Füchtbauer E-M, West M, Corydon TJ, Gregersen N, Bross P (2010) Inactivation of the hereditary spastic paraplegia-associated Hspd1 gene encoding the Hsp60 chaperone results in early embryonic lethality in mice. Cell Stress Chaperones 15:851–863
Cohen-Sfady M, Pevsner-Fischer M, Margalit R, Cohen IR (2009) Heat shock protein 60, via MyD88 innate signaling, protects B cells from apoptosis, spontaneous and induced. J Immunol Baltim Md 1950 183:890–896
Cong Y, Baker ML, Jakana J, Woolford D, Miller EJ, Reissmann S, Kumar RN, Redding-Johanson AM, Batth TS, Mukhopadhyay A et al (2010) 4.0-A resolution cryo-EM structure of the mammalian chaperonin TRiC/CCT reveals its unique subunit arrangement. Proc Natl Acad Sci U S A 107:4967–4972
Cromartie WJ, Craddock JG, Schwab JH, Anderle SK, Yang CH (1977) Arthritis in rats after systemic injection of streptococcal cells or cell walls. J Exp Med 146:1585–1602
Czarnecka AM, Campanella C, Zummo G, Cappello F (2006) Heat shock protein 10 and signal transduction: a "capsula eburnea" of carcinogenesis? Cell Stress Chaperones 11:287–294
Dasu MR, Devaraj S, Park S, Jialal I (2010) Increased toll-like receptor (TLR) activation and TLR ligands in recently diagnosed type 2 diabetic subjects. Diabetes Care 33:861–868
de Kleer I, Vercoulen Y, Klein M, Meerding J, Albani S, van der Zee R, Sawitzki B, Hamann A, Kuis W, Prakken B (2010) CD30 discriminates heat shock protein 60-induced FOXP3+ CD4+ T cells with a regulatory phenotype. J Immunol Baltim Md 1950 185:2071–2079
Deocaris CC, Kaul SC, Wadhwa R (2006) On the brotherhood of the mitochondrial chaperones mortalin and heat shock protein 60. Cell Stress Chaperones 11:116–128
Devaraj S, Dasu MR, Park SH, Jialal I (2009) Increased levels of ligands of Toll-like receptors 2 and 4 in type 1 diabetes. Diabetologia 52:1665–1668
Di Felice V, David S, Cappello F, Farina F, Zummo G (2005) Is chlamydial heat shock protein 60 a risk factor for oncogenesis? Cell Mol Life Sci CMLS 62:4–9
Dominguez MDC, Lorenzo N, Barbera A, Darrasse-Jeze G, Hernández MV, Torres A, Hernández I, Gil R, Klatzmann D, Padrón G (2011) An altered peptide ligand corresponding to a novel epitope from heat-shock protein 60 induces regulatory T cells and suppresses pathogenic response in an animal model of adjuvant-induced arthritis. Autoimmunity 44:471–482
Dunn AY, Melville MW, Frydman J (2001) Review: cellular substrates of the eukaryotic chaperonin TRiC/CCT. J Struct Biol 135:176–184
Elias D, Meilin A, Ablamunits V, Birk OS, Carmi P, Könen-Waisman S, Cohen IR (1997) Hsp60 peptide therapy of NOD mouse diabetes induces a Th2 cytokine burst and downregulates autoimmunity to various β-cell antigens. Diabetes 46:758–765
Ellis RJ, van der Vies SM (1991) Molecular chaperones. Annu Rev Biochem 60:321–347
Fenton WA, Kashi Y, Furtak K, Horwich AL (1994) Residues in chaperonin GroEL required for polypeptide binding and release. Nature 371:614–619

Flohé SB, Brüggemann J, Lendemans S, Nikulina M, Meierhoff G, Flohé S, Kolb H (2003) Human heat shock protein 60 induces maturation of dendritic cells versus a Th1-promoting phenotype. J Immunol Baltim Md 1950 170:2340–2348

Frydman J, Nimmesgern E, Erdjument-Bromage H, Wall JS, Tempst P, Hartl FU (1992) Function in protein folding of TRiC, a cytosolic ring complex containing TCP-1 and structurally related subunits. EMBO J 11:4767–4778

Galdiero M, de l'Ero GC, Marcatili A (1997) Cytokine and adhesion molecule expression in human monocytes and endothelial cells stimulated with bacterial heat shock proteins. Infect Immun 65:699–707

Gao B, Tsan M-F (2003a) Recombinant human heat shock protein 60 does not induce the release of tumor necrosis factor α from murine macrophages. J Biol Chem 278:22523–22529

Gao B, Tsan M-F (2003b) Endotoxin contamination in recombinant human heat shock protein 70 (Hsp70) preparation is responsible for the induction of tumor necrosis factor α release by murine macrophages. J Biol Chem 278:174–179

Ghosh JC, Dohi T, Kang BH, Altieri DC (2008) Hsp60 regulation of tumor cell apoptosis. J Biol Chem 283:5188–5194

Ghosh JC, Siegelin MD, Dohi T, Altieri DC (2010) Heat shock protein 60 regulation of the mitochondrial permeability transition pore in tumor cells. Cancer Res 70:8988–8993

Goh YC, Yap CT, Huang BH, Cronshaw AD, Leung BP, Lai PBS, Hart SP, Dransfield I, Ross JA (2011) Heat-shock protein 60 translocates to the surface of apoptotic cells and differentiated megakaryocytes and stimulates phagocytosis. Cell Mol Life Sci CMLS 68:1581–1592

Goloubinoff P, Gatenby AA, Lorimer GH (1989) GroE heat-shock proteins promote assembly of foreign prokaryotic ribulose bisphosphate carboxylase oligomers in *Escherichia coli*. Nature 337:44–47

Goloubinoff P, Diamant S, Weiss C, Azem A (1997) GroES binding regulates GroEL chaperonin activity under heat shock. FEBS Lett 407:215–219

Green DR (2006) At the gates of death. Cancer Cell 9:328–330

Grundtman C, Wick G (2011) The autoimmune concept of atherosclerosis. Curr Opin Lipidol 22:327–334

Gülden E, Mollérus S, Brüggemann J, Burkart V, Habich C (2008) Heat shock protein 60 induces inflammatory mediators in mouse adipocytes. FEBS Lett 582:2731–2736

Gülden E, Märker T, Kriebel J, Kolb-Bachofen V, Burkart V, Habich C (2009) Heat shock protein 60: evidence for receptor-mediated induction of proinflammatory mediators during adipocyte differentiation. FEBS Lett 583:2877–2881

Gupta S, Knowlton AA (2002) Cytosolic heat shock protein 60, hypoxia, and apoptosis. Circulation 106:2727–2733

Gupta S, Knowlton AA (2007) HSP60 trafficking in adult cardiac myocytes: role of the exosomal pathway. Am J Physiol Heart Circ Physiol 292:H3052–H3056

Gupta RS, Ramachandra NB, Bowes T, Singh B (2008) Unusual cellular disposition of the mitochondrial molecular chaperones Hsp60, Hsp70 and Hsp10. Novartis Found Symp 291:59–68 discussion 69–73, 137–140

Habich C, Burkart V (2007) Heat shock protein 60: regulatory role on innate immune cells. Cell Mol Life Sci CMLS 64:742–751

Habich C, Baumgart K, Kolb H, Burkart V (2002) The receptor for heat shock protein 60 on macrophages is saturable, specific, and distinct from receptors for other heat shock proteins. J Immunol Baltim Md 1950 168:569–576

Habich C, Kempe K, van der Zee R, Rümenapf R, Akiyama H, Kolb H, Burkart V (2005) Heat shock protein 60: specific binding of lipopolysaccharide. J Immunol Baltim Md 1950 174:1298–1305

Hansen JJ, Bross P, Westergaard M, Nielsen MN, Eiberg H, Børglum AD, Mogensen J, Kristiansen K, Bolund L, Gregersen N (2003) Genomic structure of the human mitochondrial chaperonin genes: HSP60 and HSP10 are localised head to head on chromosome 2 separated by a bidirectional promoter. Hum Genet 112:71–77

Hayoun D, Kapp T, Edri-Brami M, Ventura T, Cohen M, Avidan A, Lichtenstein RG (2012) HSP60 is transported through the secretory pathway of 3-MCA-induced fibrosarcoma tumour cells and undergoes N-glycosylation. FEBS J 279:2083–2095

Hogervorst EJM, Boog CJP, Wagenaar JPA, Wauben MHM, van der Zee R, van Eden W (1991) T cell reactivity to an epitope of the mycobacterial 65-kDa heat-shock protein (hsp 65) corresponds with arthritis susceptibility in rats and is regulated by hsp 65-specific cellular responses. Eur J Immunol 21:1289–1296

Hunt JF, Weaver AJ, Landry SJ, Gierasch L, Deisenhofer J (1996) The crystal structure of the GroES co-chaperonin at 2.8 A resolution. Nature 379:37–45

Jindal S, Dudani AK, Singh B, Harley CB, Gupta RS (1989) Primary structure of a human mitochondrial protein homologous to the bacterial and plant chaperonins and to the 65-kilodalton mycobacterial antigen. Mol Cell Biol 9:2279–2283

Juwono J, Martinus RD (2016) Does Hsp60 provide a link between mitochondrial stress and inflammation in diabetes mellitus? J Diabetes Res 2016:e8017571

Kaufman BA, Kolesar JE, Perlman PS, Butow RA (2003) A function for the mitochondrial chaperonin Hsp60 in the structure and transmission of mitochondrial DNA nucleoids in *Saccharomyces cerevisiae*. J Cell Biol 163:457–461

Kim HS, Kim EM, Lee J, Yang WH, Park TY, Kim YM, Cho JW (2006) Heat shock protein 60 modified with O-linked N-acetylglucosamine is involved in pancreatic β-cell death under hyperglycemic conditions. FEBS Lett 580:2311–2316

Kim S-C, Stice JP, Chen L, Jung JS, Gupta S, Wang Y, Baumgarten G, Trial J, Knowlton AA (2009) Extracellular heat shock protein 60, cardiac myocytes, and apoptosis. Circ Res 105:1186–1195

Kol A, Bourcier T, Lichtman AH, Libby P (1999) Chlamydial and human heat shock protein 60s activate human vascular endothelium, smooth muscle cells, and macrophages. J Clin Invest 103:571–577

Kotlo K, Xing Y, Lather S, Grillon JM, Johnson K, Skidgel RA, Solaro RJ, Danziger RS (2014) PR65A phosphorylation regulates PP2A complex signaling. PLoS One 9:e85000

Leach MD, Stead DA, Argo E, Brown AJP (2011) Identification of sumoylation targets, combined with inactivation of SMT3, reveals the impact of sumoylation upon growth, morphology, and stress resistance in the pathogen *Candida albicans*. Mol Biol Cell 22:687–702

Leavenworth JW, Tang X, Kim H-J, Wang X, Cantor H (2013) Amelioration of arthritis through mobilization of peptide-specific CD8+ regulatory T cells. J Clin Invest 123:1382–1389

Lehnardt S, Schott E, Trimbuch T, Laubisch D, Krueger C, Wulczyn G, Nitsch R, Weber JR (2008) A vicious cycle involving release of heat shock protein 60 from injured cells and activation of toll-like receptor 4 mediates neurodegeneration in the CNS. J Neurosci 28:2320–2331

Lewthwaite J, Owen N, Coates A, Henderson B, Steptoe A (2002) Circulating human heat shock protein 60 in the plasma of British civil servants: relationship to physiological and psychosocial stress. Circulation 106:196–201

Lin KM, Lin B, Lian IY, Mestril R, Scheffler IE, Dillmann WH (2001) Combined and individual mitochondrial HSP60 and HSP10 expression in cardiac myocytes protects mitochondrial function and prevents apoptotic cell deaths induced by simulated ischemia-reoxygenation. Circulation 103:1787–1792

Lin Z, Madan D, Rye HS (2008) GroEL stimulates protein folding through forced unfolding. Nat Struct Mol Biol 15:303–311

Llorca O, Carrascosa JL, Valpuesta JM (1996) Biochemical characterization of symmetric GroEL-GroES complexes. Evidence for a role in protein folding. J Biol Chem 271:68–76

Llorca O, Martín-Benito J, Ritco-Vonsovici M, Grantham J, Hynes GM, Willison KR, Carrascosa JL, Valpuesta JM (2000) Eukaryotic chaperonin CCT stabilizes actin and tubulin folding intermediates in open quasi-native conformations. EMBO J 19:5971–5979

Lund PA, Large AT, Kapatai G (2003) The chaperonins: perspectives from the Archaea. Biochem Soc Trans 31:681–685

Magnoni R, Palmfeldt J, Christensen JH, Sand M, Maltecca F, Corydon TJ, West M, Casari G, Bross P (2013) Late onset motoneuron disorder caused by mitochondrial Hsp60 chaperone deficiency in mice. Neurobiol Dis 54:12–23

Märker T, Kriebel J, Wohlrab U, Habich C (2010) Heat shock protein 60 and adipocytes: characterization of a ligand-receptor interaction. Biochem Biophys Res Commun 391:1634–1640

Märker T, Sell H, Zillessen P, Glöde A, Kriebel J, Ouwens DM, Pattyn P, Ruige J, Famulla S, Roden M et al (2012) Heat shock protein 60 as a mediator of adipose tissue inflammation and insulin resistance. Diabetes 61:615–625

Märker T, Kriebel J, Wohlrab U, Burkart V, Habich C (2014) Adipocytes from New Zealand obese mice exhibit aberrant proinflammatory reactivity to the stress signal heat shock protein 60. J Diabetes Res 2014:187153

Mendoza JA, Rogers E, Lorimer GH, Horowitz PM (1991) Chaperonins facilitate the in vitro folding of monomeric mitochondrial rhodanese. J Biol Chem 266:13044–13049

Merendino AM, Bucchieri F, Campanella C, Marcianò V, Ribbene A, David S, Zummo G, Burgio G, Corona DFV, Conway de Macario E et al (2010) Hsp60 is actively secreted by human tumor cells. PLoS One 5:e9247

Murai N, Makino Y, Yoshida M (1996) GroEL locked in a closed conformation by an interdomain cross-link can bind ATP and polypeptide but cannot process further reaction steps. J Biol Chem 271:28229–28234

Nativel B, Marimoutou M, Thon-Hon VG, Gunasekaran MK, Andries J, Stanislas G, Planesse C, Da Silva CR, Césari M, Iwema T et al (2013) Soluble HMGB1 is a novel adipokine stimulating IL-6 secretion through RAGE receptor in SW872 preadipocyte cell line: contribution to chronic inflammation in fat tissue. PLoS One 8:e76039

Nielsen KL, Cowan NJ (1998) A single ring is sufficient for productive chaperonin-mediated folding in vivo. Mol Cell 2:93–99

Nielsen KL, McLennan N, Masters M, Cowan NJ (1999) A single-ring mitochondrial chaperonin (Hsp60-Hsp10) can substitute for GroEL-GroES in vivo. J Bacteriol 181:5871–5875

Ohashi K, Burkart V, Flohé S, Kolb H (2000) Cutting edge: heat shock protein 60 is a putative endogenous ligand of the toll-like receptor-4 complex. J Immunol Baltim Md 1950 164:558–561

Ohue R, Hashimoto K, Nakamoto M, Furukawa Y, Masuda T, Kitabatake N, Tani F (2011) Bacterial heat shock protein 60, GroEL, can induce the conversion of naïve T cells into a CD4 CD25(+) Foxp3-expressing phenotype. J Innate Immun 3:605–613

Osterloh A, Meier-Stiegen F, Veit A, Fleischer B, von Bonin A, Breloer M (2004) Lipopolysaccharide-free heat shock protein 60 activates T cells. J Biol Chem 279:47906–47911

Osterloh A, Veit A, Gessner A, Fleischer B, Breloer M (2008) Hsp60-mediated T cell stimulation is independent of TLR4 and IL-12. Int Immunol 20:433–443

Pace A, Barone G, Lauria A, Martorana A, Piccionello AP, Pierro P, Terenzi A, Almerico AM, Buscemi S, Campanella C et al (2013) Hsp60, a novel target for antitumor therapy: structure-function features and prospective drugs design. Curr Pharm Des 19:2757–2764

Pearson CM (1956) Development of arthritis, periarthritis and periostitis in rats given adjuvants. Proc Soc Exp Biol Med Soc Exp Biol Med N Y N 91:95–101

Pei W, Tanaka K, Huang SC, Xu L, Liu B, Sinclair J, Idol J, Varshney GK, Huang H, Lin S et al (2016) Extracellular HSP60 triggers tissue regeneration and wound healing by regulating inflammation and cell proliferation. Npj Regen Med 1:16013

Peng C, Lu Z, Xie Z, Cheng Z, Chen Y, Tan M, Luo H, Zhang Y, He W, Yang K et al (2011) The first identification of lysine malonylation substrates and its regulatory enzyme. Mol Cell Proteomics MCP 10:M111.012658

Pfister G, Stroh CM, Perschinka H, Kind M, Knoflach M, Hinterdorfer P, Wick G (2005) Detection of HSP60 on the membrane surface of stressed human endothelial cells by atomic force and confocal microscopy. J Cell Sci 118:1587–1594

Planesse C, Nativel B, Iwema T, Gasque P, Robert-Da Silva C, Viranaïcken W (2015) Recombinant human HSP60 produced in ClearColi™ BL21(DE3) does not activate the NFκB pathway. Cytokine 73:190–195

Pockley AG, Multhoff G (2008) Cell stress proteins in extracellular fluids: friend or foe? Novartis Found Symp 291:86–95 discussion 96–100, 137–140

Pockley AG, Bulmer J, Hanks BM, Wright BH (1999) Identification of human heat shock protein 60 (Hsp60) and anti-Hsp60 antibodies in the peripheral circulation of normal individuals. Cell Stress Chaperones 4:29–35

Pockley AG, Muthana M, Calderwood SK (2008) The dual immunoregulatory roles of stress proteins. Trends Biochem Sci 33:71–79

Priya S, Sharma SK, Sood V, Mattoo RUH, Finka A, Azem A, De Los Rios P, Goloubinoff P (2013) GroEL and CCT are catalytic unfoldases mediating out-of-cage polypeptide refolding without ATP. Proc Natl Acad Sci U S A 110:7199–7204

Ranford JC, Coates ARM, Henderson B (2000) Chaperonins are cell-signalling proteins: the unfolding biology of molecular chaperones. Expert Rev Mol Med 2:1–17

Rea IM, McNerlan S, Pockley AG (2001) Serum heat shock protein and anti-heat shock protein antibody levels in aging. Exp Gerontol 36:341–352

Retzlaff C, Yamamoto Y, Okubo S, Hoffman PS, Friedman H, Klein TW (1996) Legionella pneumophila heat-shock protein-induced increase of interleukin-1 beta mRNA involves protein kinase C signalling in macrophages. Immunology 89:281

Roseman AM, Chen S, White H, Braig K, Saibil HR (1996) The chaperonin ATPase cycle: mechanism of allosteric switching and movements of substrate-binding domains in GroEL. Cell 87:241–251

Rosenberger K, Dembny P, Derkow K, Engel O, Krüger C, Wolf SA, Kettenmann H, Schott E, Meisel A, Lehnardt S (2015) Intrathecal heat shock protein 60 mediates neurodegeneration and demyelination in the CNS through a TLR4- and MyD88-dependent pathway. Mol Neurodegener 10:5

Samali A, Cai J, Zhivotovsky B, Jones DP, Orrenius S (1999) Presence of a pre-apoptotic complex of pro-caspase-3, Hsp60 and Hsp10 in the mitochondrial fraction of jurkat cells. EMBO J 18:2040–2048

Schett G, Metzler B, Kleindienst R, Amberger A, Recheis H, Xu Q, Wick G (1999) Myocardial injury leads to a release of heat shock protein (hsp) 60 and a suppression of the anti-hsp65 immune response. Cardiovasc Res 42:685–695

Shamaei-Tousi A, Stephens JW, Bin R, Cooper JA, Steptoe A, Coates ARM, Henderson B, Humphries SE (2006) Association between plasma levels of heat shock protein 60 and cardiovascular disease in patients with diabetes mellitus. Eur Heart J 27:1565–1570

Shamaei-Tousi A, Steptoe A, O'Donnell K, Palmen J, Stephens JW, Hurel SJ, Marmot M, Homer K, D'Aiuto F, Coates ARM et al (2007a) Plasma heat shock protein 60 and cardiovascular disease risk: the role of psychosocial, genetic, and biological factors. Cell Stress Chaperones 12:384–392

Shamaei-Tousi A, D'Aiuto F, Nibali L, Steptoe A, Coates ARM, Parkar M, Donos N, Henderson B (2007b) Differential regulation of circulating levels of molecular chaperones in patients undergoing treatment for periodontal disease. PLoS One 2:e1198

Sigal LH, Williams S, Soltys B, Gupta R (2001) H9724, a monoclonal antibody to Borrelia burgdorferi's flagellin, binds to heat shock protein 60 (HSP60) within live neuroblastoma cells: a potential role for HSP60 in peptide hormone signaling and in an autoimmune pathogenesis of the neuropathy of Lyme disease. Cell Mol Neurobiol 21:477–495

Sigler PB, Xu Z, Rye HS, Burston SG, Fenton WA, Horwich AL (1998) Structure and function in GroEL-mediated protein folding. Annu Rev Biochem 67:581–608

Singh B, Patel HV, Ridley RG, Freeman KB, Gupta RS (1990) Mitochondrial import of the human chaperonin (HSP60) protein. Biochem Biophys Res Commun 169:391–396

Stefano L, Racchetti G, Bianco F, Passini N, Gupta RS, Panina Bordignon P, Meldolesi J (2009) The surface-exposed chaperone, Hsp60, is an agonist of the microglial TREM2 receptor. J Neurochem 110:284–294

Tang H, Tian E, Liu C, Wang Q, Deng H (2013) Oxidative stress induces monocyte necrosis with enrichment of cell-bound albumin and overexpression of endoplasmic reticulum and mitochondrial chaperones. PLoS One 8:e59610

Terry DF, McCormick M, Andersen S, Pennington J, Schoenhofen E, Palaima E, Bausero M, Ogawa K, Perls TT, Asea A (2004) Cardiovascular disease delay in centenarian offspring: role of heat shock proteins. Ann N Y Acad Sci 1019:502–505

Trayhurn P, Wood IS (2005) Signalling role of adipose tissue: adipokines and inflammation in obesity. Biochem Soc Trans 33:1078–1081

Tsan M-F, Gao B (2004) Endogenous ligands of Toll-like receptors. J Leukoc Biol 76:514–519

Tsan M-F, Gao B (2009) Heat shock proteins and immune system. J Leukoc Biol 85:905–910

Vabulas RM, Ahmad-Nejad P, da Costa C, Miethke T, Kirschning CJ, Häcker H, Wagner H (2001) Endocytosed HSP60s use toll-like receptor 2 (TLR2) and TLR4 to activate the toll/interleukin-1 receptor signaling pathway in innate immune cells. J Biol Chem 276:31332–31339

Vachharajani V, Granger DN (2009) Adipose tissue: a motor for the inflammation associated with obesity. IUBMB Life 61:424–430

van Eden W, Tholet JER, van der Zee R, Noordzij A, van Embden JDA, Hensen EJ, Cohen IR (1988) Cloning of the mycobacterial epitope recognized by T lymphocytes in adjuvant arthritis. Nature 331:171–173

van Roon JA, van Eden W, van Roy JL, Lafeber FJ, Bijlsma JW (1997) Stimulation of suppressive T cell responses by human but not bacterial 60-kD heat-shock protein in synovial fluid of patients with rheumatoid arthritis. J Clin Invest 100:459–463

Viitanen PV, Gatenby AA, Lorimer GH (1992) Purified chaperonin 60 (groEL) interacts with the nonnative states of a multitude of *Escherichia coli* proteins. Protein Sci Publ Protein Soc 1:363–369

Wang Y, Chen L, Hagiwara N, Knowlton AA (2010) Regulation of heat shock protein 60 and 72 expression in the failing heart. J Mol Cell Cardiol 48:360–366

Wick G, Kleindienst R, Dietrich H, Xu Q (1992) Is atherosclerosis an autoimmune disease? Trends Food Sci Technol 3:114–119

Wu C (1995) Heat shock transcription factors: structure and regulation. Annu Rev Cell Dev Biol 11:441–469

Xie J, Zhu H, Guo L, Ruan Y, Wang L, Sun L, Zhou L, Wu W, Yun X, Shen A et al (2010) Lectin-like oxidized low-density lipoprotein receptor-1 delivers heat shock protein 60-fused antigen into the MHC class I presentation pathway. J Immunol Baltim Md 1950 185:2306–2313

Xu Q (2002) Role of heat shock proteins in atherosclerosis. Arterioscler Thromb Vasc Biol 22:1547–1559

Xu Q, Schett G, Perschinka H, Mayr M, Egger G, Oberhollenzer F, Willeit J, Kiechl S, Wick G (2000) Serum soluble heat shock protein 60 is elevated in subjects with atherosclerosis in a general population. Circulation 102:14–20

Yaffe MB, Farr GW, Miklos D, Horwich AL, Sternlicht ML, Sternlicht H (1992) TCP1 complex is a molecular chaperone in tubulin biogenesis. Nature 358:245–248

Yam AY, Xia Y, Lin H-TJ, Burlingame A, Gerstein M, Frydman J (2008) Defining the TRiC/CCT interactome links chaperonin function to stabilization of newly made proteins with complex topologies. Nat Struct Mol Biol 15:1255–1262

Zhang X, He M, Cheng L, Chen Y, Zhou L, Zeng H, Pockley AG, Hu FB, Wu T (2008) Elevated heat shock protein 60 levels are associated with higher risk of coronary heart disease in Chinese. Circulation 118:2687–2693

Zhang D, Sun L, Zhu H, Wang L, Wu W, Xie J, Gu J (2012) Microglial LOX-1 reacts with extracellular HSP60 to bridge neuroinflammation and neurotoxicity. Neurochem Int 61:1021–1035

Chapter 4
Heat Shock Protein 60: A Mediator of Atherosclerosis and Its Potential Therapeutic Role

Cecilia Wick

Abstract Heat shock proteins (HSP) are highly conserved families of proteins expressed by a number of cell types following exposure to stressful environmental conditions in order to protect the cells from injury. The ability of a cell to withstand these environmental stresses is critical to its capacity to adapt and remain viable. Loss of this ability may lead to a pathological state. Abnormal localization, structure or function of HSP has been associated with different pathologies, including atherosclerosis. HSP60 has in particular been identified as playing important roles in inflammatory pathways and immune reactions in atherosclerosis. This chapter will highlight the role of HSP60 in atherosclerosis and the potential to target HSP60 or peptides thereof to alter the progression or even treat atherosclerosis.

Keywords Atherosclerosis · Autoimmunity · Heat shock protein 60 · Inflammation · Regulatory T cells · Tolerization

Abbreviations

ACE	Acute coronary syndrome
AGE	Advanced glycation end product
APC	Antigen-presenting cell
$ApoE^{-/-}$	Apolipoprotein E deficient
cHSP60	*Chlamydia pneumonia* heat shock protein 60
CVD	Cardiovascular disease
DC	Dendritic cell
GARP	Glycoprotein A representations predominant

C. Wick (✉)
Department of Medicine, Unit of Rheumatology, Karolinska Institutet, Karolinska University Hospital Solna, Center for Molecular Medicine (CMM), Stockholm, Sweden
e-mail: Cecilia.Wick@ki.se

A. A. A. Asea, P. Kaur (eds.), *Chaperokine Activity of Heat Shock Proteins*, Heat Shock Proteins 16, https://doi.org/10.1007/978-3-030-02254-9_4

GroEL	Human heat shock protein 60 produced in *Escherichia coli*
HCD	High cholesterol diet
hHSP60	Human heat shock protein 60
HSP	Heat shock protein
ICAM-1	Intracellular adhesion molecule-1
IFN	Interferon
IL	Interleukin
IMT	Inter media thickness
LAP	Latency associated peptide
LDL	Low-density lipoprotein
$LDLr^{-/-}$	Low lipoprotein receptor deficient
LPS	Lipopolysaccharides
mbHSP65	*Mycobacterial tuberculosis* heat shock protein 65
MHC	Major histocompatibility complex
NO	Nitric oxid
oLDL	Oxidized low-density lipoprotein
PBMC	Peripheral blood mononuclear cells
RA	Rheumatoid arthritis
$RAG1^{-/-}$	V(D)J recombination activation gene 1 deficient
rCTD	Rat C-terminal determinants
rHSP65	Rat heat shock protein 65
SCID	Severe combined immunodeficiency
sHSP60	Soluble heat shock protein 60
SLE	Systemic lupus erythematous
SMC	Smooth muscle cell
TGF	Transforming growth factor
TLR	Toll-like receptor
TNF	Tumor necrosis factor
Treg	Regulatory T cell
VCAM-1	Vascular cell adhesion molecule-1

4.1 Introduction

Both microbial and human HSP60 have been proposed to have a role in the pathogenesis of autoimmune diseases. The mechanism by which HSP60 may act as an autoantigen is based on molecular mimicry due to the high sequence similarity between human and foreign HSP60 from bacteria and parasites that colonize humans, which leads to anti-HSP60 antibody cross-reactivity. Expression of HSP60 can be triggered by various stressors typical of chronic inflammatory diseases, as shown for example for atherosclerosis (Grundtman et al. 2011; Grundtman and Wick 2011; Wick 2016; Wick et al. 2004, 2014). Importantly, the expression of

HSP60 in stressed arterial endothelial cells is always accompanied by the simultaneous expression of adhesion molecules. This situation then allows for the attachment of HSP60-reactive T cells to target endothelial cells leading to a harmful autoimmune attack on the endothelial. If atherosclerosis risk factors persist, this still reversible inflammatory stage of atherosclerosis proceeds to complicated and larger lesions (plaques) with deleterious consequences. The role for both pro- and anti-atherogenic innate and adaptive immune responses in atherosclerosis has been proven in several studies. Moreover, common HSP60 autoantigens, against which an immune response with an activation of atheroprotective or atherogenic adaptive immune responses occurs, have been identified in animal and human models of atherosclerosis. Therefore, an induction of immune tolerance through the activation of cellular and humoral immune reactions to these antigens is hypothesized being atheroprotective. The success of using the recently identified specific immunoreactive antigenic HSP60 epitopes for tolerization further supports the idea that active vaccination may emerge as a novel immune-modulating atheroprotective strategy (Wick 2016).

4.2 Heat Shock Proteins

Heat shock proteins (HSP) are highly conserved proteins found in prokaryotes and eukaryotes (Lindquist and Craig 1988). The heat shock response was initially described over 55 years ago as the appearance of puffs in *Drosophila melanogaster* chromosomes induced by heat or treatment with respiration uncouples (Ashburner 1970; Ritossa 1962, 1964). Later, the first gene and protein products of this morphological puffing pattern were identified and the term "heat shock proteins" was created (McKenzie et al. 1975; Moran et al. 1978; Spradling et al. 1975; Tissieres et al. 1974). Increasing expressions of HSP inside the cells provides stabilization and recovery of damage proteins as well as optimal balance between the synthesis, structure formation, and degradation of proteins. It results in enhanced cell resistance to cellular stress. At the same time, HSP can release into the extracellular medium or be expressed on cell surfaces (Basu et al. 2000; Tytell et al. 1986) and in that case, their specific protective role is to control the inflammatory immune response. This, from its original description as solely an intracellular molecular chaperone, HSP have now been shown to function as initiators of the hosts immune response by binding to specific surface receptors. These bindings may activate intracellular signaling cascades, which in turn exert immunoregulatory effector functions; a process known as the chapertokine activity. Therefore, HSP are also important in the buildup of tolerance and cytoprotection, in addition, HSP has also been discussed to play important roles in the pathogenesis, prognosis, and treatment of many diseases. However, their exact mechanisms remain elusive in most cases.

4.3 Heat Shock Protein 60 (HSP60)

HSP60 was first shown to reside in the mitochondria where it can transport target proteins from the cytoplasm into the mitochondria (Czarnecka et al. 2006; Martin 1997). Accumulating evidence also suggests that HSP60 participates in apoptosis by interactions with mortalin (mitochondrial HSP70), p53, and surviving (Deocaris et al. 2006; Ghosh et al. 2008; Wadhwa et al. 2005). Moreover, recent data shows that HSP60 is not just a mitochondrial protein, as it also resides in the cytoplasm and unlike other HSP that mostly have pro-survival functions, HSP60 has either pro-survival or pro-apoptotic functions (Chandra et al. 2007). It has also been found on the cell surface where it is involved in the activation of immune system (Feng et al. 2002; Osterloh et al. 2004) and in the extracellular matrix where it has pro-inflammatory functions (Anraku et al. 2012; Lewthwaite et al. 2002). HSP60 derived peptides could be presented via both major histocompatibility complex (MHC) class I and II molecules (Anderton et al. 1993; Michaelsson et al. 2002; Zugel et al. 1995).

HSP60 is an immune-dominant molecule and peptide sequences thereof are one of the major epitopes, which can stimulate a range of different immune responses. Eukaryotic (mammalian HSP60 (hHSP60)) and prokaryotic HSP60 (the *mycobacterial tuberculosis* homologue HSP65 (mbHSP65), *Chlamydia pneumonia* homologue (cHSP60) and the *Escherichia coli* homologue (GroEL)) has a high degree of homology reaching an overall of 50–60% or even 72% in certain domains (Kaufmann 1990). Due to this fact, T cells failing to ignore autologous HSP60 may lead to a cross-recognition and eventually to an autoimmune reaction. This is the reason why HSP60 is a potential candidate for molecular mimicry and can be recognized by the immune system as being potentially pathogenic antigen(s), as a result, they can play a role in the development of both anti-infective responses and autoimmunity (Lydyard and van Eden 1990; Young and Elliott 1989). This hypothesis is evidenced by the increased serum levels and antibodies of HSP60 in several autoimmune diseases such as rheumatoid arthritis (RA), systemic lupus erythematous (SLE), diabetes, atherosclerosis, and other cardiovascular disease (CVD), just to mention a few (Gruber et al. 1996; Hoppichler et al. 2000; Horvath et al. 2001; Oda et al. 1994; Panchapakesan et al. 1992; Veres et al. 2002; Weitgasser et al. 2003; Xu et al. 1999). Moreover, T and B cell responses to self-HSP60 and HSP65 are present in healthy individuals without widespread inflammation or autoimmunity (Aalberse et al. 2011; Fischer et al. 1992). This could be due to the presence of regulatory immune responses.

4.4 Initiation of Immune Responses: The Chaperokine Activity of HSP60

Extracellular HSP60 are considered as molecules with immunomodulatory functions, either as cross-presenters of immunogenic peptides via MHC antigens or in a peptide-free version as stimulators of innate immune responses or as chaperokines.

Chaperokine is a recent term coined to better describe the unique function of a molecule to work both as a chaperone and cytokine. Since HSP60 has been shown to activate inflammatory cells that are induced to produce cytokines and other mediators of inflammation it can also function as a chaperokine. In this context, HSP60 has been found to activate mononuclear cells via binding to CD14 and activation of p38MAPK leading to the production of pro-inflammatory mediators including tumor necrosis factor (TNF), nitric oxide (NO), interleukin-1 (IL-1), IL-6, IL-12, and IL-15 promoting Th1 type responses (Kol et al. 2000; Retzlaff et al. 1994; Chen et al. 1999). HSP60 have also been reported to stimulate antigen-dependent T cell activation and production of pro-inflammatory interferon-γ (IFN-γ) (Breloer et al. 2001; More et al. 2001), and T cell adhesion to fibronectin via toll-like receptor-2 (TLR-2) (Zanin-Zhorov et al. 2003). Because of its ability to either bind to TLR-2 or TLR-4 on T cells, HSP60 controls adaptive and innate immune responses at the level of immune cells (Zanin-Zhorov and Cohen 2013). Moreover, HSP60 can bind to the neutrophils plasma membrane, enhancing production of oxidants and release of proteases (Osterloh et al. 2009). Thus, HSP60 can induce a plethora of immune responses and the list continues to grow.

4.5 Early Atherosclerosis

Atherosclerosis is the leading cause for cardiovascular mortality. Progressive growth of atherosclerotic plaques obstructing the blood flow in the arteries either by constricting the lumen or by formation of thrombus, resulting in a clinical disease manifested as coronary artery disease, stroke or peripheral arterial diseases (Bentzon et al. 2014). The disease pathogenesis involves retention of lipoproteins and recruitment of immune-inflammatory cells. Plaques may remain asymptomatic in a subclinical disease, they may get obstructive as stable angina or rupture causing acute coronary syndrome (ACS), which is the most overwhelming cause of mortality worldwide (Horvath et al. 2016; Osborn and Jaffer 2013; Roth et al. 2015).

The arterial wall is a heterogeneous three-layered structure composed of an intima, media, and adventitia. Each layer exhibits specific histologic, biochemical, and functional characteristics, and as such, each contributes in unique ways to maintaining vascular homeostasis and to regulating the vascular response to stress or injury. Endothelial- and smooth muscle cells (SMCs) are the principal cellular constituents of the intima and media. The vascular adventitia is a complex compartment of the vessel wall composed of fibroblasts, dendritic cells (DCs), macrophages, progenitor cells, vasa vasorum endothelial cells, pericytes, and adrenergic nerves. In response to vascular stress or injury, resident adventitial cells are often the first to be activated and re-programmed to then influence tone and structure of the vessel wall, to initiate and perpetuate chronic vascular inflammation, and to act to stimulate expansion of the vasa vasorum, which can act as a conduit for continued inflammatory and progenitor cell delivery to the vessel wall. A "outside-in" theory whereby vascular inflammation begins in the adventitia and advances to the media and intima was created 10 years ago (Maiellaro and Taylor 2007). This theory opposes the long-held "inside-out" concept that monocytes infiltrate the vessel wall from the luminal side. Among the arguments favoring the outside-in theory are the presence of resident immune cells in the adventitia and the homing of macrophages to that side (Galkina et al. 2006). The presence of resident T cells, B cells, macrophages, and DCs in the adventitia has been documented in humans (Houtkamp et al. 2001; Schwartz and Mitchell 1962; Watanabe et al. 2007) and experimental animal models (Galkina et al. 2006; Moos et al. 2005; Okamoto et al. 2001; Zhao et al. 2004) often at predisposed sites. The expression of these cells is often early and aortas of children, who have not developed atherosclerosis, have been shown to display leukocytes surrounding the adventitial vasa vasorum (Wick et al. 1997). T cells are suggested to play an important role in mediate pro-inflammatory and regulatory immune responses in early atherosclerosis. For example, the V(D)J recombination activation gene deficient ($Rag1^{-/-}$) crossed with low lipoprotein receptor deficient ($LDLr^{-/-}$) mice develop significantly reduced atherosclerotic lesions compared to immunocompetent $LDLr^{-/-}$ mice (Song et al. 2001). Similar findings has also been described in apolipoprotein E deficient ($ApoE^{-/-}$) mice with severe combined immunodeficiency (SCID) (Zhou et al. 2000). Importantly, transfer of $CD4^{+}$ T cells from $ApoE^{-/-}$ mice to immunodeficient $ApoE^{-/-}$ mice aggravates atherosclerosis (Zhou et al. 2000), strongly suggesting involvement of $CD4^{+}$ T cells. Coronary segments from patients with and without atherosclerotic plaque demonstrated expression of the leukocyte adhesion molecules vascular cell adhesion molecule-1 (VCAM-1), intracellular adhesion molecule-1 (ICAM-1), and E-selectin predominantly on the intimal vasa vasorum rather than the atrial luminal endothelium, which provides a means for inflammatory cells trafficking in and out of the adventitia (O'Brien et al. 1996). Correlative data suggests that inhibition of the angiogenic vasa vasorum limits the infiltration of inflammatory cells into the plaque (Moulton et al. 2003). In atherosclerotic lesions, DCs often form clusters with T cells, suggesting that direct activation of DCs might occur within the arterial wall and may therefore function as antigen-presenting cells (APCs) presenting potentially harmful exogenous or autoantigens to T cells and/or macrophages within the arterial lesion (Bobryshev and Lord 1998). Accumulation of DCs in arterial lesions is associated with plaque growth and inflammation (Liu et al. 2008). Thus, the adventitial compartment may act as the principal "injury-sensing tissue" of the vessel wall.

4.6 The "Autoimmune Concept of Atherosclerosis"

The autoimmune concept of atherosclerosis was coined by us already in the early nineties (Grundtman et al. 2011; Grundtman and Wick 2011; Wick 2016; Wick et al. 2004, 2014). The basis of the concept is that endogenous HSP60 can be expressed together with adhesion molecules (Amberger et al. 1997; Seitz et al. 1996) on the surface of endothelial cells when these are exposed to classical atherosclerosis risk factors, such as lipopolysaccharides (LPS) (Wick et al. 2008),

Chlamydia pneumoniae (Kreutmayer et al. 2013), cigarette smoke extract (Kreutmayer et al. 2011), pro-inflammatory cytokines, oxidized low-density lipoprotein (oLDL) (Amberger et al. 1997), advanced glycation end products (AGE) (Grundtman and Wick 2011), shear blood flow (Hochleitner et al. 2000; Zou et al. 1998), sodium chloride (Jakic et al. 2017), and heat shock (Pfister et al. 2005; Schett et al. 1995). This mechanism provides the prerequisite for potentially bacterial/human HSP60 cross-reactive antibodies and destruction of endothelial cells by pre-existing cellular and humoral immunity against HSP60. The pathological endothelial cell surface expression of HSP60 in predisposed atherosclerotic sites could act as a "danger signal" for pre-existing HSP60 reactive DCs (Bobryshev 2010; Bobryshev and Lord 1996, 1998; Cybulsky and Jongstra-Bilen 2010; Liu et al. 2008; Millonig et al. 2001a, b, c; Waltner-Romen et al. 1998), which could prime T cells (Almanzar et al. 2012; Kleindienst et al. 1993; Millonig et al. 2002; Xu et al. 1990) and cross-present these antigens to the T cells, leading to a harmful autoimmune attack on arterial endothelial cells maltreated by atherosclerotic risk factors. These early lesions do often consist of an increased number of γδ T cells (Kleindienst et al. 1993), which indicates that HSP60 recognition might also be possible in a non-MHC restricted fashion, since γδ T cells do not require MHC for activation (Fisch et al. 1990). However, the early inflammatory stage of atherosclerosis is still reversible, but if atherosclerosis risk factors persist, the inflammatory stage proceeds to plaque formation with deleterious consequences. At later stages of atherosclerosis, intralesional T cells, macrophages, DCs, and SMCs can also express HSP60 and the anti-HSP60 cellular immune reaction could therefore be perpetuated in situ and accelerate the disease process even further. HSP60 has during the last 25 years been recognized as one of the most important antigens in early stages of atherosclerosis and its antigenic mimicry has been thoroughly investigated in both clinical settings and in different experimental animal models.

4.7 HSP60 in Human Atherosclerosis

Early autoreactive intralesional T cells have been demonstrated to specifically react to certain hHSP60 epitopes (Almanzar et al. 2012). This is similar to what has been shown in late complicated plaques were also cross-reactive epitopes between cHSP60 and hHSP60 where described (Benagiano et al. 2005). Interestingly, some of these potentially atherogenic hHSP60 epitopes were only found in early lesions or in late plaques, while others were shared (Almanzar et al. 2012), indicating that certain hHSP60 epitopes recognized by early lesional T cells may play a pathogenic role throughout the disease. Concordant results in rabbits demonstrates that lesional T cells can specifically proliferate in response to mbHSP65 (Xu et al. 1993a, b). Furthermore, hHSP60 and 90 has been shown to play a role in modified low-density lipoprotein (LDLx) activation of plaque T cells (Liu et al. 2015). Lesional-derived T cells display a oligoclonally restricted repertoire in contrast to the polyclonal pattern of peripheral blood mononuclear cells (PBMCs), indicating that oligoclonal T cell expansion can take place in the human atherosclerotic lesions (Rossmann et al. 2008). In humans, peripheral HSP60-reactive T cells from young clinically healthy persons correlates with increased intima-media thickness (IMT) at different vascular territories (Knoflach et al. 2003, 2009). This is however not the case in elderly persons, suggesting that HSP60-specific cellular immunity play an important role at the very early stages of the disease (Knoflach et al. 2007).

Several studies have shown a correlation between high anti-hHSP60 antibody titers and/or elevated soluble HSP60 (sHSP60) levels in individuals suffering from CVD (Almanzar et al. 2012; Hoppichler et al. 2000; Pockley et al. 2000; Willeit and Kiechl 1993; Xu et al. 1993a, b; Zhang et al. 2008), suggesting that anti-hHSP60 antibody titer can be used to mirror morbidity and mortality in atherosclerosis (Xu et al. 1999). Moreover, IMT correlates with elevated sHSP60 levels in individuals with prevalent carotid atherosclerosis (Xiao et al. 2005; Xu et al. 2000). As already mentioned, due to the high degree of sequence homology between bacterial and human HSP60 cross-reactive autoantigens may incite an autoimmune response. For example, immunodominant T- and B cell epitopes from *Porphyromonas gingivalis* HSP60 has been defined and strongly linked to atherosclerosis (Choi et al. 2001, 2002, 2004a, b). Antibodies to microbial HSP60 and HSP65 can recognize specific epitopes on hHSP60 and could serve as autoimmune targets in incipient atherosclerosis (Perschinka et al. 2003). Moreover, cross-reactivity between bacterial and human anti-HSP60 antibodies can induce pathogenic events, such as cytotoxic damage of stressed endothelial cells in atherosclerosis (Mayr et al. 1999; Schett et al. 1995). Lifelong infectious load has also been discussed to correlate with antimicrobial HSP60 antibody titers and atherosclerosis (Burian et al. 2001; Ford et al. 2005; Mayr et al. 2000). Thus, these data not only indicate a pathogenic role for HSP60 but also its potential as a biomarker with applications for diagnosis, assessing prognosis, and response to treatment, as well as for preventing and treating CVD.

Overexpression of HSP60 has also been described tissues of other various inflammatory diseases suggesting a possible pathogenic role of this antigen in their pathogenesis. For example, increased expression of hHSP60 is observed in intestinal ulcer of Becheet's disease (Imamura et al. 2005), in inflamed bowel of Crohn's disease patients (Puga Yung et al. 2009), in muscle tissue of juvenile dermatomyositis patients (Elst et al. 2008), and in synovial fluid and synovial tissue of patients with juvenile idiopathic arthritis and RA (Boog et al. 1992; Huang et al. 2009). Human HSP60 is also overexpressed in inflamed lesions of atopic dermatitis where it also co-localized with $CD3^+$ T cells (Marengo et al. 2009). Moreover, PBMCs from patients with inflammatory diseases can be activated by HSP60. For instance, HSP60 induces IFN-γ secretion from atopic dermatitis patients PBMCs (Kapitein et al. 2013) and PBMCs of patients with Becheet's disease, stimulated with various epitopes of HSP60, resulted in excess amounts of Th1 cytokine production (Imamura et al. 2005). Moreover, HSP60 peptide specific T cell clones have been shown to play a significant role in the perpetuation of Chron's disease and tissue-specific T cell clones from diabetes children recognize hHSP60 as an autoantigen (Jiang et al. 2010; Puga Yung et al. 2009; Sobel and Creswell 2006). Humoral responses to HSP have also been observed in autoimmune diseases, as antibody responses to multiple

HSP families were detected in sera from RA and juvenile idiopathic arthritis patients (Huang et al. 2010; Wu et al. 2011).

4.8 HSP60-Induced Experimental Atherosclerosis

Early atherosclerotic lesions show a strong up-regulation of hHSP60 and the stress-inducible form hHSP70 in ApoE$^{-/-}$ mice (Kanwar et al. 2001). The increased expression can already be found before plaque formation is visually detected (3-weeks-old mice) at lesion-prone sites, followed by a strong and heterogeneous expression in early to advanced fibrofatty plaques (8–20-week-old mice) in lesional endothelial cells, macrophages, SMCs, and CD3^{+} T cells, with levels correlating to disease severity (Kanwar et al. 2001). However, in advanced collagenous, acellular, calcified plaques (40–69-week-old mice) the expression is markedly down-regulated and was absent in control aortas of normocholesterolemic wild-type ApoE$^{+/+}$ mice (Kanwar et al. 2001), indicating that HSP60 might be a good marker for progression stages of atherosclerosis. The mechanisms responsible for the decline of HSP expression in late-stage aortic lesions of aged mice are unknown. However, the acellular nature of large necrotic cores due to apoptosis (Crisby et al. 1997) and the cytolytic effects of anti-HSP65/60 autoantibodies (Schett et al. 1997) provide an obvious explanation.

One of the first studies demonstrating that mbHSP65 immunization could aggravate atherosclerosis was performed in normocholesterolemic rabbits (Xu et al. 1992). The immunized rabbits developed atherosclerotic plaques irrespectively of their cholesterol levels and T cells isolated from these lesions specifically responded to mbHSP65 in vitro (Metzler and Wraith 1999; Xu et al. 1992, 1993a) similar to what has been described in humans (Almanzar et al. 2012; Benagiano et al. 2005; Rossmann et al. 2008). These findings were later confirmed in mice where both C57BL/6J fed high cholesterol diet (HCD) and LDLr$^{-/-}$ mice fed normal chow diet revealed enhanced early atherosclerotic lesions consisting of an increased number of CD3^{+} or CD4^{+} T cells after mbHSP65 immunizations (Afek et al. 2000; George et al. 1999). Transfer of CD3^{+} mbHSP65 reactive lymphocytes to syngenic mice led to an enhancement of fatty streak formation (Shoenfeld et al. 2000), supporting the fact that early mbHSP65 reactive T cells may initiate atherosclerosis as also described in humans (Almanzar et al. 2012). Enhanced fatty streak formation and increased IFN-γ/IL-4 ratio was also found in C57BL/6NJcl mice given HCD together with hHSP60 immunization (Mori et al. 2000). Concordantly, rabbits immunized with P277, a 24-residue peptide of hHSP60, showed an aggravation of the disease (Xiong et al. 2016). Surface expression of P277 was observed on stressed endothelial cells and was suggested to mediate the autoimmune attack and promote the disease (Xiong et al. 2016). We have recently shown that immunization with mbHSP65 in ApoE$^{-/-}$ mice or control (C57BL/6J) mice (14-weeks-old) resulted in a significant aggravation of the disease irrespectively of their diet (Grundtman et al. 2015), Figs. 4.1a, b and 4.2a–c. Similar results were recently described in aged

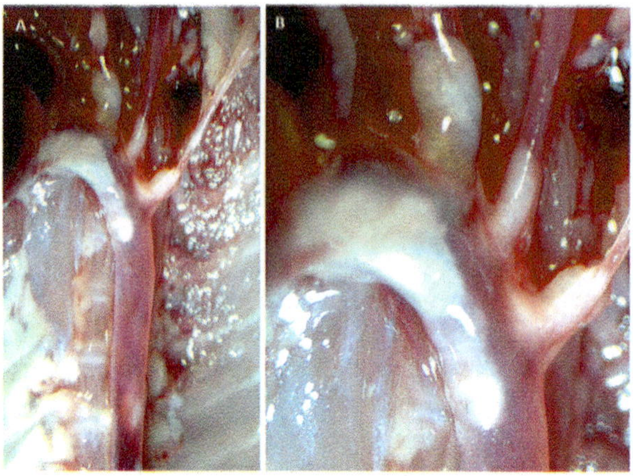

Fig. 4.1 Lesions in the aortic arch of an mbHSP65 immunized ApoE$^{-/-}$ mouse. (**a**) Severe stenosis can be seen in the aortic lumen in a 22-weeks-old ApoE$^{-/-}$ mouse after HCD for 14 weeks and four mbHSP65 immunizations (first immunization at week 8, second at week 11, third at week 14, and the last immunization at week 17). (**b**) Close up of (**a**)

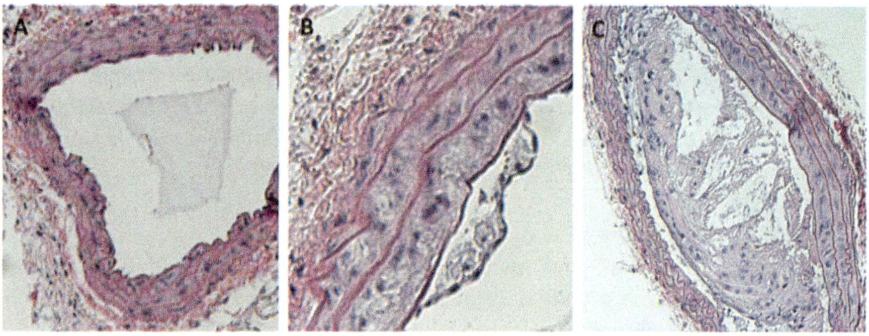

Fig. 4.2 Haematoxylin and eosin stained cross-sectional of the brachiocephalic artery from a 22-week-old ApoE$^{-/-}$ mouse (**a**) unimmunized and fed normal chow diet, (**b**) unimmunized and fed HCD for 14 weeks, and (**c**) mbHSP65 immunized four times (first immunization at week 8, second at week 11, third at week 14, and the last immunization at week 17) and fed HCD for 14 weeks. Plaque formation is notably greater in the immunized mouse compared to the unimmunized ones

ApoE$^{-/-}$ mice (65-weeks-old) with long-established atherosclerosis (Wick et al. 2018). Early inflammatory stages of atherosclerotic lesions induced by mbHSP65 immunization can regress in the absence of additional risk factors for atherosclerosis and T cell activation (Sun et al. 2010; Xu et al. 1996). In contrast, mbHSP65 immunization in alum and anti-CD45RB monoclonal antibody reduced atherosclerotic lesions in ApoE$^{-/-}$ mice, IgG1 antibody titer (a Th2 driven IgG) against mbHSP65 was also elevated (Klingenberg et al. 2012). The discrepancy of these results with earlier studies is possibly due to the sort of adjuvant used or the stage of atherosclerosis (Klingenberg et al. 2012). In fact, some studies indicate that adjuvant itself can modulate the immune response, usually leading to a reduced atherosclerosis (Khallou-Laschet et al. 2006). Moreover, rats immunized with mbHSP65 showed a brisk and sustained humoral response with increased anti-mbHSP65 IgG titers and pronounced splenocyte proliferation to mbHSP65 together with an enhanced neointimal growth (George et al. 2003). Similarly, high-titer Ig treatment with anti-hHSP60 autoantibodies can accelerate atherosclerosis in ApoE$^{-/-}$ mice (Foteinos et al. 2005). Administration of a specific mouse monoclonal antibody (II–13) recognizing amino acid residues 288–366 of hHSP60 could effectively induce atherosclerotic lesions in these mice. II–13 injection resulted in endothelial cell damage followed by increased leukocyte attachment and accumulation of macrophages and SMCs in lesions, which could be prevented by F(ab)2 segments derived from these antibodies (Foteinos et al. 2005). However, the monoclonal antibody (ML-30), reactive to mbHSP65, lacked cytotoxic effects against cells in vitro (Foteinos et al. 2005). As already described, cross-reactive antibodies between bacterial/human HSP60 can induce cytotoxic damage of stressed endothelial cells (Mayr et al. 1999; Schett et al. 1995), indicating that humoral immune reactions to bacterial HSP may play an important role in the process of vascular endothelial injury, one of the key events in the pathogenesis of atherosclerosis. Thus, these results strongly indicate that HSP60 is a protein, which is involved in the pathogenesis of atherosclerosis. However, it is important to remember that HSP60 is a rather large protein which, when processed, give rise to a multitude of potential epitopes of which only a few are dominant. Different epitopes from the same HSP in the same disease may therefore have very different functional effects on the immune response, some being pro-inflammatory and others tolerogenic (Wick et al. 2014). We have recently performed a study where we functionally characterized novel atherogenic and atheroprotective mbHSP65 epitopes (Grundtman et al. 2015). These epitopes may be potential candidates for the development of a tolerizing vaccine to prevent and treat atherosclerosis, while keeping protective immunity to non-atherogenic domains of mbHSP65 intact. Thus, disrupting or blocking certain spots in the monomer will have an impact on its functions, including oligomerization and interacting with other molecules, and thus interfere with one or more of the roles played by the molecule inside and outside the cell. This dependency of intrinsic and overall functions of the integrity of specific structural sites opens the road to developing compounds that will bind those sites and, thereby, block HSP60 when it is a pathologic factor.

4.9 HSP60 Tolerization in Atherosclerosis

Contrary to its role as a target for pathogenic autoimmune inflammatory processes, HSP60 has been shown to activate immunoregulatory pathways that may lead to suppression of the disease. Thus, several research groups have shown that immune modulation by restoring tolerance to self-antigens such as HSP60 is a promising therapeutic model for atherosclerosis. Mucosal administration of antigen is known to generate tolerance and suppressions Th1 immune responses through production of IL-4, IL-10, and transforming growth factor-β (TGF-β), which induces unresponsiveness of immune system against a given antigen (Faria and Weiner 1999). Oral administration with mbHSP65 significantly attenuated atherogenesis. The reduction in lesion size was accompanied with reduced lymphocyte reactivity and specific mbHSP65 reactivity in spleenoctes in orally treated mice immunized either with mbHSP65 or *Mycobacterium tuberculosis* (Harats et al. 2002). Lymph node cells of treated mice produced more IL-4 compared with non-tolerized cells. However, no suppressive effect was seen on Th2 cytokine secretion, as evidenced by the unaltered IFN-γ production (Harats et al. 2002). These results are consistent with previous reports suggesting that IL-4 plays a protective role in atherosclerosis (Huber et al. 2001). Furthermore, treatment with a combination $ApoB_{661-680}$ and $hHSP60_{153-163}$ peptides induced a significant reduction of atherosclerosis compared to single immunization (Lu et al. 2010), possibly due to a reduced production of IFN-γ and an increased production of IL-10. Similarly, oral administration of both mbHSP60 and $HSP60_{253-268}$ peptide reduced atherosclerosis (van Puijvelde et al. 2007). In the latter study, although the number of $CD4^+CD25^+Foxp3^+$ regulatory T cells (Tregs) and the expression of IL-10 and TGF-β were increased, casuality was not tested. Oral administration of mbHSP65 delivered by *Lactococcus lactis* reduced atherosclerotic plaque in aortas of $LDLr^{-/-}$ mice, accompanied by elevation of IL-10 production (Jing et al. 2011). We have recently shown that atherosclerosis is attenuated, pro-inflammatory cytokines are decreased, and anti-inflammatory cytokines increased in the aorta in both $ApoE^{-/-}$ and control C57BL/6J mice after oral mbHSP65 administration (Grundtman et al. 2015). This was accompanied by increased numbers of $CD4^+CD25^+FoxP3^+$ Tregs (Grundtman et al. 2015). These results were recently confirmed in aged $ApoE^{-/-}$ mice with established atherosclerosis, suggesting that oral mbHSP65 tolerization could be used both to prevent and to treat chronic atherosclerosis (Wick et al. 2018). Both nasal and oral mbHSP65 tolerization has been investigated in $LDLr^{-/-}$ mice maintained on HCD. A significant decrease in lesion size, a reduction in macrophages-positive area in the aortic arch, decreased IFN-γ expression, increased IL-10 expression, and a reduced number of $CD4^+$ T cells were found in nasally treated mice (Maron et al. 2002). The titer of IgG1 antibody against mbHSP65 was increased in this study (Maron et al. 2002). Nasal administration of mbHSP65 reduced atherosclerosis also in HCD fed rabbits (Xiong et al. 2009).

As already described, activated $CD4^+$ T cells reactive to HSP60 mainly secrete Th1 cytokines in atherosclerotic lesions, leading to an imbalance between Th1 and the Tregs specific for self- and non-self-antigens (Mallat et al. 2005). Restoring this balance between Th1 cells and Tregs may be effective in ameliorating atherosclerosis, suggesting that Tregs play a crucial role in controlling the initiation and progression of atherosclerosis (Ait-Oufella et al. 2006). Mounting evidence supports that the newly identified $CD4^+LAP^+$ Tregs that express latency associated peptide (LAP) on their surface is associated with oral tolerance induction and following inhibition of atherosclerosis (Ait-Oufella et al. 2006; Sasaki et al. 2009; Taleb et al. 2008; Weiner et al. 2011). It was just recently shown that nasal administration of *Helicobacter pylori* HSP60 inhibits atherosclerotic formation through immune tolerance, which is established by $CD4^+LAP^+$ and $CD4^+CD25^+FoxP3^+$ Tregs, and that TGF-β, mainly induced by $CD4^+LAP^+$ Tregs, played a crucial role in the prolonged atheroprotective mechanisms (Li et al. 2012). In a similar study, nasal *Helicobacter pylori* HSP60 caused a significant reduction in plaque size at the aortic root in the early stages of atherosclerosis (Zhong et al. 2016). Notably, a significant increase in activated $CD4^+CD25^+$ glycoprotein A representations predominant $(GARP)^+$ Tregs type 1 and $CD4^+CD25^+Foxp3^+$ Tregs as well as a marked decrease in the numbers of Th1 and Th17 cells was detected in spleens and in cervical lymph nodes of *Helicobacter pylori* HSP60 treated mice (Zhong et al. 2016). Moreover, nasal *Helicobacter pylori* HSP60 increased the production of TGF-β and IL-10 and decreased the secretion of IFN-γ and IL-17. Interestingly, the atheroprotective role of nasal HSP60 treatment was abrogated partly by the neutralization of IL-10 (Zhong et al. 2016). Taken together, these data strongly indicate that the HSP60 antigen or peptides thereof may be used to ameliorate or even treat atherosclerosis.

4.10 HSP60 Tolerization in Other Autoimmune Diseases

HSP60-peptides have been shown to prevent autoimmune disease in multiple experimental animal models. For instance, HSP60/65 tolerization has been demonstrated to be beneficial in different arthritis models (van Eden et al. 2005, 2007; Zonneveld-Huijssoon et al. 2013). A dose dependent protection or exacerbation of pristine-induced arthritis was seen after immunization with *Mycobacterium vaccae* (a mycobacterial stain in cattle which contain high amounts of HSP65) (Thompson et al. 1991). A T cell response has also been found in adjuvant arthritis after DNA vaccination with human HSP70 and HSP90 (Quintana et al. 2004). The T cell immunity to HSP70 and HSP90 induced an HSP60-specific T cell response, which could modulate the arthritogenic response in adjuvant arthritis. Similarly, DNA vaccination with HSP60 induced HSP70-speciic T cell immunity (Quintana et al. 2004). Epitope mapping studies showed that these HSP60-specific T cells induced by human HSP70 vaccination could react with known regulatory HSP60 epitopes. This suggests that the regulatory mechanisms induced by HSP60, HSP70, and HSP90 are reinforced by an immune network that connects their reactivity (Quintana et al. 2004). The rat HSP65 (rHSP65) C-terminal determinants (rCTD) are displayed as the region whereas the dominant epitopes following the processing and presentation of native rHSP65 is located (Durai et al. 2004). Furthermore, peptides comprising rCTD, when administrated into Lewis rats in a synthetic adjuvant, also induced a T cell response that is protective against adjuvant arthritis (Durai et al. 2004). Thus, anti-self-HSP65 immune response is protective against adjuvant arthritis. Studies using DNA vaccination approach have revealed the disease-regulating activity of human HSP60 in adjuvant arthritis. Human HSP60 was more effective than mbHSP65 in suppressing adjuvant arthritis (Quintana et al. 2002). This protective effect was associated with increased IFN-γ and TGF-β to hHSP60, but reduced IFN-γ to epitope 180–188 of mbHSP65 (Quintana et al. 2002). Subsequent experiments have identified Hu3 as a regulatory epitope of hHSP60 ad immunization with this epitope, but not its mycobacterial homolog, was protective against adjuvant arthritis (Quintana et al. 2003). Interesting, presentation of HSP60 by activated T cells can turn them into targets for anti-ergotypic T cells specific for HSP60 (Quintana et al. 2008). However, the direct interaction between the anti-ergotypic T regulators (anti-HSP60) and the activated T cells can also down-regulate the regulators. Thus, by functioning as an ergotope, HSP60 can control both the effector T cells and the regulatory HSP60-epcific T cells that control them.

HSP65 has also been investigated in different diabetes models. For example, HSP65 has been found to serve as a carrier for the diabetogenic peptide P277-based vaccine and intranasal administration with P277 carried by HSP65 has been shown to decrease the incidence of diabetes, inhibit insulitis, reduce IgG_{2a} isotype antibodies to P277, and pro-inflammatory cytokines (Liang et al. 2010). Intranasal vaccination with P277, in tandem repeat sequences carried by HSP65, indicates that HSP65 may serve as a particularly advantageous carrier for P277-based vaccines and mucosal administration may therefore be a useful therapeutic approach for treatment of type I diabetes (Liang et al. 2010). Moreover, there are increasing numbers of reports suggesting that HSP could also be beneficial in the treatment of various forms of cancer (Saini and Sharma 2017).

4.11 Conclusions

HSP60 has been identified as a target for both pathogenic as well as protective immune mechanisms in many experimental and human inflammatory diseases. The up-regulation of this protein at sites of inflammation on one hand and its ability to skew the immune system towards suppression of disease on the other hand may explain the spontaneous remission of many experimental models of these inflammatory diseases. Atheroprotection has been demonstrated with the induction of immune tolerance through activation of cellular and humoral immune reactions to HSP60. Recent identification of specific immunoreactive antigenic epitopes of HSP60 antigens and the induction of atheroprotection by using them in immunization supports the idea that active vaccination may emerge as a novel immune-modulating atheroprotective strategy. However, understanding of putative mechanisms of this multifactorial disease is important and increased knowledge of the intricate networks governing these tolerizations is highly warranted.

Acknowledgements I grateful to Professor *em.* Georg Wick, Laboratory of Autoimmunity, Section of Experimental Pathophysiology and Immunology, Biocenter, Innsbruck Medical University, Innsbruck, Austria, for his kind help and valuable comments.

References

Aalberse JA, Kapitein B, de Roock S, Klein MR, de Jager W, van der Zee R, Hoekstra MO, van Wijk F, Prakken BJ (2011) Cord blood CD4+ T cells respond to self heat shock protein 60 (HSP60). PLoS One 6:e24119

Afek A, George J, Gilburd B, Rauova L, Goldberg I, Kopolovic J, Harats D, Shoenfeld Y (2000) Immunization of low-density lipoprotein receptor deficient (LDL-RD) mice with heat shock protein 65 (HSP-65) promotes early atherosclerosis. J Autoimmun 14:115–121

Ait-Oufella H, Salomon BL, Potteaux S, Robertson AK, Gourdy P, Zoll J, Merval R, Esposito B, Cohen JL, Fisson S et al (2006) Natural regulatory T cells control the development of atherosclerosis in mice. Nat Med 12:178–180

Almanzar G, Ollinger R, Leuenberger J, Onestingel E, Rantner B, Zehm S, Cardini B, van der Zee R, Grundtman C, Wick G (2012) Autoreactive HSP60 epitope-specific T-cells in early human atherosclerotic lesions. J Autoimmun 39:441–450

Amberger A, Maczek C, Jurgens G, Michaelis D, Schett G, Trieb K, Eberl T, Jindal S, Xu Q, Wick G (1997) Co-expression of ICAM-1, VCAM-1, ELAM-1 and Hsp60 in human arterial and venous endothelial cells in response to cytokines and oxidized low-density lipoproteins. Cell Stress Chaperones 2:94–103

Anderton SM, van der Zee R, Goodacre JA (1993) Inflammation activates self hsp60-specific T cells. Eur J Immunol 23:33–38

Anraku I, Rajasuriar R, Dobbin C, Brown R, Lewin SR, Suhrbier A (2012) Circulating heat shock protein 60 levels are elevated in HIV patients and are reduced by anti-retroviral therapy. PLoS One 7:e45291

Ashburner M (1970) Patterns of puffing activity in the salivary gland chromosomes of Drosophila. V. Responses to environmental treatments. Chromosoma 31:356–376

Basu S, Binder RJ, Suto R, Anderson KM, Srivastava PK (2000) Necrotic but not apoptotic cell death releases heat shock proteins, which deliver a partial maturation signal to dendritic cells and activate the NF-kappa B pathway. Int Immunol 12:1539–1546

Benagiano M, D'Elios MM, Amedei A, Azzurri A, van der Zee R, Ciervo A, Rombola G, Romagnani S, Cassone A, Del Prete G (2005) Human 60-kDa heat shock protein is a target autoantigen of T cells derived from atherosclerotic plaques. J Immunol 174:6509–6517

Bentzon JF, Otsuka F, Virmani R, Falk E (2014) Mechanisms of plaque formation and rupture. Circ Res 114:1852–1866

Bobryshev YV (2010) Dendritic cells and their role in atherogenesis. Lab Investig 90:970–984

Bobryshev YV, Lord RS (1996) Structural heterogeneity and contacting interactions of vascular dendritic cells in early atherosclerotic lesions of the human aorta. J Submicrosc Cytol Pathol 28:49–60

Bobryshev YV, Lord RS (1998) Mapping of vascular dendritic cells in atherosclerotic arteries suggests their involvement in local immune-inflammatory reactions. Cardiovasc Res 37:799–810

Boog CJ, de Graeff-Meeder ER, Lucassen MA, van der Zee R, Voorhorst-Ogink MM, van Kooten PJ, Geuze HJ, van Eden W (1992) Two monoclonal antibodies generated against human hsp60 show reactivity with synovial membranes of patients with juvenile chronic arthritis. J Exp Med 175:1805–1810

Breloer M, Domer B, Moré SH, Roderian T, Fleischer B, von Bonin A (2001) Heat shock proteins as "danger signals": eukaryotic Hsp60 enhanses and accelerates antigen-specific IFN-gamma production in T cells. Eur J Immunol 31:2051–2059

Burian K, Kis Z, Virok D, Endresz V, Prohaszka Z, Duba J, Berencsi K, Boda K, Horvath L, Romics L et al (2001) Independent and joint effects of antibodies to human heat-shock protein 60 and Chlamydia pneumoniae infection in the development of coronary atherosclerosis. Circulation 103:1503–1508

Chandra D, Choy G, Tang DG (2007) Cytosolic accumulation of HSP60 during apoptosis with or without apparent mitochondrial release: evidence that its pro-apoptotic or pro-survival functions involve differential interactions with caspase-3. J Biol Chem 282:31289–31301

Chen W, Syldath U, Bellmann K, Burkart V, Kolb H (1999) Human 60-kDa heat-shock protein: a danger signal to the innate immune system. J Immunol 162:3212–3219

Choi J, Chung SW, Kim SJ, Kim SJ (2001) Establishment of Porphyromonas gingivalis-specific T-cell lines from atherosclerosis patients. Oral Microbiol Immunol 16:316–318

Choi JI, Chung SW, Kang HS, Rhim BY, Kim SJ, Kim SJ (2002) Establishment of Porphyromonas gingivalis heat-shock-protein-specific T-cell lines from atherosclerosis patients. J Dent Res 81:344–348

Choi JI, Chung SW, Kang HS, Rhim BY, Park YM, Kim US, Kim SJ (2004a) Epitope mapping of Porphyromonas gingivalis heat-shock protein and human heat-shock protein in human atherosclerosis. J Dent Res 83:936–940

Choi JI, Kang HS, Park YM, Kim SJ, Kim US (2004b) Identification of T-cell epitopes of Porphyromonas gingivalis heat-shock-protein 60 in periodontitis. Oral Microbiol Immunol 19:1–5

Crisby M, Kallin B, Thyberg J, Zhivotovsky B, Orrenius S, Kostulas V, Nilsson J (1997) Cell death in human atherosclerotic plaques involves both oncosis and apoptosis. Atherosclerosis 130:17–27

Cybulsky MI, Jongstra-Bilen J (2010) Resident intimal dendritic cells and the initiation of atherosclerosis. Curr Opin Lipidol 21:397–403

Czarnecka AM, Campanella C, Zummo G, Cappello F (2006) Mitochondrial chaperones in cancer: from molecular biology to clinical diagnostics. Cancer Biol Ther 5:714–720

Deocaris CC, Kaul SC, Wadhwa R (2006) On the brotherhood of the mitochondrial chaperones mortalin and heat shock protein 60. Cell Stress Chaperones 11:116–128

Durai M, Gupta RS, Moudgil KD (2004) The T cells specific for the carboxyl-terminal determinants of self (rat) heat-shock protein 65 escape tolerance induction and are involved in regulation of autoimmune arthritis. J Immunol 172:2795–2802

Elst EF, Klein M, de Jager W, Kamphuis S, Wedderburn LR, van der Zee R, Albani S, Kuis W, Prakken BJ (2008) Hsp60 in inflamed muscle tissue is the target of regulatory autoreactive T cells in patients with juvenile dermatomyositis. Arthritis Rheum 58:547–555

Faria AM, Weiner HL (1999) Oral tolerance: mechanisms and therapeutic applications. Adv Immunol 73:153–264

Feng H, Zeng Y, Graner MW, Katsanis E (2002) Stressed apoptotic tumor cells stimulate dendritic cells and induce specific cytotoxic T cells. Blood 100:4108–4115

Fisch P, Malkovsky M, Braakman E, Sturm E, Bolhuis RL, Prieve A, Sosman JA, Lam VA, Sondel PM (1990) Gamma/delta T cell clones and natural killer cell clones mediate distinct patterns of non-major histocompatibility complex-restricted cytolysis. J Exp Med 171:1567–1579

Fischer HP, Sharrock CE, Panayi GS (1992) High frequency of cord blood lymphocytes against mycobacterial 65-kDa heat-shock protein. Eur J Immunol 22:1667–1669

Ford PJ, Gemmell E, Hamlet SM, Hasan A, Walker PJ, West MJ, Cullinan MP, Seymour GJ (2005) Cross-reactivity of GroEL antibodies with human heat shock protein 60 and quantification of pathogens in atherosclerosis. Oral Microbiol Immunol 20:296–302

Foteinos G, Afzal AR, Mandal K, Jahangiri M, Xu Q (2005) Anti-heat shock protein 60 autoantibodies induce atherosclerosis in apolipoprotein E-deficient mice via endothelial damage. Circulation 112:1206–1213

Galkina E, Kadl A, Sanders J, Varughese D, Sarembock IJ, Ley K (2006) Lymphocyte recruitment into the aortic wall before and during development of atherosclerosis is partially L-selectin dependent. J Exp Med 203:1273–1282

George J, Shoenfeld Y, Afek A, Gilburd B, Keren P, Shaish A, Kopolovic J, Wick G, Harats D (1999) Enhanced fatty streak formation in C57BL/6J mice by immunization with heat shock protein-65. Arterioscler Thromb Vasc Biol 19:505–510

George J, Greenberg S, Barshack I, Goldberg I, Keren G, Roth A (2003) Immunity to heat shock protein 65 – an additional determinant in intimal thickening. Atherosclerosis 168:33–38

Ghosh JC, Dohi T, Kang BH, Altieri DC (2008) Hsp60 regulation of tumor cell apoptosis. J Biol Chem 283:5188–5194

Gruber R, Lederer S, Bechtel U, Lob S, Riethmuller G, Feucht HE (1996) Increased antibody titers against mycobacterial heat-shock protein 65 in patients with vasculitis and arteriosclerosis. Int Arch Allergy Immunol 110:95–98

Grundtman C, Wick G (2011) The autoimmune concept of atherosclerosis. Curr Opin Lipidol 22:327–334

Grundtman C, Kreutmayer SB, Almanzar G, Wick MC, Wick G (2011) Heat shock protein 60 and immune inflammatory responses in atherosclerosis. Arterioscler Thromb Vasc Biol 31:960–968

Grundtman C, Jakic B, Buszko M, Onestingel E, Almanzar G, Demetz E, Dietrich H, Cappellano G, Wick G (2015) Mycobacterial heat shock protein 65 (mbHSP65)-induced atherosclerosis: preventive oral tolerization and definition of atheroprotective and atherogenic mbHSP65 peptides. Atherosclerosis 242:303–310

Harats D, Yacov N, Gilburd B, Shoenfeld Y, George J (2002) Oral tolerance with heat shock protein 65 attenuates Mycobacterium tuberculosis-induced and high-fat-diet-driven atherosclerotic lesions. J Am Coll Cardiol 40:1333–1338

Hochleitner BW, Hochleitner EO, Obrist P, Eberl T, Amberger A, Xu Q, Margreiter R, Wick G (2000) Fluid shear stress induces heat shock protein 60 expression in endothelial cells in vitro and in vivo. Arterioscler Thromb Vasc Biol 20:617–623

Hoppichler F, Koch T, Dzien A, Gschwandtner G, Lechleitner M (2000) Prognostic value of antibody titre to heat-shock protein 65 on cardiovascular events. Cardiology 94:220–223

Horvath L, Czirjak L, Fekete B, Jakab L, Prohaszka Z, Cervenak L, Romics L, Singh M, Daha MR, Fust G (2001) Levels of antibodies against C1q and 60 kDa family of heat shock proteins in the sera of patients with various autoimmune diseases. Immunol Lett 75:103–109

Horvath M, Hajek P, Stechovsky C, Honek J, Spacek M, Veselka J (2016) The role of near-infrared spectroscopy in the detection of vulnerable atherosclerotic plaques. Arch Med Sci 12:1308–1316

Houtkamp MA, de Boer OJ, van der Loos CM, van der Wal AC, Becker AE (2001) Adventitial infiltrates associated with advanced atherosclerotic plaques: structural organization suggests generation of local humoral immune responses. J Pathol 193:263–269

Huang QQ, Sobkoviak R, Jockheck-Clark AR, Shi B, Mandelin AM 2nd, Tak PP, Haines GK 3rd, Nicchitta CV, Pope RM (2009) Heat shock protein 96 is elevated in rheumatoid arthritis and activates macrophages primarily via TLR2 signaling. J Immunol 182:4965–4973

Huang MN, Yu H, Moudgil KD (2010) The involvement of heat-shock proteins in the pathogenesis of autoimmune arthritis: a critical appraisal. Semin Arthritis Rheum 40:164–175

Huber SA, Sakkinen P, David C, Newell MK, Tracy RP (2001) T helper-cell phenotype regulates atherosclerosis in mice under conditions of mild hypercholesterolemia. Circulation 103:2610–2616

Imamura Y, Kurokawa MS, Yoshikawa H, Nara K, Takada E, Masuda C, Tsukikawa S, Ozaki S, Matsuda T, Suzuki N (2005) Involvement of Th1 cells and heat shock protein 60 in the pathogenesis of intestinal Behcet's disease. Clin Exp Immunol 139:371–378

Jakic B, Buszko M, Cappellano G, Wick G (2017) Elevated sodium leads to the increased expression of HSP60 and induces apoptosis in HUVECs. PLoS One 12:e0179383

Jiang H, Canfield SM, Gallagher MP, Jiang HH, Jiang Y, Zheng Z, Chess L (2010) HLA-E-restricted regulatory CD8(+) T cells are involved in development and control of human autoimmune type 1 diabetes. J Clin Invest 120:3641–3650

Jing H, Yong L, Haiyan L, Yanjun M, Yun X, Yu Z, Taiming L, Rongyue C, Liang J, Jie W et al (2011) Oral administration of Lactococcus lactis delivered heat shock protein 65 attenuates atherosclerosis in low-density lipoprotein receptor-deficient mice. Vaccine 29:4102–4109

Kanwar RK, Kanwar JR, Wang D, Ormrod DJ, Krissansen GW (2001) Temporal expression of heat shock proteins 60 and 70 at lesion-prone sites during atherogenesis in ApoE-deficient mice. Arterioscler Thromb Vasc Biol 21:1991–1997

Kapitein B, Aalberse JA, Klein MR, de Jager W, Hoekstra MO, Knol EF, Prakken BJ (2013) Recognition of self-heat shock protein 60 by T cells from patients with atopic dermatitis. Cell Stress Chaperones 18:87–95

Kaufmann SH (1990) Heat shock proteins and the immune response. Immunol Today 11:129–136

Khallou-Laschet J, Tupin E, Caligiuri G, Poirier B, Thieblemont N, Gaston AT, Vandaele M, Bleton J, Tchapla A, Kaveri SV et al (2006) Atheroprotective effect of adjuvants in apolipoprotein E knockout mice. Atherosclerosis 184:330–341

Kleindienst R, Xu Q, Willeit J, Waldenberger FR, Weimann S, Wick G (1993) Immunology of atherosclerosis. Demonstration of heat shock protein 60 expression and T lymphocytes bearing alpha/beta or gamma/delta receptor in human atherosclerotic lesions. Am J Pathol 142:1927–1937

Klingenberg R, Ketelhuth DF, Strodthoff D, Gregori S, Hansson GK (2012) Subcutaneous immunization with heat shock protein-65 reduces atherosclerosis in Apoe(−)/(−) mice. Immunobiology 217:540–547

Knoflach M, Kiechl S, Kind M, Said M, Sief R, Gisinger M, van der Zee R, Gaston H, Jarosch E, Willeit J et al (2003) Cardiovascular risk factors and atherosclerosis in young males: ARMY study (Atherosclerosis Risk-Factors in Male Youngsters). Circulation 108:1064–1069

Knoflach M, Kiechl S, Mayrl B, Kind M, Gaston JS, van der Zee R, Faggionato A, Mayr A, Willeit J, Wick G (2007) T-cell reactivity against HSP60 relates to early but not advanced atherosclerosis. Atherosclerosis 195:333–338

Knoflach M, Kiechl S, Penz D, Zangerle A, Schmidauer C, Rossmann A, Shingh M, Spallek R, Griesmacher A, Bernhard D et al (2009) Cardiovascular risk factors and atherosclerosis in young women: atherosclerosis risk factors in female youngsters (ARFY study). Stroke 40:1063–1069

Kol A, Lichtman AH, Finberg RW, Libby P, Kurt-Jones EA (2000) Cutting edge: heat shock protein (HSP) 60 activates the innate immune response: CD14 is an essential receptor for HSP60 activation of mononuclear cells. J Immunol 164:13–17

Kreutmayer SB, Messner B, Knoflach M, Henderson B, Niederegger H, Bock G, Van der Zee R, Wick G, Bernhard D (2011) Dynamics of heat shock protein 60 in endothelial cells exposed to cigarette smoke extract. J Mol Cell Cardiol 51:777–780

Kreutmayer S, Csordas A, Kern J, Maass V, Almanzar G, Offterdinger M, Ollinger R, Maass M, Wick G (2013) Chlamydia pneumoniae infection acts as an endothelial stressor with the potential to initiate the earliest heat shock protein 60-dependent inflammatory stage of atherosclerosis. Cell Stress Chaperones 18:259–268

Lewthwaite J, Owen N, Coates A, Henderson B, Steptoe A (2002) Circulating human heat shock protein 60 in the plasma of British civil servants: relationship to physiological and psychosocial stress. Circulation 106:196–201

Li H, Ding Y, Yi G, Zeng Q, Yang W (2012) Establishment of nasal tolerance to heat shock protein-60 alleviates atherosclerosis by inducing TGF-beta-dependent regulatory T cells. J Huazhong Univ Sci Technol Med Sci 32:24–30

Liang J, Aihua Z, Yu W, Yong L, Jingjing L (2010) HSP65 serves as an immunogenic carrier for a diabetogenic peptide P277 inducing anti-inflammatory immune response in NOD mice by nasal administration. Vaccine 28:3312–3317

Lindquist S, Craig EA (1988) The heat-shock proteins. Annu Rev Genet 22:631–677

Liu P, Yu YR, Spencer JA, Johnson AE, Vallanat CT, Fong AM, Patterson C, Patel DD (2008) CX3CR1 deficiency impairs dendritic cell accumulation in arterial intima and reduces atherosclerotic burden. Arterioscler Thromb Vasc Biol 28:243–250

Liu A, Ming JY, Fiskesund R, Ninio E, Karabina SA, Bergmark C, Frostegard AG, Frostegard J (2015) Induction of dendritic cell-mediated T-cell activation by modified but not native low-density lipoprotein in humans and inhibition by annexin a5: involvement of heat shock proteins. Arterioscler Thromb Vasc Biol 35:197–205

Lu X, Chen D, Endresz V, Xia M, Faludi I, Burian K, Szabo A, Csanadi A, Miczak A, Gonczol E et al (2010) Immunization with a combination of ApoB and HSP60 epitopes significantly reduces early atherosclerotic lesion in Apobtm2SgyLdlrtm1Her/J mice. Atherosclerosis 212:472–480

Lydyard PM, van Eden W (1990) Heat shock proteins: immunity and immunopathology. Immunol Today 11:228–229

Maiellaro K, Taylor WR (2007) The role of the adventitia in vascular inflammation. Cardiovasc Res 75:640–648

Mallat Z, Ait-Oufella H, Tedgui A (2005) Regulatory T cell responses: potential role in the control of atherosclerosis. Curr Opin Lipidol 16:518–524

Marengo EB, Commodaro AG, Peron JP, de Moraes LV, Portaro FC, Belfort R Jr, Rizzo LV, Sant'Anna OA (2009) Administration of Mycobacterium leprae rHsp65 aggravates experimental autoimmune uveitis in mice. PLoS One 4:e7912

Maron R, Sukhova G, Faria AM, Hoffmann E, Mach F, Libby P, Weiner HL (2002) Mucosal administration of heat shock protein-65 decreases atherosclerosis and inflammation in aortic arch of low-density lipoprotein receptor-deficient mice. Circulation 106:1708–1715

Martin J (1997) Molecular chaperones and mitochondrial protein folding. J Bioenerg Biomembr 29:35–43

Mayr M, Metzler B, Kiechl S, Willeit J, Schett G, Xu Q, Wick G (1999) Endothelial cytotoxicity mediated by serum antibodies to heat shock proteins of Escherichia coli and Chlamydia pneumoniae: immune reactions to heat shock proteins as a possible link between infection and atherosclerosis. Circulation 99:1560–1566

Mayr M, Kiechl S, Willeit J, Wick G, Xu Q (2000) Infections, immunity, and atherosclerosis: associations of antibodies to Chlamydia pneumoniae, Helicobacter pylori, and cytomegalovirus with immune reactions to heat-shock protein 60 and carotid or femoral atherosclerosis. Circulation 102:833–839

McKenzie SL, Henikoff S, Meselson M (1975) Localization of RNA from heat-induced polysomes at puff sites in *Drosophila melanogaster*. Proc Natl Acad Sci U S A 72:1117–1121

Metzler B, Wraith DC (1999) Inhibition of T-cell responsiveness by nasal peptide administration: influence of the thymus and differential recovery of T-cell-dependent functions. Immunology 97:257–263

Michaelsson J, Teixeira de Matos C, Achour A, Lanier LL, Karre K, Soderstrom K (2002) A signal peptide derived from hsp60 binds HLA-E and interferes with CD94/NKG2A recognition. J Exp Med 196:1403–1414

Millonig G, Niederegger H, Rabl W, Hochleitner BW, Hoefer D, Romani N, Wick G (2001a) Network of vascular-associated dendritic cells in intima of healthy young individuals. Arterioscler Thromb Vasc Biol 21:503–508

Millonig G, Niederegger H, Wick G (2001b) Analysis of the cellular composition of the arterial intima with modified en face techniques. Lab Investig 81:639–641

Millonig G, Schwentner C, Mueller P, Mayerl C, Wick G (2001c) The vascular-associated lymphoid tissue: a new site of local immunity. Curr Opin Lipidol 12:547–553

Millonig G, Malcom GT, Wick G (2002) Early inflammatory-immunological lesions in juvenile atherosclerosis from the Pathobiological Determinants of Atherosclerosis in Youth (PDAY)-study. Atherosclerosis 160:441–448

Moos MP, John N, Grabner R, Nossmann S, Gunther B, Vollandt R, Funk CD, Kaiser B, Habenicht AJ (2005) The lamina adventitia is the major site of immune cell accumulation in standard chow-fed apolipoprotein E-deficient mice. Arterioscler Thromb Vasc Biol 25:2386–2391

Moran L, Mirault ME, Arrigo AP, Goldschmidt-Clermont M, Tissieres A (1978) Heat shock of *Drosophila melanogaster* induces the synthesis of new messenger RNAs and proteins. Philos Trans R Soc Lond Ser B Biol Sci 283:391–406

More SH, Breloer M, von BA (2001) Eukaryotic heat shock proteins as molecular links in innate and adaptive immune responses: Hsp60-mediated activation of cytotoxic T cells. Int Immunol 13:1121–1127

Mori Y, Kitamura H, Song QH, Kobayashi T, Umemura S, Cyong JC (2000) A new murine model for atherosclerosis with inflammation in the periodontal tissue induced by immunization with heat shock protein 60. Hypertens Res 23:475–481

Moulton KS, Vakili K, Zurakowski D, Soliman M, Butterfield C, Sylvin E, Lo KM, Gillies S, Javaherian K, Folkman J (2003) Inhibition of plaque neovascularization reduces macrophage accumulation and progression of advanced atherosclerosis. Proc Natl Acad Sci U S A 100:4736–4741

O'Brien KD, McDonald TO, Chait A, Allen MD, Alpers CE (1996) Neovascular expression of E-selectin, intercellular adhesion molecule-1, and vascular cell adhesion molecule-1 in human atherosclerosis and their relation to intimal leukocyte content. Circulation 93:672–682

Oda A, Miyata M, Kodama E, Satoh H, Sato Y, Nishimaki T, Nomaguchi H, Kasukawa R (1994) Antibodies to 65Kd heat-shock protein were elevated in rheumatoid arthritis. Clin Rheumatol 13:261–264

Okamoto E, Couse T, De Leon H, Vinten-Johansen J, Goodman RB, Scott NA, Wilcox JN (2001) Perivascular inflammation after balloon angioplasty of porcine coronary arteries. Circulation 104:2228–2235

Osborn EA, Jaffer FA (2013) Imaging atherosclerosis and risk of plaque rupture. Curr Atheroscler Rep 15:359

Osterloh A, Meier-Stiegen F, Veit A, Fleischer B, von Bonin A, Breloer M (2004) Lipopolysaccharide-free heat shock protein 60 activates T cells. J Biol Chem 279:47906–47911

Osterloh A, Geisinger F, Pie'davent M, Fleischer B, Brattig N, Breloer M (2009) Heat shock protein 60 (HSP60) stimulates neutrophil effector functions. J Leukoc Biol 86:423–434

Panchapakesan J, Daglis M, Gatenby P (1992) Antibodies to 65 kDa and 70 kDa heat shock proteins in rheumatoid arthritis and systemic lupus erythematosus. Immunol Cell Biol 70(Pt 5):295–300

Perschinka H, Mayr M, Millonig G, Mayerl C, van der Zee R, Morrison SG, Morrison RP, Xu Q, Wick G (2003) Cross-reactive B-cell epitopes of microbial and human heat shock protein 60/65 in atherosclerosis. Arterioscler Thromb Vasc Biol 23:1060–1065

Pfister G, Stroh CM, Perschinka H, Kind M, Knoflach M, Hinterdorfer P, Wick G (2005) Detection of HSP60 on the membrane surface of stressed human endothelial cells by atomic force and confocal microscopy. J Cell Sci 118:1587–1594

Pockley AG, Wu R, Lemne C, Kiessling R, de Faire U, Frostegard J (2000) Circulating heat shock protein 60 is associated with early cardiovascular disease. Hypertension 36:303–307

Puga Yung GL, Fidler M, Albani E, Spermon N, Teklenburg G, Newbury R, Schechter N, van den Broek T, Prakken B, Billetta R et al (2009) Heat shock protein-derived T-cell epitopes contribute to autoimmune inflammation in pediatric Crohn's disease. PLoS One 4:e7714

Quintana FJ, Carmi P, Mor F, Cohen IR (2002) Inhibition of adjuvant arthritis by a DNA vaccine encoding human heat shock protein 60. J Immunol 169:3422–3428

Quintana FJ, Carmi P, Mor F, Cohen IR (2003) DNA fragments of the human 60-kDa heat shock protein (HSP60) vaccinate against adjuvant arthritis: identification of a regulatory HSP60 peptide. J Immunol 171:3533–3541

Quintana FJ, Carmi P, Mor F, Cohen IR (2004) Inhibition of adjuvant-induced arthritis by DNA vaccination with the 70-kd or the 90-kd human heat-shock protein: immune cross-regulation with the 60-kd heat-shock protein. Arthritis Rheum 50:3712–3720

Quintana FJ, Mimran A, Carmi P, Mor F, Cohen IR (2008) HSP60 as a target of anti-ergotypic regulatory T cells. PLoS One 3:e4026

Retzlaff C, Yamamoto Y, Hoffman PS, Friedman H, Klein TW (1994) Bacterial heat shock proteins directly induce cytokine mRNA and interleukin-1 secreation in macrophage cultures. Infect Immun 62:5689–5693

Ritossa FM (1962) A new puffing pattern induced by temperature shock and DNP in Drosophila. Experientia 18:571–573

Ritossa FM (1964) Experimental activation of specific loci in polytene chromosomes of drosophila. Exp Cell Res 35:601–607

Rossmann A, Henderson B, Heidecker B, Seiler R, Fraedrich G, Singh M, Parson W, Keller M, Grubeck-Loebenstein B, Wick G (2008) T-cells from advanced atherosclerotic lesions recognize hHSP60 and have a restricted T-cell receptor repertoire. Exp Gerontol 43:229–237

Roth GA, Huffman MD, Moran AE, Feigin V, Mensah GA, Naghavi M, Murray CJ (2015) Global and regional patterns in cardiovascular mortality from 1990 to 2013. Circulation 132:1667–1678

Saini J, Sharma PK (2017) Clinical, prognostic and therapeutic significance of heat shock proteins in cancer. Curr Drug Targets 18:1478–1490. E-pub Ahead of Print

Sasaki N, Yamashita T, Takeda M, Shinohara M, Nakajima K, Tawa H, Usui T, Hirata K (2009) Oral anti-CD3 antibody treatment induces regulatory T cells and inhibits the development of atherosclerosis in mice. Circulation 120:1996–2005

Schett G, Xu Q, Amberger A, Van der Zee R, Recheis H, Willeit J, Wick G (1995) Autoantibodies against heat shock protein 60 mediate endothelial cytotoxicity. J Clin Invest 96:2569–2577

Schett G, Metzler B, Mayr M, Amberger A, Niederwieser D, Gupta RS, Mizzen L, Xu Q, Wick G (1997) Macrophage-lysis mediated by autoantibodies to heat shock protein 65/60. Atherosclerosis 128:27–38

Schwartz CJ, Mitchell JR (1962) Cellular infiltration of the human arterial adventitia associated with atheromatous plaques. Circulation 26:73–78

Seitz CS, Kleindienst R, Xu Q, Wick G (1996) Coexpression of heat-shock protein 60 and intercellular-adhesion molecule-1 is related to increased adhesion of monocytes and T cells to aortic endothelium of rats in response to endotoxin. Lab Investig 74:241–252

Shoenfeld Y, Harats D, George J (2000) Heat shock protein 60/65, beta 2-glycoprotein I and oxidized LDL as players in murine atherosclerosis. J Autoimmun 15:199–202

Sobel DO, Creswell K (2006) Characterization of anti-islet cytotoxic human T-cell clones from patients with type 1 diabetes mellitus. Autoimmunity 39:323–332

Song L, Leung C, Schindler C (2001) Lymphocytes are important in early atherosclerosis. J Clin Invest 108:251–259

Spradling A, Penman S, Pardue ML (1975) Analysis of drosophila mRNA by in situ hybridization: sequences transcribed in normal and heat shocked cultured cells. Cell 4:395–404

Sun J, Hartvigsen K, Chou MY, Zhang Y, Sukhova GK, Zhang J, Lopez-Ilasaca M, Diehl CJ, Yakov N, Harats D et al (2010) Deficiency of antigen-presenting cell invariant chain reduces atherosclerosis in mice. Circulation 122:808–820

Taleb S, Tedgui A, Mallat Z (2008) Regulatory T-cell immunity and its relevance to atherosclerosis. J Intern Med 263:489–499

Thompson SJ, Butcher PD, Patel VK, Rook GA, Stanford J, van der Zee R, Elson CJ (1991) Modulation of pristane-induced arthritis by mycobacterial antigens. Autoimmunity 11:35–43

Tissieres A, Mitchell HK, Tracy UM (1974) Protein synthesis in salivary glands of *Drosophila melanogaster*: relation to chromosome puffs. J Mol Biol 84:389–398

Tytell M, Greenberg SG, Lasek RJ (1986) Heat shock-like protein is transferred from glia to axon. Brain Res 363:161–164

van Eden W, van der Zee R, Prakken B (2005) Heat-shock proteins induce T-cell regulation of chronic inflammation. Nat Rev Immunol 5:318–330

van Eden W, Wick G, Albani S, Cohen I (2007) Stress, heat shock proteins, and autoimmunity: how immune responses to heat shock proteins are to be used for the control of chronic inflammatory diseases. Ann N Y Acad Sci 1113:217–237 van Puijvelde GH, van Es T, van Wanrooij EJ, Habets KL, de Vos P, van der Zee R, van Eden W, van Berkel TJ, Kuiper J (2007) Induction of oral tolerance to HSP60 or an HSP60-peptide activates T cell regulation and reduces atherosclerosis. Arterioscler Thromb Vasc Biol 27:2677–2683

Veres A, Szamosi T, Ablonczy M, Szamosi T Jr, Singh M, Karadi I, Romics L, Fust G, Prohaszka Z (2002) Complement activating antibodies against the human 60 kDa heat shock protein as a new independent family risk factor of coronary heart disease. Eur J Clin Investig 32:405–410

Wadhwa R, Takano S, Kaur K, Aida S, Yaguchi T, Kaul Z, Hirano T, Taira K, Kaul SC (2005) Identification and characterization of molecular interactions between mortalin/mtHsp70 and HSP60. Biochem J 391:185–190

Waltner-Romen M, Falkensammer G, Rabl W, Wick G (1998) A previously unrecognized site of local accumulation of mononuclear cells. The vascular-associated lymphoid tissue. J Histochem Cytochem 46:1347–1350

Watanabe M, Sangawa A, Sasaki Y, Yamashita M, Tanaka-Shintani M, Shintaku M, Ishikawa Y (2007) Distribution of inflammatory cells in adventitia changed with advancing atherosclerosis of human coronary artery. J Atheroscler Thromb 14:325–331

Weiner HL, da Cunha AP, Quintana F, Wu H (2011) Oral tolerance. Immunol Rev 241:241–259

Weitgasser R, Lechleitner M, Koch T, Galvan G, Muhlmann J, Steiner K, Hoppichler F (2003) Antibodies to heat-shock protein 65 and neopterin levels in patients with type 1 diabetes mellitus. Exp Clin Endocrinol Diabetes 111:127–131

Wick C (2016) Tolerization against atherosclerosis using heat shock protein 60. Cell Stress Chaperones 21:201–211

Wick G, Romen M, Amberger A, Metzler B, Mayr M, Falkensammer G, Xu Q (1997) Atherosclerosis, autoimmunity, and vascular-associated lymphoid tissue. FASEB J 11:1199–1207

Wick G, Knoflach M, Xu Q (2004) Autoimmune and inflammatory mechanisms in atherosclerosis. Annu Rev Immunol 22:361–403

Wick MC, Mayerl C, Backovic A, van der Zee R, Jaschke W, Dietrich H, Wick G (2008) In vivo imaging of the effect of LPS on arterial endothelial cells: molecular imaging of heat shock protein 60 expression. Cell Stress Chaperones 13:275–285

Wick G, Jakic B, Buszko M, Wick MC, Grundtman C (2014) The role of heat shock proteins in atherosclerosis. Nat Rev Cardiol 11:516–529

Wick C, Onestingel E, Demetz E, Dietrich H, Wick G (2018) Oral tolerization with mycobacterial heat shock protein 65 reduces chronic experimental atherosclerosis in aged mice. Gerontology 64:36–48

Willeit J, Kiechl S (1993) Prevalence and risk factors of asymptomatic extracranial carotid artery atherosclerosis. A population-based study. Arterioscler Thromb 13:661–668

Wu CT, Ou LS, Yeh KW, Lee WI, Huang JL (2011) Serum heat shock protein 60 can predict remission of flare-up in juvenile idiopathic arthritis. Clin Rheumatol 30:959–965

Xiao Q, Mandal K, Schett G, Mayr M, Wick G, Oberhollenzer F, Willeit J, Kiechl S, Xu Q (2005) Association of serum-soluble heat shock protein 60 with carotid atherosclerosis: clinical significance determined in a follow-up study. Stroke 36:2571–2576

Xiong Q, Li J, Jin L, Liu J, Li T (2009) Nasal immunization with heat shock protein 65 attenuates atherosclerosis and reduces serum lipids in cholesterol-fed wild-type rabbits probably through different mechanisms. Immunol Lett 125:40–45

Xiong Q, Feng J, Zhang Y, Sun Y, Lu Y, Li T, Zhang X, Cao R, Jin L, Wu J (2016) Promotion of atherosclerosis in high cholesterol diet-fed rabbits by immunization with the P277 peptide. Immunol Lett 170:80–87

Xu QB, Oberhuber G, Gruschwitz M, Wick G (1990) Immunology of atherosclerosis: cellular composition and major histocompatibility complex class II antigen expression in aortic intima, fatty streaks, and atherosclerotic plaques in young and aged human specimens. Clin Immunol Immunopathol 56:344–359

Xu Q, Dietrich H, Steiner HJ, Gown AM, Schoel B, Mikuz G, Kaufmann SH, Wick G (1992) Induction of arteriosclerosis in normocholesterolemic rabbits by immunization with heat shock protein 65. Arterioscler Thromb 12:789–799

Xu Q, Kleindienst R, Waitz W, Dietrich H, Wick G (1993a) Increased expression of heat shock protein 65 coincides with a population of infiltrating T lymphocytes in atherosclerotic lesions of rabbits specifically responding to heat shock protein 65. J Clin Invest 91:2693–2702

Xu Q, Willeit J, Marosi M, Kleindienst R, Oberhollenzer F, Kiechl S, Stulnig T, Luef G, Wick G (1993b) Association of serum antibodies to heat-shock protein 65 with carotid atherosclerosis. Lancet 341:255–259

Xu Q, Kleindienst R, Schett G, Waitz W, Jindal S, Gupta RS, Dietrich H, Wick G (1996) Regression of arteriosclerotic lesions induced by immunization with heat shock protein 65-containing material in normocholesterolemic, but not hypercholesterolemic, rabbits. Atherosclerosis 123:145–155

Xu Q, Kiechl S, Mayr M, Metzler B, Egger G, Oberhollenzer F, Willeit J, Wick G (1999) Association of serum antibodies to heat-shock protein 65 with carotid atherosclerosis: clinical significance determined in a follow-up study. Circulation 100:1169–1174

Xu Q, Schett G, Perschinka H, Mayr M, Egger G, Oberhollenzer F, Willeit J, Kiechl S, Wick G (2000) Serum soluble heat shock protein 60 is elevated in subjects with atherosclerosis in a general population. Circulation 102:14–20

Young RA, Elliott TJ (1989) Stress proteins, infection, and immune surveillance. Cell 59:5–8

Zanin-Zhorov A, Cohen IR (2013) Signaling via TLR2 and TLR4 directly down-regulates T cell effector functions: the regulatory face of danger signals. Front Immunol 4:211

Zanin-Zhorov A, Nussbaum G, Franitza S, Cohen IR, Lider O (2003) T cells respond to heat shock protein 60 via TLR2: activation of adhesion and inhibition of chemokine receptors. FASEB J 17:1567–1569

Zhang X, He M, Cheng L, Chen Y, Zhou L, Zeng H, Pockley AG, Hu FB, Wu T (2008) Elevated heat shock protein 60 levels are associated with higher risk of coronary heart disease in Chinese. Circulation 118:2687–2693

Zhao L, Moos MP, Grabner R, Pedrono F, Fan J, Kaiser B, John N, Schmidt S, Spanbroek R, Lotzer K et al (2004) The 5-lipoxygenase pathway promotes pathogenesis of hyperlipidemia-dependent aortic aneurysm. Nat Med 10:966–973

Zhong Y, Tang H, Wang X, Zeng Q, Liu Y, Zhao XI, Yu K, Shi H, Zhu R, Mao X (2016) Intranasal immunization with heat shock protein 60 induces CD4(+) CD25(+) GARP(+) and type 1 regulatory T cells and inhibits early atherosclerosis. Clin Exp Immunol 183:452–468

Zhou X, Nicoletti A, Elhage R, Hansson GK (2000) Transfer of CD4(+) T cells aggravates atherosclerosis in immunodeficient apolipoprotein E knockout mice. Circulation 102:2919–2922

Zonneveld-Huijssoon E, Albani S, Prakken BJ, van Wijk F (2013) Heat shock protein bystander antigens for peptide immunotherapy in autoimmune disease. Clin Exp Immunol 171:20–29

Zou Y, Dietrich H, Hu Y, Metzler B, Wick G, Xu Q (1998) Mouse model of venous bypass graft arteriosclerosis. Am J Pathol 153:1301–1310

Zugel U, Schoel B, Yamamoto S, Hengel H, Morein B, Kaufmann SH (1995) Crossrecognition by CD8 T cell receptor alpha beta cytotoxic T lymphocytes of peptides in the self and the mycobacterial hsp60 which share intermediate sequence homology. Eur J Immunol 25:451–458

Chapter 5
Role of Heat Shock Protein 90 in Autoimmune Inflammatory Rheumatic Diseases

Hana Storkanova and Michal Tomcik

Abstract Hsp90 is the most studied member of the heat shock proteins family (HSP), which are characterized by induction by increased temperature and various other types of stress. It is a highly conserved molecular chaperone that plays a significant role in many cellular processes. Hsp90 is required for the proper conformation and activation of a number of client cellular proteins, including protein kinases, transcription factors and steroid receptors that play an important role in signal transduction. It also regulates activation of innate immunity, antigen presentation, and the induction of proinflammatory cytokines and chemokines by macrophages and dendritic cells. These properties predispose Hsp90 to a potential role in the pathogenesis of autoimmune inflammatory rheumatic diseases. This article provides an overview of the available knowledge about the potential role of Hsp90 in currently studied rheumatic diseases as a promising candidate for targeted therapy or biomarker of disease activity and severity or a predictor of therapeutic response.

Keywords Autoimmunity · Chaperone · Cytokine · Hsp90 · Inflammation · Rheumatic diseases

Abbreviations

17-DMAG	17-dimethylaminoethylamino-17-demethoxygeldanamycin
AAV	Adeno-associated virus
APC	Antigen presenting cells

H. Storkanova
Institute of Rheumatology, 1st Faculty of Medicine, Charles University, Prague, Czech Republic

Department of Rheumatology, 1st Faculty of Medicine, Charles University, Prague, Czech Republic

M. Tomcik (✉)
Institute of Rheumatology and Department of Rheumatology, 1st Faculty of Medicine, Charles University, Prague, Czech Republic
e-mail: tomcik@revma.cz

A. A. A. Asea, P. Kaur (eds.), *Chaperokine Activity of Heat Shock Proteins*, Heat Shock Proteins 16, https://doi.org/10.1007/978-3-030-02254-9_5

AS	Ankylosing spondylitis
ATP	Adenosine triphosphate
CD	Cluster of differentiation
Cdc37	Cell division cycle 37
CK2	Casein kinase 2
CRP	C-reactive protein
DAMP	Danger activated molecular pattern
DAS-28	Disease activity score-28
DC	Dendritic cell
DM	Dermatomyositis
ERK	Extracellular signal-regulated kinase
GM-CSF	Granulocyte-macrophage colony-stimulating factor
HLA-B27	Human leukocyte antigen B27
HSP	Heat shock protein family
Hsp	Heat shock proteins
IBM	Inclusion body myositis
ICAM-1	Intracellular adhesion molecule-1
Ig	Immunoglobulin
IIM	Idiopathic inflammatory myopathies
IKK	I-κB kinase
IL	Interleukin
ILD	Interstitial lung disease
IMNM	Immune-mediated necrotizing myopathy
iNOS	Inducible nitric oxide synthase
kDa	KiloDalton
LOX-1	Lectin-like oxidized low-density lipoprotein receptor-1
MAPK	Mitogen-activated protein kinase
MCP-1	Monocyte chemoattractant protein-1
MHC	Major histocompatibility complex
MIP-1	Macrophage inflammatory protein-1
MyoD	Myogenic regulatory protein
NF-κB	Nucelar factor κB
NK	Natural killer cells
nr-axSpA	Non-radiographic axial spondyloarthritis
OA	Osteoarthritis
p23	Prostaglandin E synthase 3
PBMC	Peripheral blood mononuclear cells
PM	Polymyositis
PRR	Pattern recognition receptors
PsA	Psoriatic arthritis
RA	Rheumatoid arthritis
RANTES	Regulated on activation, normal T cell expressed and secreted
SLE	Systemic lupus erythematosus
SpA	Spondyloarthritis
SS	Sjögren's syndrome
SSc	Systemic sclerosis

TGF	Transforming growth factor
TLR	Toll-like receptor
TNF	Tumor necrosis factor
TRAP1	TNF receptor-associated protein 1
TβRI	TGF-β type I receptor
VCAM-1	Vascular cell adhesion molecule

5.1 Introduction

The first mention of the existence of heat shock proteins (Hsp) was reported in 1974 in *Drosophila* salivary glands, in which the synthesis of a small number of proteins occurred immediately after the exposure of the cells to higher temperatures (Schlesinger 1990). This has led to the discovery of a universal protective mechanism that prokaryotic and eukaryotic cells use to maintain cellular homeostasis and biological processes (Lindquist and Craig 1988). In eukaryotic cells, the synthesis of Hsp can be induced not only by hyperthermia but also by a variety of stressful situations that lead to the accumulation of non-native proteins, including changes in the intracellular environment, the presence of viral infection (Santoro 2000), processes involved in tissue healing and remodelling (Laplante et al. 1998) or exposition to UV radiation (Cao et al. 1999). Transcription of Hsp genes is also activated under physiological conditions during cell cycle, cell development, differentiation and proliferation. The primary function of these proteins is to recognize and bind improperly folded proteins, and to stabilize them by creating the right three dimensional conformation. In this way, the proteins are protected against degradation in the proteasome. They also play an important role during protein transport via membranes. Hsp are so called molecular chaperones, which are divided into groups according to their molecular weight (Hsp100, Hsp90, Hsp70, Hsp60, Hsp40 and small Hsp) (Lindquist and Craig 1988).

Ninety-kDa molecular weight Hsp90 is a highly conserved molecular chaperone present in bacteria and eukaryotic organisms. It plays an important role in many cell processes, including control of cell cycle and survival (Burrows et al. 2004; Echeverria and Picard 2010; Pearl and Prodromou 2000). Hsp90 is required for the correct folding and activation of a number of cellular proteins and protein complexes. Proteins that interact with Hsp90 are collectively called clients, and many of them are protein kinases, transcription factors and steroid receptors that play an important role in signal transduction (McClellan et al. 2007). Hsp90 is an important part of the cell response to stress situations and has a key function in inducing cell homeostasis (Taipale et al. 2010).

In eukaryotic organisms, more Hsp90 protein homologues can be found in cytosol (Hsp90-α, Hsp90-β) or in various cellular structures such as endoplasmic reticulum (endoplasmin/GRP-94) and mitochondria (TRAP1, TNF Receptor-Associated Protein 1). Cytosolic Hsp90 isoforms are the result of gene duplication that occurred early during eukaryotic evolution (Chen et al. 2006). Hsp90-α is encoded by the HSP90AA1 gene on human chromosome 14q and belongs to inducible proteins that are expressed only under non-physiological conditions (Zuehlke et al. 2015). Hsp90-β is encoded by the HSP90AB1 gene on the human chromosome 6p and is constitutively expressed. In spite of sharing 85% of identical sequences, these isoforms have completely different properties (Millson et al. 2007; Taherian et al. 2008). Bacterial form Hsp90 is called HtpG and is usually non-essential. Unlike bacteria, eukaryotes require functional cytoplasmic Hsp90 for viability under all physiological and non-physiological conditions (Lund 2001). The necessary presence of Hsp90 has thus enabled the use of yeast to provide a more detailed analysis of the structural and functional properties of Hsp90 (Meyer et al. 2003; Obermann et al. 1998). However, the molecular nature of Hsp90 underlying binding and selection of its clients is still unknown. To date, the Hsp90 relationship with the intracellular steroid hormone receptor has been studied most extensively. The Hsp90 interaction with the receptor is preceded by the association with the Hsp70/sp40 chaperone complex that is linked to Hsp90 by the Hop70 co-chaperone (Hsp70/Hsp90 organizing protein) (Kimmins and MacRae 2000). However, it is not clear which molecule from the mentioned complex of cooperating chaperones mediates specificity in protein selection. The specificity for protein kinases, which are the most frequent clients, is similarly not given by Hsp90 itself, but probably by the Cdc37 (cell division cycle 37) adapter protein, which mediates the mutual binding of both molecules (Pearl 2005).

Hsp90 is an ATP (adenosine triphosphate)-dependent homodimer which is comprised of three flexible parts: (i) the N-terminal domain binding the molecule of ATP; (ii) the middle domain, a binding site for clients with a catalytic loop; and (iii) the C-terminal domain mediating the dimerization of Hsp90 (Li and Buchner 2013). The functional purpose of Hsp90 is to bind and subsequently hydrolyse ATP. After binding ATP, the conformational change of Hsp90 from the hydrophobic open position to a closed one occurs with a transitory link of the N-terminal domains (Csermely et al. 1993). Through the dimerization of N-terminal domains, ATP hydrolysis and Hsp90 activation occurs, allowing efficient binding, stabilization and subsequent release of the client proteins.

The mechanism of stabilization and proper arrangement by which Hsp90 acts on its protein clients is mediated through a large complex of five core proteins: (i) Hsp90 itself; (ii, iii) a co-chaperone Hop70 or Cdc37 which serves as an adapter and mediator of the binding between Hsp70 and Hsp90; (iv) the p23 protein (prostaglandin E synthase 3) that mediates the binding of the client; and (v) the Hsp70-Hsp40 chaperone complex, which allows the Hsp90-client interaction (Pratt et al. 2004).

However, recent studies have described another possible mechanism by which Hsp interfere with biological processes, that is by activating innate and adaptive immunity due to their cytokine activity, which has been only marginally studied to date (Tsan and Gao 2004a, b; Wallin et al. 2002). In association with the above-mentioned chaperone activity of Hsp, such as maintaining cellular functions by preventing misfolding and aggregation of proteins, Hsp were usually described as intracellular proteins. However, if they are released into the extracellular environment, it indicates non-physiological changes and leads to induction of pro-inflammatory responses (Tsan and Gao 2004a, b). Extracellular Hsp, specifically

Hsp60, Hsp70, Hsp90 and gp96, are capable of inducing pro-inflammatory cytokines such as IL-1β, IL-12, TNF-α and GM-CSF (Asea et al. 2000; Basu et al. 2000). Hsp have also been reported to increase the activity of inducible nitric oxide synthase (iNOS) and to release chemokines including MCP-1, MIP-1 and RANTES by macrophages, monocytes and dendritic cells (Lehner et al. 2000; Panjwani et al. 2002). Hsp60, Hsp70 and gp96 have an ability to induce the maturation of dendritic cells due to the upregulation of MHC class I and class II molecules and co-stimulatory molecules CD80, CD86 and CD40 (Somersan et al. 2001). Hsp also affect the proliferation, migration, and activity of natural killer (NK) cells (Multhoff 2002; Multhoff et al. 1999) and the activation of macrophages and T cells via Toll-like receptor 4 (TLR4) (Zanin-Zhorov et al. 2003). Unfortunately, to date there are very few studies that could clarify the exact mechanism or signalling pathway underlying the cytokine activity of Hsp. However, it has been described that cytokine activity of Hsp60, Hsp70 and gp96 is mediated by the interaction with the complex of TLR2 and TLR4 and their cofactor CD14 leading to the activation of signal transduction pathway of NF-κB and MAPK (Asea et al. 2000, 2002; Basu et al. 2000; Swaroop et al. 2016). Hsp60 is known for its ability to induce the IL-6 secretion from macrophages, endothelial cells and smooth muscle cells and the expression of the adhesion molecules (intracellular adhesion molecule (ICAM)-1, vascular cell adhesion molecule (VCAM)-1 and endothelial-leukocyte adhesion molecule E-selectin) on the surface of endothelial cells (Ko et al. 2000, 1999). Extracellular Hsp70 is capable to bind with high affinity to the plasma membrane of antigen presenting cells (APC) and to elicit release of intracellular Ca^{2+}. Recently, it has been reported that Hsp70 induces the nuclear translocation and the activation of NF-κB axis and stimulates the upregulation of pro-inflammatory cytokines (IL-1β, IL-6, IL-12 and TNF-α) in monocytes (Asea 2003; Somensi et al. 2017). Furthermore, many researchers were interested in the specific interaction between Hsp70 and a variety of cells (macrophages, NK cells, monocytes and B cells), and subsequently determined a number of receptors in addition to previously mentioned TLR2 or TLR4, such as receptors from scavenger superfamily Lox-1 (Delneste et al. 2002; Nakamura et al. 2002), and the co-stimulatory molecule CD40 (Becker et al. 2002). Gp96 is another member of the Hsp90 family that is known for its ability to release nitric oxide from APC (Panjwani et al. 2002) and to mediate DC maturation via upregulation of MHC class II and CD86 molecules and secretion of IL-12, TNF-α (Singh-Jasuja et al. 2000).

Hsp also participate in antigen presentation. Hsp bind antigenic peptides and create a Hsp-peptide complex that is presented to APC, such as DC and macrophages. These interactions lead to presentation of antigenic peptides to T cells and to non-antigen-specific signalling that activates APC and stimulates immune responses, including secretion of inflammatory cytokines, chemokines, and maturation of DC and other cell types (Tsan and Gao 2004a, b). Hsp released into the extracellular environment may be recognized by immune cells as a danger signal (danger activated molecular pattern (DAMP)) (Matzinger 2002). Thus, they are able to interact with pattern recognition receptors (PRR) and to activate the pro-inflammatory signalling pathways (Asea et al. 2002; Vabulas et al. 2002).

Hsp90 in particular is involved in autoimmune and tumour immune responses and participates in antigen presentation (Tsan and Gao 2004a, b). The stimulation of macrophages with Hsp90 leads to cytokine secretion, especially of IL-1β, and also of TNF-α and IL-12 in a small extent (Basu et al. 2000). Extracellular Hsp90 induces an increased expression and secretion of IL-6 and IL-8 in a signalling pathway partially dependent upon NF-κB via TLR4 and stimulates ERK activation (Bohonowych et al. 2014; Chung et al. 2009). However, the cytokine activity of Hsp90 is still insufficiently understood. Hsp90 has been investigated mostly in association with its chaperone activity or the ability to create the Hsp90-antigenic peptide complex that is subsequently presented to APC through the process known as antigen cross-presentation (Murshid et al. 2010; Srivastava 2002). In addition, extracellular Hsp90 is also considered as DAMP (Wheeler 2011). The association of Hsp90 with the innate immunity system is further demonstrated by its interaction with lipopolysaccharides (LPS) (Byrd et al. 1999; Triantafilou et al. 2001). Given the lack of studies elucidating the mechanisms underlying the cytokine activity of Hsp90, there is an unmet need for further investigation on the functional roles of extracellular Hsp90.

In recent years, Hsp90 has become a major target for the treatment of tumour diseases, (Neckers and Workman 2012), due to its ability to bind to regulatory factors and due to its role in the regulation of mitogenesis and cell cycle progression (Neckers and Ivy 2003). It was further investigated as a therapeutic target for neurodegenerative diseases and the development of antivirals (Geller et al. 2012; Kalia et al. 2010).

5.2 Rheumatoid Arthritis

Rheumatoid arthritis (RA) is a chronic inflammatory joint disease that affects about 1% of the population. Women are affected more commonly (up to 5:1). Typically, it is a chronic, symmetric progressive polyarthritis with a dominant involvement of small hand or foot joints. The disease begins most often in middle age with a maximum incidence of 50–60 years of age. In most patients, rheumatoid factors and in significant proportion of the patients antibodies against citrullinated peptides can be found in serum. Both types of autoantibodies are associated with worse prognosis of arthritis and more frequent extra-articular manifestations.

Only one study on Hsp90 in RA has investigated the expression of this protein in serum of RA patients. Despite the fact that higher concentrations of Hsp90 in RA patients compared to healthy subjects did not achieve statistical significance, the study showed a positive correlation with higher CRP values and negative correlation with IgM antibodies. In addition, there was a significant decrease in serum Hsp90 concentrations in RA patients treated with anti-TNF therapy compared to patients without this therapy (Procházková et al. 2013). Our recent analysis of plasma Hsp90 showed statistically significantly higher Hsp90 concentrations compared to healthy controls and positive correlation with CRP and disease activity assessed by DAS-28 (unpublished data). The link between Hsp90 and RA was further investigated by detection of antibodies against Hsp90. Elevated concentrations of anti-Hsp90 antibodies were measured in serum of RA patients compared to control sera. Patients with longer duration of disease had higher anti-Hsp90 levels than patients with early RA (Hayem et al. 1999). In RA associated with interstitial lung disease (RA-ILD), autoantibodies against the citrullinated Hsp90 molecules have been detected. The formation of these autoantibodies precedes the development of RA-ILD, which highlights the important role of autoantibodies against citrullinated Hsp90 in the initiation of interstitial lung disease (Deane and Nicolls 2013). Hsp90, due to its prominent role in inflammatory processes, has become the focus of a group of scientists attempting to inhibit the function of this protein in preclinical models of RA using the small molecule Hsp90 inhibitor SNX-7081. They have shown that inhibition of Hsp90 can positively influence processes in inflammatory diseases such as blocking of the NF-κB pathway, subsequent production of proinflammatory cytokines (IL-1, TNF-α) and NO production (Rice et al. 2008). A recent study has described that Hsp90 plays an important chaperone role with the co-chaperone Cdc37 in stabilization of I-κB kinase (IKK), that is important for the activation NF-κB signalling (Chen et al. 2002). These data contribute to confirming the important role of Hsp90 in the pathogenesis of this disease and its potential as a candidate therapeutic target and a biomarker of inflammation and disease activity in RA. However, no currently available studies can clarify the underlying mechanism of Hsp90 chaperone and cytokine activity in RA. There is still an unmet need to describe the client proteins, interactions, receptors, signalling pathways and specific cell types involved in the chaperone and cytokine activity of Hsp90 in RA, that would elucidate its role in local joint and systemic inflammation and disease activity.

5.3 Systemic Lupus Erythematosus

Systemic lupus erythematosus (SLE) is a systemic autoimmune disease with various clinical manifestations that is associated with production of organ non-specific autoantibodies and immune-mediated involvement of several organs. The most commonly seen are skin and joint inflammatory manifestations and haematological involvement, but also vital organs (kidneys, lungs, nervous system) are often affected. This is a relatively rare disease (prevalence of approximately 0.1%), which affects women more commonly (in a ratio of 5–10:1) than men. Disease typically begins in young women of childbearing age, usually during the two to three decades.

The first data on increased Hsp90 expression in patients with SLE was demonstrated in peripheral blood mononuclear cells (PBMC), particularly in patients with more active form of disease compared to patients with disease in remission and healthy control subjects or patients with RA (Norton et al. 1989). Increased surface expression of Hsp90 on lymphocytes and monocytes was further demonstrated by flow cytometry in SLE patients, with elevated values significantly different from healthy controls only for lymphocytes. Increased Hsp90 expression was also detected in patients with higher disease activity (Erkeller-Yüksel et al. 1992). Another study demonstrated higher Hsp90 expression in PBMC in patients with thrombocytopenia, which is a common manifestation of antiphospholipid syndrome in SLE (Deguchi et al. 1987). Twomey et al. subsequently pointed out that elevated Hsp90 expression in PBMC depended on increased HSP90AB gene transcription, which was only seen in SLE patients compared to healthy subjects or patients with lower Hsp90 levels (Twomey et al. 1993). Different concentrations of Hsp90 were detected in neuropsychiatric, cardio-respiratory or renal involvement in this disease. Autoantibodies and Hsp90-reactive cells were detected not only in RA but also in SLE (Panchapakesan et al. 1992). However, the principles of autoimmune reactions and antibody production against Hsp90 have not yet been fully elucidated. The first data on autoantibodies against Hsp90 in SLE have been reported by Minota et al. who detected elevated levels of anti-Hsp90 antibodies of the IgG class in patients with SLE compared to SSc or RA patients and the healthy control group (Minota et al. 1988). A recent study has highlighted the potential association of Hsp90 with IL-6 in patients with SLE, as autoantibody levels were significantly correlated with increasing concentrations of proinflammatory cytokines (Ripley et al. 2001). Earlier studies have also demonstrated the ability of IL-6 to increase the expression of HSP90AB1 gene since in vitro stimulation of lymphocytes with IL-6 led to significant overproduction of Hsp90 (Stephanou et al. 1997). The association of IL-6 and Hsp90 is further confirmed by a positive correlation of these two molecules, and by their increased levels in patients with higher disease activity (Hu et al. 2006). Not only IL-6 plays an important role in activating gene expression but also IL-10, which directly activates transcription of the Hsp90-β and Hsp90-α genes (Ripley et al. 1999). Even though the function of Hsp90 in pathogenesis of SLE is still not understood, it has been reported that Hsp90 is capable to induce the secretion of IL-6 via TLR4 and the NF-κB pathway (Bohonowych et al. 2014). Hsp90 as a chaperone has become the target of SLE research based on its important stabilizing interactions with receptors for steroid molecules, a predicted key factor in regulating the effectiveness of glucocorticoids, which represent one of the most common therapies for patients with SLE (Stephanou et al. 1998). Recent studies have revealed that nucleotide replacement in the HSP90AA1 gene may affect the response of patients to glucocorticoid therapy (Zou et al. 2016). Our laboratory was interested in the expression of extracellular Hsp90 in the plasma of patients with SLE. However, the difference in Hsp90 concentrations between SLE patients and healthy controls did not reach statistical significance. We also did not find any associations between Hsp90 and individual clinical parameters. However, the pilot cohort analysed by us, due to the heterogeneous nature of SLE, might have been insufficiently large to find significant differences in the systemic concentrations of Hsp90 or to detect any significant clinical associations (unpublished data). In addition, the above-mentioned effects of Hsp90 in SLE have still not been properly understood, and more studies are needed to describe the client proteins, receptors and signalling pathways involved in the chaperone and/or cytokine activity of Hsp90 in this heterogenous disease.

5.4 Idiopathic Inflammatory Myopathies

Idiopathic inflammatory myopathies (IIM) are a heterogenous group of chronic autoimmune inflammatory diseases, which comprise diseases such as polymyositis (PM), dermatomyositis (DM), inclusion body myositis (IBM) and immune-mediated necrotizing myopathy (IMNM). IIM is an autoimmune inflammatory disease of striated muscles, manifested by symmetrical muscle weakness, especially of the proximal muscle groups of the limbs. In addition, characteristic dermatitis is present in DM. Also other organs, such as the lungs, digestive tract, heart and joints can be affected in IIM. IIM is a rare disease with a prevalence of approximately 0.001%. It is more likely to affect women (the ratio is 5:2) than men, the peak incidence is in the fifth decade of age, but the disease can manifest at any age. DM has paraneoplastic origin in approximately 20% of cases.

Local Hsp90 expression was investigated in several muscle diseases, and found specifically in PM and DM in regenerating muscle fibers, and then compared with normal and non-regenerating muscle fibers. In the case of regenerating and developing muscle fibers of DM and PM, increased expression of Hsp90 and ubiquitin was found in comparison to control samples. Here, however, the authors believe that the increased expression of Hsp90 and ubiquitin is more related to myogenesis, where their secretion is to a large extent regulated by activation of the developmental mechanisms rather than by the disease itself (Bornman et al. 1996). Their findings are supported by the chaperone interaction of Hsp90 with myogenic regulatory protein (MyoD) and casein kinase (CK2) which has been implicated as a regulator of MyoD activity during myogenesis (Johnson et al. 1996; Shaknovich et al. 1992). The chaperone activity of Hsp90 is also believed to be in activating the mechanisms, which prevent tissue damage, since elevated levels of both Hsp90 and Hsp70 in damaged muscle fibers in muscle dystrophy and IIM have been detected. Hsp90 was significantly elevated in cytotoxic macrophages invading non-necrotic fibers in PM and IBM. The Hsp90 expression correlated with the degree of muscle damage (De Paepe et al. 2009). DM is associated with higher expression of Hsp90 by endothelial cells in the blood vessels, which is likely to act as a danger signal in case of vascular damage. Immune cells, macrophages and T cells that invade non-necrotic fibers in PM and IBM expressed a significant amount of Hsp90 as well (De Paepe et al. 2012). In this case one could hypothesize that Hsp90 could have possible cytokine activity in the autocrine and/or paracrine activation of macrophages and T cells via stimulation of NF-κB pathway or antigen cross-presentation (Bohonowych et al. 2014; Chung et al. 2009; Murshid et al. 2010). Our group detected increased local expression of Hsp90 in muscle tissues of PM, DM and IMNM patients compared to healthy controls, particularly in inflammatory cells, degenerating, regenerating and/or necrotic fibers. In addition, we have demonstrated elevated plasma concentrations of Hsp90 in patients with IIM, which correlate with disease activity, extent and severity of damage and the involvement of proximal skeletal muscles, the heart and lungs (unpublished data). However, the role of Hsp90 or its therapeutic potential in IIM has not yet been fully elucidated.

5.5 Systemic Sclerosis

Systemic sclerosis (SSc) is a systemic autoimmune disease characterized by fibroproliferative changes in the skin and some internal organs (such as the lungs, heart, and gastrointestinal tract), vasculopathy and activation of the immune system. Very characteristic is skin fibrosis (scleroderma), which may be diffuse (diffuse cutaneous SSc) or restricted to distal limbs, face and neck (limited cutaneous SSc). Disease occurs rarely, with a prevalence of about 0.01%, and is about four times more likely to affect women with usual onset at middle or older age.

The role of Hsp90 in the pathogenesis of SSc is not fully elucidated, as there are very few studies to address this topic to date. Earlier studies have described the important chaperone activity of Hsp90 in the stabilization and proper spatial conformation of the TGF-β receptor (Wrighton et al. 2008) and Src kinases that are an essential part of the TGF-β signalling cascade (Beyer and Distler 2013; Koga et al. 2006; Skhirtladze et al. 2008). TGF-β is considered to be the major profibrotic mediator in SSc. Inhibition of Hsp90 increased the degradation of the TGF-β receptor and Src kinases in the proteasome which led to a decrease in the activation of the TGF-β signalling (Wrighton et al. 2008). In our earlier study we detected an increased expression of Hsp90 in the fibrotic skin of SSc patients compared to healthy subjects, particularly in fibroblasts and endothelial cells in the upper dermis layer and in keratinocytes of the epidermis. Increased expression of Hsp90 was also detected in experimental mouse models of skin fibrosis. In vitro experiments showed elevated basal Hsp90 expression at mRNA and protein level in cultured SSc fibroblasts compared to healthy fibroblasts. The stimulatory effect of TGF-β on Hsp90 production was noticeable in cultures of healthy fibroblasts in comparison to SSc fibroblasts since the TGF-β cascade is continuously activated in SSc fibroblasts (Tomcik et al. 2014). In mouse models, skin fibrosis induced by injection of replication-deficient adeno-associated virus (AAV) encoding a constitutively active TGF-β type I receptor (TβRI) was accompanied by an increased expression of Hsp90. Thus, the TGF-β signal cascade can significantly amplify the effects of Hsp90 in tissue fibrosis. We further described reduced pro-fibrotic effects of TGF-β by inactivating the Hsp90 molecule both in vitro (using dermal fibroblasts) and in vivo (using experimental murine models of dermal fibrosis). Inhibition of Hsp90 with 17-DMAG (17-Dimethylaminoethylamino-17-demethoxygeldanamycin) significantly reduced histological fibrotic findings, the accumulation of collagen, and prevented the activation of fibroblasts in three preclinical experimental mouse models that mimicked various stages of SSc (Tomcik et al. 2014). Based on the above-mentioned data, we subsequently analyzed extracellular Hsp90 in the plasma of SSc patients and healthy controls. In SSc patients, we detected significantly higher Hsp90 concentrations than in healthy individuals. Increased Hsp90 was associated with higher inflammatory activity, impaired lung function, and the extent and severity of skin involvement (unpublished data). These data highlight the important chaperone activity of Hsp90 in the signalling pathway of TGF-β, its role as an amplifier of abundant TGF-β signalling, a potential molecular target of anti-fibrotic therapy, and as a novel biomarker of lung and skin involvement and inflammation in SSc. Even though no data is available on the cytokine activity of Hsp90 in SSc, based on our published and yet unpublished data one can speculate, that extracellular Hsp90, possibly released by cells with increased intracellular expression of Hsp90, such as endothelial cells and dermal fibroblasts, could exert its cytokine effects in an autocrine and/or paracrine manner on perivascular inflammatory cells and induce secretion of pro-inflammatory and pro-fibrotic cytokines, which further amplify the pro-inflammatory and pro-fibrotic signalling pathways involved in the pathogenesis of SSc.

5.6 Other Rheumatic Diseases

A recent study assessed the link between the expression of extracellular Hsp90 and the chornic fatigue level of primary Sjögren's syndrome patients (SS). Sjögren's syndrome is a chronic autoimmune disease characterized by lymphocytic infiltration of exocrine, especially salivary and lacrimal glands, dry eyes and mouth or other mucous membranes and the presence of circulating autoantibodies (anti-Ro (SSA), anti-La (SSB), and rheumatoid factors). Patients with higher fatigue levels showed significantly higher Hsp90 concentrations than those with lower scores evaluating chronic fatigue. Based on this result, the authors investigated the relationship of depression to patient fatigue and the presence of Hsp90. However, there were no significant differences in Hsp90 concentrations in patients with more frequent depression compared to those without it (Bårdsen et al. 2016).

Our research has also focused on Hsp90 analysis in patients with spondyloarthritis (SpA) – specifically in patients with ankylosing spondylitis (AS) and non-radiographic axial spondyloarthritis (nr-axSpA). SpA is a group of inflammatory rheumatic diseases affecting the backbone structures, peripheral joints, tendons, entheses, and extraarticular structures as well. Diseases belonging to the group of SpA share similar clinical features and have a common genetic basis in HLA-B27 antigen, which is differently expressed in different units. Our group detected significantly increased plasma Hsp90 in SpA patients compared to the control group. Increased concentrations were measured in patients with more extensive axial involvement (unpublished data). Similarly, we detected elevated plasma Hsp90 concentrations in patients with psoriatic arthritis (PsA), yet another disease from the group of SpA characterized by typical skin involvement (unpublished data).

We also detected increased plasma Hsp90 levels in osteoarthritis (OA) patients when compared to healthy controls and a possible association with some parameters of the metabolic syndrome (unpublished data). OA is the most common joint disease affecting approximately 12% of the general population. It is a multifactorial process leading to cartilage degradation and productive changes of the subchondral bone. A detailed analysis of associations with disease-specific clinical parameters in the aforementioned rheumatic diseases is currently ongoing. To our knowledge, Hsp90 has not yet been studied in other rheumatic diseases. Unfortunately, there are only a few studies available presenting only limited data on the role of Hsp90 in the previously mentioned diseases, which provides only description of possible associations with some clinical aspects. To date, no conclusions can be drawn from these findings as to whether Hsp90 expression is a mere epiphenomenon or a significant contributor to the pathogenesis of these diseases based on its potential chaperone and cytokine activity.

5.7 Conclusions

Heat shock proteins are the family of molecular chaperones whose expression increases after exposure to high temperature. In eukaryotic cells, however, their synthesis is also induced due to other stressful situations, such as low temperatures, viral infection, changes in pH or intracellular environment, UV exposure, starvation, immobility, strenuous physical load, but also during tissue healing and remodelling. Hsp selectively recognize and bind improperly folded protein clients, and eventually stabilize them in proper spatial conformation to protect them against protease degradation. Recently, a cytokine activity of Hsp has been reported in activating the innate and adaptive immunity. Hsp are capable to stimulate the secretion of pro-inflammatory cytokines, chemokines and to induce maturation of DC or activation of macrophages, T cells and NK cells. Furthermore, extracellular Hsp are also classified as DAMPs. Hsp90 is essential for stabilizing many receptors, protein kinases and transcription factors that regulate a number of biological processes important for cell growth and survival, but also for regulating tissue fibrosis, angiogenesis, or antigen presentation, induction of cytokine secretion and activation of the immune system. These processes contribute in varying degrees to the pathogenesis of a heterogeneous group of autoimmune inflammatory rheumatic diseases, making Hsp90 an interesting target for both basic and applied research. However, the role of Hsp90 in rheumatic diseases has only sporadically been investigated and its exact position, function and mechanism in the pathogenesis of these diseases are yet to be elucidated. From research in SSc, Hsp90 can be perceived as a chaperone, mediator and amplificator of TGF-β induced tissue fibrosis, a potential target of anti-fibrotic therapy, and a biomarker of inflammation and fibrotic involvement of the skin and lungs. Increased local expression documented in affected skeletal muscle in IIM and fibrotic skin in SSc confirms a possible association with inflammation (IIM and SSc), regeneration (IIM) and fibrosis (SSc). In these conditions with increased local expression of Hsp90 one could hypothesize on possible cytokine activity of Hsp90 in the autocrine and/or paracrine activation of inflammatory cells in the perivascular (IIM and SSc), endomysial (IIM) and dermal (SSc) milieu via stimulation of pro-inflammatory (IIM and SSc) and/or pro-fibrotic (SSc) signalling pathways and cytokines or via antigen cross-presentation. Predominantly increased systemic expression in several other autoimmune inflammatory rheumatic diseases (RA, SpA, PsA) supports the proposed association with inflammation or disease activity or the extent of tissue involvement, nonetheless, these findings do not provide conclusions as to whether Hsp90 expression is only an epiphenomenon or a significant molecule in the pathogenesis. However, extensive studies demonstrating the chaperokine role of Hsp90 in pathogenesis by in vitro or in vivo experiments are yet to be performed in most rheumatic diseases. Due to the development of a number of new generation Hsp90 inhibitors and an increasing number of clinical trials investigating their efficacy predominantly in oncological indications, new findings on Hsp90 function in the pathogenesis of rheumatic diseases could also have direct translational potential. Another unmet need in rheumatology is the search for new appropriate biomarkers of development of the disease, disease activity, extent of organ involvement, or prediction of treatment response, which would outweigh the benefits of the well-established currently used markers with their known limitations. Thus, there is a room for exploration and use of this molecule, whose increased systemic expression can be found in many diseases characterized by activation of the immune system and local or systemic chronic inflammation. However, in order to clarify the role, chaperokine activity, and potential of Hsp90 in rheumatic disorders, a number of in-depth studies and validation of already demonstrated data in independent cohorts will be needed.

Acknowledgements This chapter was supported by grant projects AZV 16-33542A, AZV 16-33574A, SVV 260263, PRVOUK, UNCE 204022, and the Ministry of Health of the Czech Republic [Research Project No. 00023728].

References

Asea A (2003) Chaperokine-induced signal transduction pathways. Exerc Immunol Rev 9:25–33

Asea A, Kraeft S-K, Kurt-Jones EA, Stevenson MA, Chen LB, Finberg RW, Koo GC, Calderwood SK (2000) HSP70 stimulates cytokine production through a CD14-dependant pathway, demonstrating its dual role as a chaperone and cytokine. Nat Med 6(4):435–442

Asea A, Rehli M, Kabingu E, Boch JA, Baré O, Auron PE, Stevenson MA, Calderwood SK (2002) Novel signal transduction pathway utilized by extracellular HSP70. J Biol Chem 277(17):15028–15034

Bårdsen K, Nilsen MM, Kvaløy JT, Norheim KB, Jonsson G, Omdal R (2016) Heat shock proteins and chronic fatigue in primary Sjögren's syndrome. Innate Immun 22(3):162–167

Basu S, Binder RJ, Suto R, Anderson KM, Srivastava PK (2000) Necrotic but not apoptotic cell death releases heat shock proteins, which deliver a partial maturation signal to dendritic cells and activate the NF-kappa B pathway. Int Immunol 12(11):1539–1546

Becker T, Hartl F-U, Wieland F (2002) CD40, an extracellular receptor for binding and uptake of Hsp70-peptide complexes. J Cell Biol 158(7):1277–1285

Beyer C, Distler JHW (2013) Tyrosine kinase signaling in fibrotic disorders. Biochim Biophys Acta (BBA) Mol Basis Dis 1832(7):897–904

Bohonowych JE, Hance MW, Nolan KD, Defee M, Parsons CH, Isaacs JS (2014) Extracellular Hsp90 mediates an NF-κB dependent inflammatory stromal program: implications for the prostate tumor microenvironment. Prostate 74(4):395–407

Bornman L, Polla BS, Gericke GS (1996) Heat-shock protein 90 and ubiquitin: developmental regulation during myogenesis. Muscle Nerve 19(5):574–580

Burrows F, Zhang H, Kamal A (2004) Hsp90 activation and cell cycle regulation. Cell Cycle (Georgetown, Tex) 3(12):1530–1536

Byrd CA, Bornmann W, Erdjument-Bromage H, Tempst P, Pavletich N, Rosen N, Nathan CF, Ding A (1999) Heat shock protein 90 mediates macrophage activation by Taxol and bacterial lipopolysaccharide. Proc Natl Acad Sci U S A 96(10):5645–5650

Cao Y, Ohwatari N, Matsumoto T, Kosaka M, Ohtsuru A, Yamashita S (1999) TGF-beta1 mediates 70-kDa heat shock protein induction due to ultraviolet irradiation in human skin fibroblasts. Proc Natl Acad Sci U S A 96(10):5645–5650

Chen G, Cao P, Goeddel DV (2002) TNF-induced recruitment and activation of the IKK complex require Cdc37 and Hsp90. Mol Cell 9(2):401–410

Chen B, Zhong D, Monteiro A (2006) Comparative genomics and evolution of the HSP90 family of genes across all kingdoms of organisms. BMC Genomics 7:156

Chung S-W, Lee J-H, Choi K-H, Park Y-C, Eo S-K, Rhim B-Y, Kim K (2009) Extracellular heat shock protein 90 induces interleukin-8 in vascular smooth muscle cells. Biochem Biophys Res Commun 378(3):444–449

Csermely P, Kajtár J, Hollósi M, Jalsovszky G, Holly S, Kahn CR, Gergely P, Söti C, Mihály K, Somogyi J (1993) ATP induces a conformational change of the 90-kDa heat shock protein (hsp90). J Biol Chem 268(3):1901–1907

De Paepe B, Creus KK, Martin J-J, Weis J, De Bleecker JL (2009) A dual role for HSP90 and HSP70 in the inflammatory myopathies: from muscle fiber protection to active invasion by macrophages. Ann N Y Acad Sci 1173:463–469

De Paepe B, Creus KK, Weis J, De Bleecker JL (2012) Heat shock protein families 70 & 90 in Duchenne muscular dystrophy and inflammatory myopathy: Balancing muscle protection & destruction. Neuromuscul Disord Neuromuscul Disord 22(1):26–33

Deane KD, Nicolls MR (2013) Developing better biomarkers for connective tissue disease-associated interstitial lung disease: citrullinated hsp90 autoantibodies in rheumatoid arthritis. Arthritis Rheum 65(4):864–868

Deguchi Y, Negoro S, Kishimoto S (1987) Heat-shock protein synthesis by human peripheral mononuclear cells from sle patients. Biochem Biophys Res Commun 148(3):1063–1068

Delneste Y, Magistrelli G, Gauchat J, Haeuw J, Aubry J, Nakamura K, Kawakami-Honda N, Goetsch L, Sawamura T, Bonnefoy J et al (2002) Involvement of LOX-1 in dendritic cell-mediated antigen cross-presentation. Immunity 17(3):353–362

Echeverria PC, Picard D (2010) Molecular chaperones, essential partners of steroid hormone receptors for activity and mobility. Biochim Biophys Acta (BBA) Mol Cell Res 1803(6):641–649

Erkeller-Yüksel FM, Isenberg DA, Dhillon VB, Latchman DS, Lydyard PM (1992) Surface expression of heat shock protein 90 by blood mononuclear cells from patients with systemic lupus erythematosus. J Autoimmun 5(6):803–814

Geller R, Taguwa S, Frydman J (2012) Broad action of Hsp90 as a host chaperone required for viral replication. Biochim Biophys Acta (BBA) Mol Cell Res 1823(3):698–706

Hayem G, De Bandt M, Palazzo E, Roux S, Combe B, Eliaou JF, Sany J, Kahn MF, Meyer O (1999) Anti-heat shock protein 70 kDa and 90 kDa antibodies in serum of patients with rheumatoid arthritis. Ann Rheum Dis 58(5):291–296

Hu S, Xu Q, Xiao W, Huang M (2006) The expression of molecular chaperone HSP90 and IL-6 in patients with systemic lupus erythematosus. J Huazhong Univ Sci Technol 26(6):664–666

Johnson SE, Wang X, Hardy S, Taparowsky EJ, Konieczny SF (1996) Casein kinase II increases the transcriptional activities of MRF4 and MyoD independently of their direct phosphorylation. Mol Cell Biol 16(4):1604–1613

Kalia SK, Kalia LV, McLean PJ (2010) Molecular chaperones as rational drug targets for Parkinson's disease therapeutics. CNS Neurol Disord Drug Targets 9(6):741–753

Kimmins S, MacRae TH (2000) Maturation of steroid receptors: an example of functional cooperation among molecular chaperones & their associated proteins. Cell Stress Chaperones 5(2):76–86

Koga F, Xu W, Karpova TS, McNally JG, Baron R, Neckers L (2006) Hsp90 inhibition transiently activates Src kinase and promotes Src-dependent Akt and Erk activation. Proc Natl Acad Sci 103(30):11318–11322

Kol A, Bourcier T, Lichtman AH, Libby P (1999) Chlamydial & human heat shock protein 60s activate human vascular endothelium, smooth muscle cells, and macrophages. J Clin Investig 103(4):571

Kol A, Lichtman AH, Finberg RW, Libby P, Kurt-Jones EA (2000) Cutting edge: heat shock protein (HSP) 60 activates the innate immune response: CD14 is an essential receptor for HSP60 activation of mononuclear cells. J Immunol (Baltimore, Md: 1950) 164(1):13–17

Laplante AF, Moulin V, Auger FA, Landry J, Li H, Morrow G, Tanguay RM, Germain L (1998) Expression of heat shock proteins in mouse skin during wound healing. J Histochem Cytochem Off J Histochem Soc 46(11):1291–1301

Lehner T, Bergmeier LA, Wang Y, Tao L, Sing M, Spallek R, van der Zee R (2000) Heat shock proteins generate β-chemokines which function as innate adjuvants enhancing adaptive immunity. Eur J Immunol 30(2):594–603

Li J, Buchner J (2013) Structure, function and regulation of the hsp90 machinery. Biomed J 36(3):106–117

Lindquist S, Craig EA (1988) The heat-shock proteins. Annu Rev Genet 22:631–677

Lund PA (2001) Microbial molecular chaperones. Adv Microb Physiol 44:93–140

Matzinger P (2002) The danger model: a renewed sense of self. Science 296(5566):301–305

McClellan AJ, Xia Y, Deutschbauer AM, Davis RW, Gerstein M, Frydman J (2007) Diverse cellular functions of the Hsp90 molecular chaperone uncovered using systems approaches. Cell 131(1):121–135

Meyer P, Prodromou C, Hu B, Vaughan C, Roe SM, Panaretou B, Piper PW, Pearl LH (2003) Structural & functional analysis of the middle segment of Hsp90: implications for ATP hydrolysis and client protein and cochaperone interactions. Mol Cell 11(3):647–658

Millson SH, Truman AW, Rácz A, Hu B, Panaretou B, Nuttall J, Mollapour M, Söti C, Piper PW (2007) Expressed as the sole Hsp90 of yeast, the alpha and beta isoforms of human Hsp90 differ with regard to their capacities for activation of certain client proteins, whereas only Hsp90beta generates sensitivity to the Hsp90 inhibitor radicicol. FEBS J 274(17):4453–4463

Minota S, Koyasu S, Yahara I, Winfield J (1988) Autoantibodies to the heat-shock protein hsp90 in systemic lupus erythematosus. J Clin Invest 81(1):106–109

Multhoff G (2002) Activation of natural killer cells by heat shock protein 70. Int J Hyperth 18(6):576–585

Multhoff G, Mizzen L, Winchester CC, Milner CM, Wenk S, Eissner G, Kampinga HH, Laumbacher B, Johnson J (1999) Heat shock protein 70 (Hsp70) stimulates proliferation and cytolytic activity of natural killer cells. Exp Hematol 27(11):1627–1636

Murshid A, Gong J, Calderwood SK (2010) Heat shock protein 90 mediates efficient antigen cross presentation through the scavenger receptor expressed by endothelial cells-I. J Immunol 185(5):2903–2917

Nakamura T, Hinagata J, Tanaka T, Imanishi T, Wada Y, Kodama T, Doi T (2002) HSP90, HSP70, and GAPDH directly interact with the cytoplasmic domain of macrophage scavenger receptors. Biochem Biophys Res Commun 290(2):858–864

Neckers L, Ivy SP (2003) Heat shock protein 90. Curr Opin Oncol 15(6):419–424

Neckers L, Workman P (2012) Hsp90 molecular chaperone inhibitors: are we there yet? Clin Cancer Res Off J Am Assoc Cancer Res 18(1):64–76

Norton PM, Isenberg DA, Latchman DS (1989) Elevated levels of the 90 kd heat shock protein in a proportion of SLE patients with active disease. J Autoimmun 2(2):187–195

Obermann WM, Sondermann H, Russo AA, Pavletich NP, Hartl FU (1998) In vivo function of Hsp90 is dependent on ATP binding and ATP hydrolysis. J Cell Biol 143(4):901–910

Panchapakesan J, Daglis M, Gatenby P (1992) Antibodies to 65 kDa and 70 kDa heat shock proteins in rheumatoid arthritis & systemic lupus erythematosus. Immunol Cell Biol 70(Pt 5):295–300

Panjwani NN, Popova L, Srivastava PK (2002) Heat shock proteins gp96 and hsp70 activate the release of nitric oxide by APCs. J Immunol (Baltimore, Md: 1950) 168(6):2997–3003

Pearl LH (2005) Hsp90 and Cdc37 – a chaperone cancer conspiracy. Curr Opin Genet Dev 15(1):55–61

Pearl LH, Prodromou C (2000) Structure and in vivo function of Hsp90. Curr Opin Struct Biol 10(1):46–51

Pratt WB, Galigniana MD, Harrell JM, DeFranco DB (2004) Role of hsp90 and the hsp90-binding immunophilins in signalling protein movement. Cell Signal 16(8):857–872

Procházková L, Hulejová H, Němec P, Šenolt L (2013) Cirkulující protein tepelného šoku 90 (HSP90) u pacientů s revmatoidní artritidou a axiální spondyloartritidou. Čes Revmatol 21(4):164–169

Rice JW, Veal JM, Fadden RP, Barabasz AF, Partridge JM, Barta TE, Dubois LG, Huang KH, Mabbett SR, Silinski MA et al (2008) Small molecule inhibitors of Hsp90 potently affect inflammatory disease pathways and exhibit activity in models of rheumatoid arthritis. Arthritis Rheum 58(12):3765–3775

Ripley BJ, Stephanou A, Isenberg DA, Latchman DS (1999) Interleukin-10 activates heat-shock protein 90beta gene expression. Immunology 97(2):226–231

Ripley BJ, Isenberg D, Latchman D (2001) Elevated levels of the 90kDa heat shock protein (hsp90) in SLE correlate with levels of IL-6 and autoantibodies to hsp90. J Autoimmun 17(4):341–346

Santoro MG (2000) Heat shock factors and the control of the stress response. Biochem Pharmacol 59(1):55–63

Schlesinger MJ (1990) Heat shock proteins. J Biol Chem 265(21):12111–12114

Shaknovich R, Shue G, Kohtz DS (1992) Conformational activation of a basic helix-loop-helix protein (MyoD1) by the C-terminal region of murine HSP90 (HSP84). Mol Cell Biol 12(11):5059–5068

Singh-Jasuja H, Scherer HU, Hilf N, Arnold-Schild D, Rammensee HG, Toes RE, Schild H (2000) The heat shock protein gp96 induces maturation of dendritic cells and down-regulation of its receptor. Eur J Immunol 30(8):2211–2215

Skhirtladze C, Distler O, Dees C, Akhmetshina A, Busch N, Venalis P, Zwerina J, Spriewald B, Pileckyte M, Schett G et al (2008) Src kinases in systemic sclerosis: central roles in fibroblast activation and in skin fibrosis. Arthritis Rheum 58(5):1475–1484

Somensi N, Brum PO, de Miranda Ramos V, Gasparotto J, Zanotto-Filho A, Rostirolla DC, da Silva Morrone M, Moreira JCF, Pens Gelain D (2017) Extracellular HSP70 activates ERK1/2, NF-kB and pro-inflammatory gene transcription through binding with RAGE in A549 human lung cancer cells. Cell Physiol Biochem Int J Exp Cell Physiol Biochem Pharmacol 42(6):2507–2522

Somersan S, Larsson M, Fonteneau JF, Basu S, Srivastava P, Bhardwaj N (2001) Primary tumor tissue lysates are enriched in heat shock proteins and induce the maturation of human dendritic cells. J Immunol (Baltimore, Md: 1950) 167(9):4844–4852

Srivastava P (2002) Roles of heat-shock proteins in innate and adaptive immunity. Nat Rev Immunol 2(3):185–194

Stephanou A, Amin V, Isenberg DA, Akira S, Kishimoto T, Latchman DS (1997) Interleukin 6 activates heat-shock protein 90 beta gene expression. Biochem J 321(Pt 1):103–106

Stephanou A, Latchman DS, Isenberg DA, Yellon DM, Latchman DS, Ellis RJ, Schultz DR, Arnold PI, Hickey E, Brandon SE et al (1998) The regulation of heat shock proteins and their role in systemic lupus erythematosus. Sem Arthritis Rheum 28(3):155–162

Swaroop S, Sengupta N, Suryawanshi AR, Adlakha YK, Basu A (2016) HSP60 plays a regulatory role in IL-1β-induced microglial inflammation via TLR4-p38 MAPK axis. J Neuroinflammation 13(1):27

Taherian A, Krone PH, Ovsenek N (2008) A comparison of Hsp90alpha & Hsp90beta interactions with cochaperones and substrates. Biochem Cell Biol Biochim Biol Cell 86(1):37–45

Taipale M, Jarosz DF, Lindquist S (2010) HSP90 at the hub of protein homeostasis: emerging mechanistic insights. Nat Rev Mol Cell Biol 11(7):515–528

Tomcik M, Zerr P, Pitkowski J, Palumbo-Zerr K, Avouac J, Distler O, Becvar R, Senolt L, Schett G, Distler JH (2014) Heat shock protein 90 (Hsp90) inhibition targets canonical TGF-β signalling to prevent fibrosis. Ann Rheum Dis 73(6):1215–1222

Triantafilou K, Triantafilou M, Dedrick RL (2001) A CD14-independent LPS receptor cluster. Nat Immunol 2(4):338–345

Tsan M-F, Gao B (2004a) Cytokine function of heat shock proteins. Cell Physiol 286(4):C739–C744

Tsan M-F, Gao B (2004b) Heat shock protein and innate immunity. Cell Mol Immunol 1(4):274–279

Twomey BM, Dhillon VB, McCallum S, Isenberg DA, Latchman DS (1993) Elevated levels of the 90 kD heat shock protein in patients with systemic lupus erythematosus are dependent upon enhanced transcription of the hsp90β gene. J Autoimmun 6(4):495–506

Vabulas RM, Ahmad-Nejad P, Ghose S, Kirschning CJ, Issels RD, Wagner H (2002) HSP70 as endogenous stimulus of the toll/interleukin-1 receptor signal pathway. J Biol Chem 277(17):15107–15112

Wallin RPA, Lundqvist A, Moré SH, von Bonin A, Kiessling R, Ljunggren H-G (2002) Heat-shock proteins as activators of the innate immune system. Trends Immunol 23(3):130–135

Wheeler DS (2011) Extracellular heat shock proteins: alarmins for the host immune system. Open Inflamm J 4(1):49–60

Wrighton KH, Lin X, Feng X-H (2008) Critical regulation of TGFbeta signaling by Hsp90. Proc Natl Acad Sci U S A 105(27):9244–9249

Zanin-Zhorov A, Nussbaum G, Franitza S, Cohen IR, Lider O (2003) T cells respond to heat shock protein 60 via TLR2: activation of adhesion and inhibition of chemokine receptors. FASEB J 17(11):1567–1569

Zou Y-F, Xu J-H, Gu Y-Y, Pan F-M, Tao J-H, Wang D-G, Xu S-Q, Xiao H, Chen P-L, Liu S et al (2016) Single nucleotide polymorphisms of HSP90AA1 gene influence response of SLE patients to glucocorticoids treatment. SpringerPlus 5:222

Zuehlke AD, Beebe K, Neckers L, Prince T (2015) Regulation and function of the human HSP90AA1 gene. Gene 570(1):8–16

Chapter 6
Upregulation of Cytoprotective Chaperones Mediate Better Tolerance to High Altitude

Richa Rathor and Geetha Suryakumar

Abstract Disturbance in redox homeostasis at high altitude due to hypobaric hypoxia has been implicated in several physiological as well as pathological consequences. A shift in the redox potential leads to modulation of key biological cascades which are intricately regulated by proteins. Enhanced oxidative protein modifications at high altitude trigger several protective responses such as the heat shock response or the unfolded protein response to restore the protein homeostasis thus counteracting the stress and promoting cell survival. Upregulation of heat shock response enhanced hypoxic tolerance and better survival under low oxygen tensions. Modulation of proteostasis by upregulation of heat shock proteins and concomitant protein quality control mechanisms play a very important role in differential tolerance or susceptibility to environmental hypoxia. Hence, the present work potentiate the use of proteostasis modulators either chemical or herbal which can be developed as therapeutic interventions for the treatment of a multitude of hypoxia related pathophysiologies.

Keywords Chaperones · Chaperokine · ER stress · High altitude · HSP70 · Tolerance

Abbreviations

4PBA	4-Phenyl butyric acid
AICAR	5-Aminoimidazole-4-carboxamide ribonucleotide
AMPK	AMP-activated protein kinase
mTOR	Mammalian target of rapamycin
NAFLD	Non-alcoholic fatty liver disease
PKA	Protein kinase A
TUDCA	Tauroursodeoxycholic acid
UPR	Unfolded protein response

R. Rathor · G. Suryakumar (✉)
Cellular Biochemistry Division, Defence Institute of Physiology and Allied Sciences, Delhi, India
e-mail: sgeetha@dipas.drdo.in; geetha.suryakumar@gmail.com

A. A. A. Asea, P. Kaur (eds.), *Chaperokine Activity of Heat Shock Proteins*, Heat Shock Proteins 16, https://doi.org/10.1007/978-3-030-02254-9_6

6.1 Introduction

One of the physiological conditions which affect the oxygen availability in our body is the high altitude environment. Apart from the characteristic fall in the partial pressure of oxygen, known as hypobaric hypoxia, other stressors including low temperature, high wind velocity, UV rays from the sun, dehydration and a lack of antioxidant nutrients, also affect the physiological function in the high altitude milieu. Hypobaric hypoxia has several consequences for the oxygen economy of the body and requires the body to mount adaptive coping measures for minimizing the effect of low oxygen tensions. Cells respond to stress in a variety of ways ranging from activation of pathways that promote survival to eliciting programmed cell death that eliminates damaged cells. The cell's initial response to a stressful stimulus is geared towards helping the cell to defend against and recover from the insult. Several interlinked pathways are engaged in restoring the cellular homeostasis, but one of the well-identified mechanisms that act as the first line of defense against cellular injury, is the response mediated by the heat shock family of stress proteins (heat shock proteins (HSP). Heat shock response (HSR), plays a critical role in amelioration of acute as well as severe challenges from internal or external environmental stimuli that pose a lethal threat to survival of the organism.

Enhanced free radical generation leading to oxidative stress have been implicated in the pathophysiology of hypoxic exposure under high altitude environment. Earlier studies have brought to light the effect of reactive oxygen species on proteins such as nitrosylation of tyrosine, glutathiolation, and formation of amino acid oxidation adducts. Our recent research evaluated how hypoxia, acute, chronic or intermittent, differentially modulates the protein folding homeostasis and subsequently mediates both protective and deleterious responses on the cardiac function. An intricate interplay of hypoxia induced adaptive cytoprotective responses mediated by heat shock proteins and maintenance of protein folding homeostasis may be central to the susceptibility or tolerance to adverse environments at high altitude.

The manifestation of the unfolded protein response and its complex downstream signalling molecules under low oxygen conditions may form an exciting possibility of targeting the UPR components as an effective therapeutic strategy for hypoxia induced pathophysiology. Thus the present work was designed to elucidate the activation of heat shock response and proteostasis under acute and chronic hypobaric hypoxia. Further the modulation of proteostasis for better adaptation under high altitude conditions has also been explored.

6.1.1 High Altitude, Hypobaric Hypoxia and Oxidtive Stress

More than 140 million people reside at altitudes over 2500 m above sea level worldwide. Physiological stresses, imposed by hypoxia, cold, wind, and harmful ionizing radiations, contribute to a decrease in physical and mental performance at altitude. The reduced pressure of atmospheric oxygen, known as hypobaric hypoxia, has several consequences for the oxygen economy of the body. Decreased oxygen pressure at high altitude leads to enhanced generation of reactive oxygen (ROS) and nitrogen species (RONS). The degree of altitude makes the difference of severity of oxidative protest. High altitude response is usually initiated above 3000 m (Sinha et al. 2009a; Faiss et al. 2013; Miller et al. 2013; Lewis et al. 2014; McGinnis et al. 2014). Numerous reactive species generating system are activated at high altitude such as mitochondrial electron transport chain, xanthine oxidase and nitric oxide synthase (Mohanraj et al. 1998; Radak et al. 1994; Serrano et al. 2003). High altitude associated hypoxia also leads to reductive stress. It is a well known mechanism that ROS is generated during electron transport chain at complex I and complex III stage but unavailability of appropriate oxygen causes accumulation of reducing equivalents and this condition is termed as reductive stress. This stress leads to auto-oxidation of one or more mitochondrial complexes like ubiquinone-ubiquinol redox couple and form ROS. Along with this increased cellular NADH/NAD+ ratio was also observed during hypoxia associated reductive stress.

6.1.2 *Effect of ROS on Proteins*

Reactive Oxygen Species have important role in several cell signaling processes. However, during cellular stress, they culminate into a state known as oxidative stress, in which ROS levels increase so dramatically that leads to significant structural damage. In recent years, it has become known that proteins can be modified in many ways by ROS, which include oxygen free radicals and peroxides. A number of potentially adverse reactions, commonly known as redox reactions, involve post-translational alterations of proteins and have been specifically linked to the pathology of several physiological disorders, including degenerative diseases, aging, atherosclerosis, and skeletal muscle dysfunction. There are multiple forms of post-translation oxidative modification in which the structure and function of proteins can be substantially influenced (Fig. 6.1).

Excessive ROS production leads to protein modification in different ways, either direct or indirect. Direct protein modification includes nitrosylation, carbonylation disulphide formation and glutathionylation. Contrary, protein may be modified indirectly via conjugation with breakdown products of fatty acid peroxidation. Enormous ROS production and its interaction with protein leads to site-specific amino acid modification, fragmentation of the peptide chain, aggregation of cross-linked reaction products, altered electric charge and increased susceptibility of proteins to proteolysis. Protein carbonyl content and Advance oxidation protein products (AOPPs) could be considered as biomarkers of protein oxidation (Dalle-Donne et al. 2003; Capeillere-Blandin et al. 2004). The formation of protein carbonyl could be due to oxidative cleavage of protein backbone, oxidation of amino acids and via

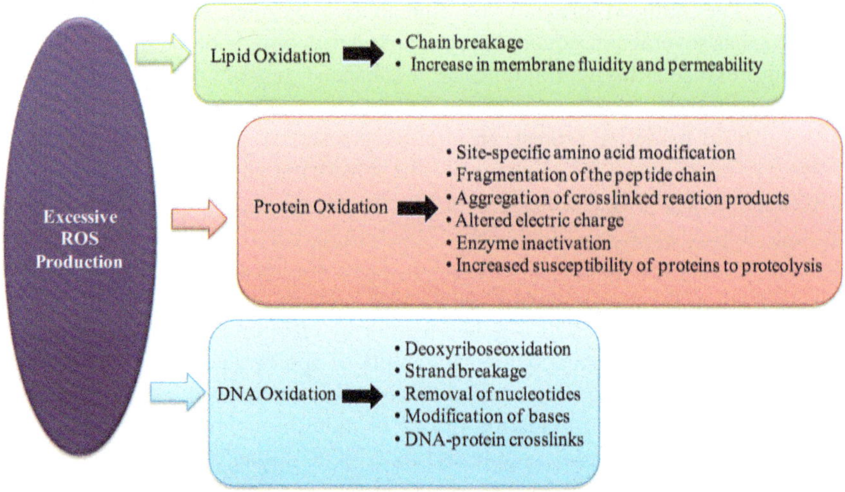

Fig. 6.1 Excessive ROS production and oxidation of biomolecules

binding of lipid peroxidation products such as hydroxynonenal (HNE) or acrolein, (Stadtman 1992; Butterfield and Stadtman 1997). Further, thiol containing and sulphur containing amino acids are sensitive to ROS as they are oxidized and form disulfide bridge. Activated oxygen also attack amino acids such as methionine and tyrosine which form methionine sulphoxide derivatives and bityrosine respectively (Brot and Weissbach 1982; Davies 1987; Sharma et al. 2012). In vivo plasma levels of AOPP also correlated with the levels of dityrosine which are also a badge of oxidized proteins (Witko-Sarsat et al. 1996, 1998).

6.1.3 Hypoxia and Post-Translation Modification

Hypoxia not only affects protein expression but also amend the function of existing proteins via post-translational modification of amino acids (Kumar and Klein 2004). Mostly phosphorylation-dephosphorylation reactions play a paramount role during hypoxia. Large number of studies have showed that acute, chronic and intermittent hypoxic condition leads to either phosphorylation or dephosphorylation of proteins which entail transcription factors (Mishra et al. 2002; Tang et al. 2002), translation initiation factors (Koumenis et al. 2002; Row et al. 2003), protein kinases (Kulisz et al. 2002), cell adhesion and cytoskeletal proteins (Weinberg et al. 2001), cytosolic and mitochondrial enzymes (Kayyali et al. 2001), cell surface receptors (Gurd et al. 2002), and neurotransmitter-related proteins and synthetic enzymes (Shi et al. 2002) (Fig. 6.2). Since the protein pool is exposed to a continuous influence of metabolic and environmental damaging agents, leading to formation of oxidized

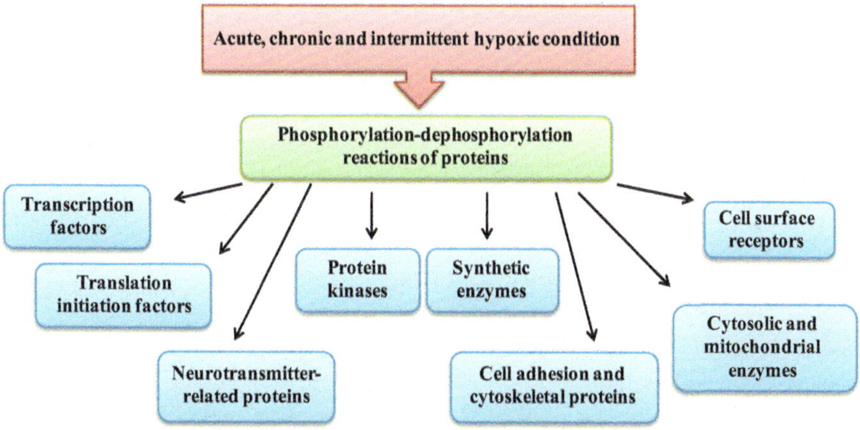

Fig. 6.2 Types of phosphorylation-dephosphorylation protein reactions

and misfolded proteins there are several sophisticated mechanisms to ensure the maintenance of protein homeostasis. The response mediated by the heat shock family of stress proteins, acts as the first line of defense against cellular injury to restore the proper folding of proteins.

6.1.4 *Chaperones: Cytosolic and ER Chaperones*

Proteins are more resourceful and structurally complex bio macromolecules which are engaged in most of the biological processes. For proper function, protein should retain its conformational flexibility and thermodynamically stable state in physiological environment. Hence, protein quality control and proteostasis (maintenance of protein homeostasis) is essential for cellular health. Proteostasis is maintained via number of networks like molecular chaperones and their regulators (involved in de novo folding and refolding), the ubiquitin-proteasome system (UPS) and autophagy system (Powers et al. 2009). Chaperones are functional proteins which are required for proper protein folding in the cell under physiological and stress conditions (Beissinger and Buchner 1998). In other definition, molecular chaperones are proteins which interact and stabilize the other protein and achieve its functional active conformation and interestingly it is not present in the final 3D structure protein (Hartl 1996; Herbst et al. 1997). Basically chaperones are divided into two categories: (i) ATP-dependent chaperones in which ATP hydrolysis is required for large conformational changes for correcting protein folding, also known as "foldases" and (ii) ATP-independent chaperones which quickly bind to misfolded proteins and later pass them to ATP-dependent chaperones for folding, also known as "holdases" (Haslbeck 2002; Horowitz et al. 2017).

Table 6.1 The location and functions of HSP families

Location	HSP class	Function
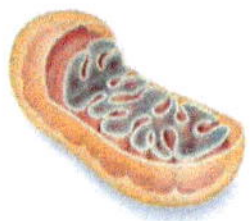 Mitochondria	HSP10	Protein folding in mitochondria
	HSP60	Protein assembly by forming hetero-oligomeric protein complex
	HSP75	Associated with reducing infarct area by ischemic brain injury and protection against apoptotic death
 Endoplasmic reticulum (ER)	HSP47	Caring protocollagen
	GRP78	Enhancing the cellular resistance to apoptosis
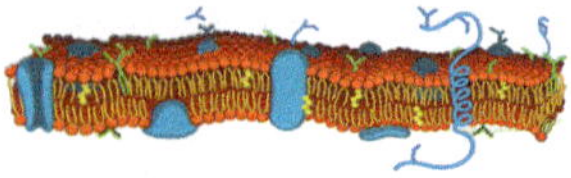 Plasma membrane	HSPb12	Embryonic cardiac muscle development, liposomal membrane protection
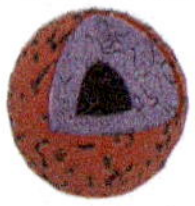 Nucleus and cytosol	HSP110	Helping immune response
Cytosol	HSP40	Chaperoning intermediate filament
Cytoplasm	HSP70	Helping protein folding (assembly & refolding) in endoplasmic reticulum
	HSP90	Helping myosin folding and sarcomere formation
	HSP100	Involved in refolding the aggregates

Heat shock proteins are a highly conserved, ubiquitiously expressed family of stress response proteins. They are expressed at low levels under normal physiological conditions. HSP function as molecular chaperones, facilitating protein folding, preventing protein aggregation and targeting improperly folded proteins to specific degradative pathways. In response to cellular stress, such as hyperthermia, hypoxia, oxidative damage, physical injury or chemical stressors the expression of HSP increases dramatically. This general, coordinated activation of HSP expression, called the heat shock response (HSR), is a ubiquitous adaptation mechanism of organisms that helps them to survive and adapt to a wide range of environmental challenges. Chaperones are classified according to their molecular weight (HSP40, HSP60, HSP70, HSP90, HSP100 and other small HSP. Basically, HSP families are ranging from 10 to more than 100 kDa in molecular weight; they are located at different compartments of cell and also perform various functions (Table 6.1).

6.1.5 Role of Chaperones in Protein Folding

Under cellular stress conditions, oxidative protein modifications lead to formation of oxidized proteins that could lose their secondary or tertiary structure which further compromise the stability, activity and function of proteins (Nedic et al. 2013; Perez et al. 2009). As proteins perform major cellular functions, cells activate several protective mechanisms to recover fully functional protein termed as the proteome quality control system or the proteostasis network (PN) (Roth and Balch 2011; Trougakos et al. 2013). Proteostasis network (PN) is complex protein machinery which controls protein quality via two ways: Proteome quality control 1 and Proteome quality control 2. Proteome quality control 1 consist of folding and unfolding module (UPR and intra- and extra chaperones) while proteome quality control 2 comprises degradation module (mitochondrial proteases, ubiquitin-proteasome system and autophagy-lysosomal system) (Table 6.2).

Molecular chaperones play a central role in proteostasis and they are tightly regulated at transcriptional and post-transcriptional level during stress conditions. Basically, chaperone gene expression is regulated by heat-shock factor (HSFs), a small group transcriptional factors and four HSFs are reported in mammals but among them HSF1 is dominating and highly conserved transcriptional factor (Akerfelt et al. 2007). Under normal condition, HSF1 is present in an inert stage in which inactive monomer, HSF1 is bound to different chaperones/co-chaperones like HSP90, HSP70, HSP40 and others (Anckar and Sistonen 2007; Niforou et al. 2014). While, under stress conditions, HSF1 containing chaperones is bound to unfolded proteins and releases monomeric HSF1 which undergoes phosphorylation and trimerization and ultimately lead to translocation into the nucleus and bind to heat shock element (HSEs). Following this process numerous types of HSP are generated for maintaining and improving proteostasis in cell (Brinkmeier and Ohlendieck 2014; Jee 2016).

6.1.6 Role of HSP in Maintaining Protein Homeostasis

The best characterized organelle chaperones for their role in quality control are those involved in maintenance of protein homeostasis in the ER. The direct involvement of the ER in protein synthesis demands a high content of chaperones in its

Table 6.2 Proteome quality control: a complex mechanism

Proteome quality control	Module	Activities
Proteome quality control 1	Folding and unfolding module	UPR
		Intra chaperones
		Extra chaperones
Proteome quality control 2	Degradation module	Mitochondrial proteases
		Ubiquitin-proteasome system
		Autophagy-lysosomal system

lumen, which sense unfolded protein products and facilitate their folding. Prime chaperones and folding sensors are glucose regulated proteins 78 (GRP78; also known as immunoglubin binding protein- BiP), GRP94, lectins, calnexin, calreticulin, thiol-disulfide oxidoreductases, protein disulfide isomerase (PDI) and ERp57; all are residing in ER in high concentration. Plethora of stressful stimulants such as oxidative stress, hypoglycemia, hyperglycemia, disturbance in calcium homeostasis, viral infection, aging, hypoxia, extreme exercise, redox imbalance, altered protein glycosylation or protein folding defects lead to activation of an adaptive signaling cascade know as the unfolded protein response (UPR). UPR is well-described signaling mechanism which are activated to restore the cellular integrity via chaperone induction, protein translation attenuation and degradation of misfolded proteins (Hetz 2012; Rutkowski and Hegde 2010; Chakrabarti et al. 2011). Any stressful condition that triggers UPR pathway immediately initiate three protective pathways: (i) up-regulation of GRP78/BiP so that refolding could start (ii) attenuation of protein translation and (iii) degradation of misfolded proteins via proteosome, the process known as ER associated degradation (ERAD). A variety of cell intrinsic or extrinsic stresses evoke perturbations in the folding environment of the endoplasmic reticulum (ER), collectively known as ER stress. In normal condition, ER homeostasis is crucial for cell survival but during stress condition, accumulation of unfolded protein stimulates a signal cascade which reduces protein synthesis and increase the capacity of protein folding. Three ER membrane embedded sensor initiate this process, namely inositol requiring enzyme 1 (IRE1), protein kinase R-like ER protein kinase (PERK) and activating transcription factor 6 (ATF6) (Hotamisligil 2010; Zhang and Kaufman 2008) (Figs. 6.3 and 6.4).

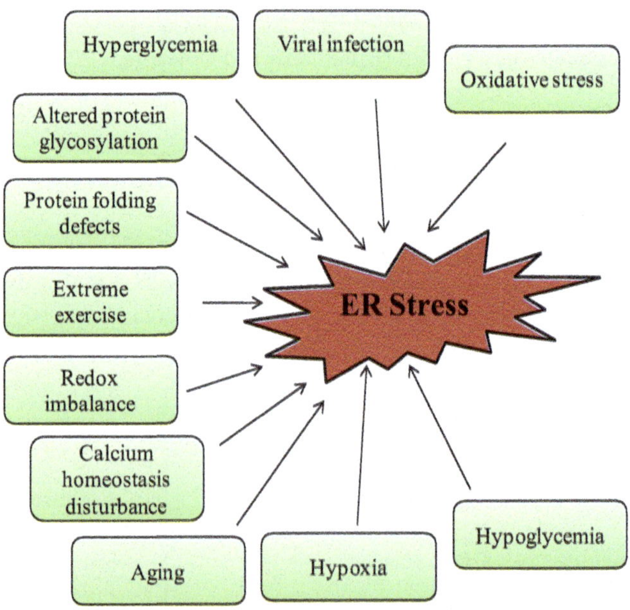

Fig. 6.3 Activation of ER stress during different stressful stimulants

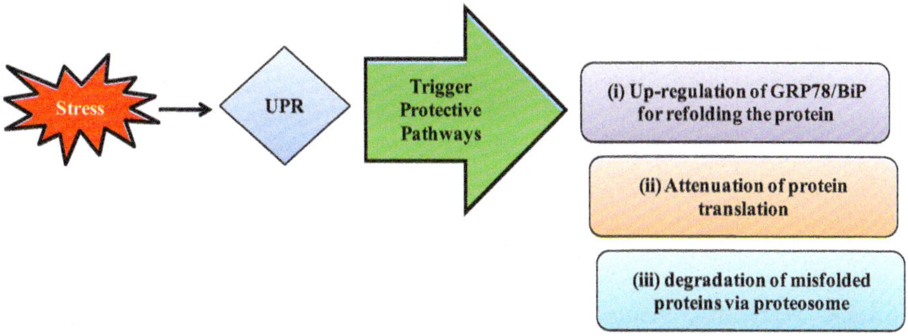

Fig. 6.4 Stress triggers UPR which initiate protective pathway

6.1.6.1 Glucose-Regulated Protein GRP78/BiP

GRP78 is 78 kDa protein and also known as BiP. In normal conditions, the above mentioned three ER sensors are bound to GRP78/BiP. However when ER is overloaded with misfolded proteins, GRP78/BiP immediately binds to the hydrophobic region of misfolded proteins and dissociates from the sensors. The dissociation leads to activation of IRE-1, PERK and ATF6. GRP78 is the first chaperones which are upregulated during UPR hence it is also considered as a key regulator of ER stress (Hotamisligil 2010).

6.1.6.2 Inositol Requiring Enzyme 1 (IRE1)

IRE1 has serine/threonine kinase activity in its cytoplasmic domain. During ER stress, kinase-kinase and nuclease-nuclease interactions, both mediate the formation of an IRE1 homodimer. ER-induced dimerization leads to the proximal position of cytosolic domains for trans-autophosphorylation of kinase activation loop. Hence, ER-induced dimerization and trans-autophosphorylation arise its endonuclease activity, thus it excise a 26-nucleotide intron from mRNA of X-box binding protein (XBP1) to generate spliced XBP1. This is a potent translational transactivator that upregulate UPR targeted genes like BiP/GRP78 and XBP1-U (Calfon et al. 2002; Yoshida et al. 2001). But persistent ER stress leads to dissociation of IRE1 cluster so that its endonuclease activity gets vanished and UPR is switched to apoptosis from adaptive phase. Few studies demonstrated that cell survival could increase if IRE1 activity is sustained artificially.

6.1.6.3 Protein Kinase R-Like ER Protein Kinase (PERK)

PERK is type I transmembrane serine/thronine kinase and second sensor of ER stress response. Like IRE1, it is also activated via transautophosphorylation and homodimerization under conditions of ER stress. Activated PERK attenuates protein translation and alleviates ER protein overload via phosphorylating eukaryotic translation initiation factor 2 α (eIF2α) that inhibits the assembly of 80S ribosome (Shi et al. 1998; Harding et al. 1999). Beside this, PERK/eIF2α also inhibit cyclinD1 expression that leads to cell cycle arrest (Brewer and Diehl 2000) and boost activating transcription factor-4 (ATF4) (Harding et al. 2000) that further the enhances transcription of C/EBP-homologous protein (CHOP) causing apoptosis (Zinszner et al. 1998).

6.1.6.4 Activating Transcription Factor 6 (ATF6)

ATF6 is considered as third major sensor of ER stress response. Once cell is accumulated with misfolded proteins, GRP78 dissociates from ATF6 and ATF6 translocates to golgi complex where it gets cleaved by site 1 and site 2 proteases (Haze et al. 1999). Then cleaved ATF6 translocates to nucleus and binds to ER stress response element CCAAT (N) 9CCACG which increase the transcription BiP/GRP78, GRP94, XBP1, CHOP and PDI to increase protein folding, secretion, and degradation (Okada et al. 2002). In conclusion, the three complementary sensors (IRE1, PERK and ATF6) regulate the adaptive and mal-adaptive response of cell to survival as well as apoptosis (Fig. 6.5). Even though the

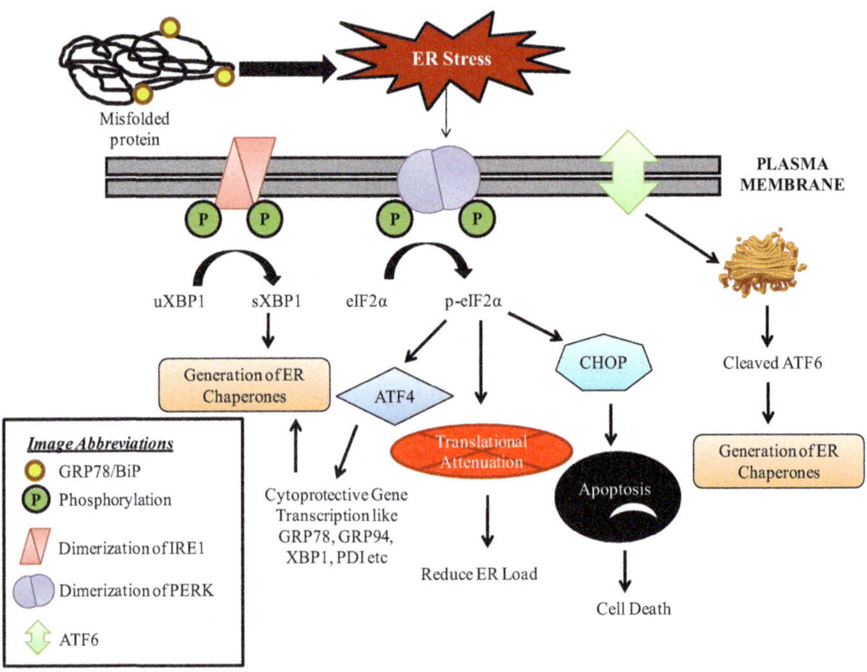

Fig. 6.5 ER Stress Signaling Pathway

oxidative protein modifications under hypobaric hypoxia are well studied, very few studies focus on the occurrence of this multifaceted ER stress response under HA milieu.

6.1.7 Chaperokine

Previously it was assumed that chaperones are mainly involved with protein folding and quality control system but recent research has described its dual function and these chaperones are termed as chaperokine. As per the scientific definition, Chaperokine is the term which describe the dual function of extracellular heat shock protein ie both chaperone and cytokine function. One of the best known chaperokine is eHSP70 (Hsp72) (Asea 2008).

The function of seventy kilodaltan HSP depends on its localization: intracellular or extracellular. Intracellular HSP70 engaged in protein chaperoning, transport and folding which used to initiate if cell meets any disease or stress conditions like atherosclerosis (Pockley et al. 2003), hypertension (Pockley et al. 2002), renal disease (Wright et al. 2000), sickle cell anemia (Adewoye et al. 2005), aging (Terry et al. 2004), heat stress (Georgopoulos and Welch 1993) or hypoxic stress (Jain et al. 2016a, b). On the other side, extracellular HSP72 (eHSP72) involved in upregulation of pro-inflammatory cytokine including TNF- α, interleukin-1 beta (IL-1β) and IL-6 by antigen presenting cells (APC).

6.1.7.1 Chaperokine-Induced Signaling

Except T-lymphocytes, other immune cells such as natural killer (NK) cells (Gross et al. 2003) and antigen presenting cells (APC) like dendritic cells (Asea et al. 2002), macrophages, peripheral blood monocytes (Sondermann et al. 2000), B-lymphocytes (Arnold-Schild et al. 1999) efficiently binds with eHSP70. Asea and his team addressed the steps involved in chaperokine induced pro-inflammatory cytokine production. They revealed that eHSP70 binds with plasma membrane of APC that lead to rapid intracellular Ca_i^{2+} flux within 10 s (Asea et al. 2000). This step is considered as an exclusive step which differentiates chaperokine induced signal mechanism from LPS-induced signalling mechanism as in LPS does not resultant into intracellular calcium flux (Ca_i^{2+}). The rapid eHSP70 induced intracellular calcium flux (Ca_i^{2+}) leads to phosphorylation of I-κBα and activation of NF-κB that further migrate to nucleus and activate promoter of target genes (Baeuerle and Baltimore 1988). Approx. 30 min used to take place from binding of eHSP70 to release and nuclear translocation of NF-κB. In vitro studies were also performed by the same group which revealed that eHSP72 induced NF-κB promoter activity is MyD88-dependant, CD14-dependant and is transduced via both Toll like receptors 2 (TLR2) and toll like receptors 4 (TLR4). Ergo, the presence of both

TLR2 and TLR4 synergisticly activate eHSP72 induced pro-inflammatory cytokine production (Asea et al. 2002). Vabulas and colleagues also made an agreement with the same study by their independent studies (Vabulas et al. 2002). Interestingly, within 2–4 h post-exposure of APC to eHSP70, a significant release of pro-inflammatory cytokines covering TNF-α, IL-1β, IL-6, IL-12 and GM-CSF (Srivastava 2002); nitric oxide (NO), a potent apoptogenic mediator and chemokines like MIP-1, MCP-1 and RANTES was observed (Panjwani et al. 2002).

6.1.7.2 Biological Significance

Circulating eHSP72 was found increased and upregulated in various disease conditions like renal disease, hypertension, atherosclerosis, sickle cell anemia and phychological stresses (Wright et al. 2000; Pockley et al. 2003; Adewoye et al. 2005). It was hypothesized that eHSP72 may act like "danger signal" to immune cells and promote immune responses. Campisi et al. 2003 reported the increment in eHSP72 which act as danger signal which further facilitate nitric oxide (NO) response and promote immune responses and also improved host defense system. Possible circulating eHSP72 used to provoke immune system to real or superficial danger. Recently, eHSP72 also proposed as a biomarker of sarcopenia (Perreault et al. 2016). However, extensive research in this field is required to decipher the exploitation of chaperokines for medicinal purpose.

6.1.8 Factors for Hypoxia Tolerance

Every individual have the different ability to respond against environmental stress and it is also true for high altitude associated stressors. Morphological and physiological differences which could be related to genetic adaptation or inherent threshold to tolerate stress also play a prime role in adaptation (Moore 2001). It is also observed that lowlanders are more susceptible to high altitude stress as compared to high altitude natives, indicating that high and low altitude population has diverse metabolic strategies to overcome the stress (Gelfi et al. 2004). Surprisingly, antioxidant enzymes were found to be lower in highlanders which indicating their better adaptation to high altitude (Sinha et al. 2009a, b; Jain et al. 2013a). Low oxygen concentration stabilized the protein hypoxia-inducible factor (HIF -1 alpha) that further regulated more than 100 hypoxia-sensitive genes like as inducible nitric oxide synthase (iNOS), glucose transporter (GLUT-1), erythropoietin (EPO), heme oxygenase-1 (HO-1), Heat shock proteins (HSP60, HSP70, HSP90 etc.) and activation of other nuclear factors.

Our recent studies indicate that the upregulation of cytoprotective mechanisms such as antioxidant levels, heat shock response and Hypoxia-responsive proteins imparts tolerance to hypobaric hypoxia. Our studies by Jain et al. (2013a) show the role of HIF-1 alpha regulated proteins in hypoxia tolerance. Animals were grouped to susceptible (<10 min), normal (10–25 min), and tolerant (>25 min) as per the gasping time. The results concluded that differential induction and stabilization of HIF-1alpha activated genes mediate tolerance to hypoxic stress. Up regulation of genes which are involved in increased oxygen transport and delivery (EPO, and HO-1), vascular tone modulators and angiogenesis (ET-1 and (VEGF), and glucose utilization (GLUT1) enhance the tolerance of an organism during acute hypoxic stress. Molecular chaperones system, HSP70 and HSP90 also played a key role in adaptive response of animal during acute hypoxic stress. Further, we also studied the function of myocardial ER chaperones due to synergistic adverse effect, cold and hypoxic stress in rats (Jain et al. 2013b). Further we (Jain et al. 2014) also reported that rats could able to sustain under sub lethal hypoxia via activation of chaperones and enhanced NO generation (Fig. 6.6).

The role of chaperones in protein folding/unfolding, repair of misfolded proteins and preventing the aggregation of such modified proteins is well established. In our study a significant increase in cytoprotective HSP70 and HSP60 was observed in cold hypoxia exposure that was also harmonized with increase HSP90 levels in cold hypoxia exposed rats (Jain et al. 2013b). Few other studies also stated the protective role of HSP70 against hypoxic insult (Mohan et al. 2001). Along with this, heme oxygenase-1 (HSP32) also showed the similar pattern of elevated level at cold

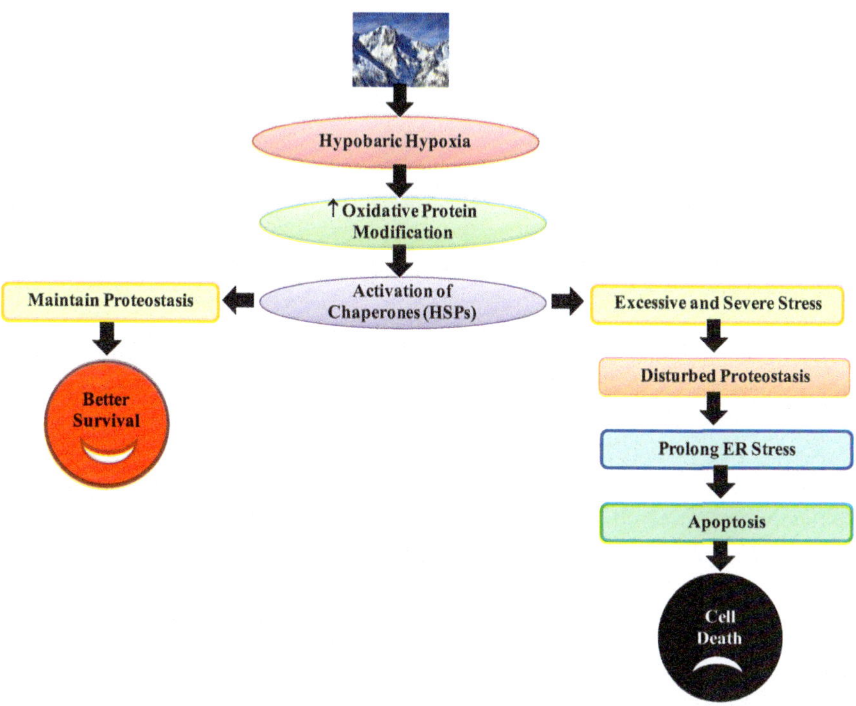

Fig. 6.6 High altitude associated hypobaric hypoxia induced cell death via chaperone activation

hypoxic stress which indicated the altered proteins load during stress (Jain et al. 2013a). Hence from our research we have comprehensively identified the role of chaperones in imparting tolerance to hypobaric hypoxia.

We observed that animals susceptible to hypoxic stress showed higher protein oxidative damage and the presence of such modified proteins led to an increase in expression of ER chaperones GRP78, PDI and calreticulin. There was a significant upregulation of pro-apoptotic CHOP and caspase-3 in susceptible animals indicating the inability to maintain proteostasis. However the animals which could better endure the reduced oxygen availability did not show a significant degree of oxidized protein damage, ER stress activation and cell death. The higher expression of pro-survival kinases may also have contributed to the better tolerance. These results indicated the role of protein quality control machinery and maintenance of proteostasis in the ability to sustain under low oxygen conditions.

In Vascular Pharmacology, our results (Jain et al. 2016a, b) demonstrated the role of ER stress in cardiac hypertrophy and injury in rats during chronic hypobaric hypoxia. The work described the right ventricular cardiac hypertrophy along with the activation of PI3K/Akt/mTOR signalling pathway during chronic hypobaric hypoxia. As the duration of hypoxic exposure is increased, there is a distinct increase in the damage to proteins, increasing the load of misfolded proteins on the ER. While the more extensive studies on the UPR pathways remain to be done, our observation suggest that prolonged hypoxic protocols are capable of inducing all the three arms of the ER stress response such as activation of ATF4 expression, eIF2α phosphorylation, ER chaperone expression and protein degradation. Although all three branches of UPR work together to counter ER stress and restore the homeostasis in the ER, paradoxically it is the very same UPR which activates cell death pathway under persistent ER stress. One major event responsible for this switches between pro-survival and proapoptotic responses are the induction of CHOP, which forms a common point of convergence for all 3 arms of the UPR. In our study we found a progressive increase in apoptotic caspase-12, 9 and 3 and CHOP expression, with the increase in duration of hypoxic exposure. Suzen et al. (2015) also demonstrated the adverse effect of ER stress in various cardiovascular diseases including myocardium I/R injury.

Not only heart our recent research focused on skeletal muscle which was also affected under acute hypobaric hypoxic stress. The hypoxic exposure enhanced the oxidative protein modifications and altered proteostasis. This further resulted into upregulation of HSP60, HSP70 and ER stress marker, GRP78 (Agrawal et al. 2017). The up regulation of chaperones consequently activated the pro-apoptotic signalling cascade which involved CHOP and caspase3. Other studies also reported increase in HSP90 and HSP70 in retinas which could lead to cell injury during hypobaric hypoxia exposure (Minet et al. 1999; Xin et al. 2017).

6.1.9 *Role of ER Stress/Chaperones in Therapeutic Interventions*

ER stress response displays a dichotomic yin-yang characteristic. Basically, yin yang (Shadow and light) is Chinese philosophy according to that a balance occurs between cell survival and cell death. During ER stress, opposing efforts of prosurvival (yang) GRP78 and proapoptotic (yin) CHOP and IRE1-regulated JNK represent prime protein for this struggle (Fig. 6.4). GRP78 protein is a key chaperone to initiate the process to pursue adaptation and cellular survival. On the other hand, persistent and severe ER stress activates the CHOP and IRE1-regulated JNK pro-apoptotic pathway which lead to cell death (Nishitoh 2012; Verma and Datta 2012). There is accumulating evidence which implicate the role of UP and ER stress response in various human diseases. Thus, it is considered as a novel therapeutic target in several diseases. Recent research have demonstrated that ER stress modulators could become novel therapeutic target against numerous diseases and these could be pharmacological or nutraceuticals. Few of them are listed in the Table 6.3 (Fig. 6.7).

Table 6.3 ER stress modulators and their effect on various diseases

Category	Drug	Effect on disease	References
mTOR inhibitors	Rapamycin	Diabetes, apoptosis, hepatic ischemia, NAFLD	Tersey et al. (2012), Marhfour et al. (2012), and Harding et al. (2001)
Chemical chaperones	4PBA	Adipogenesis, high altitude associated hypoxia	Seimon et al. (2010)
	TUDCA	Apoptosis, steatohepatitis	Puri et al. (2008)
AMPK activators	Metformin	Renal fibrosis, Apoptosis, vascular dysfunction	Qin et al. (2010), Yin et al. (2015) and Yorimitsu and Klionsky (2007)
	Salicylate/ Salsalate	Insulin resistance,	Bachar-Wikstrom et al. (2013);
	AICAR	EDR, Cardiac hypoxic injury, Apoptosis	Zhang et al. (2011) and Usui et al. (2012)
GLP-1 receptor agonists and DPP-4 inhibitors	Exenatide	Apoptosis (β-cell), NAFLD	Kim et al. (2015) and Basseri et al. (2009)
	Vildagliptin	β-cell loss	Sen et al. (2013)
	Gemigliptin	Apoptosis (cardiomyocyte)	Alhusaini et al. (2010)
PPARs agonists	Fenofibrate	NAFLD, EDV	Li et al. (2015) and Jung et al. (2014)
	Pioglitazone	β-cell dysfunction, Apoptosis (macrophage)	Yusta et al. (2006)
	GW1516	Insulin resistance (skeletal myocytes) Cardiac hypertrophy	Cunha et al. (2009) and Yamane et al. (2011)
ARBs	Valsartan	Apoptosis (cardiomyocyte, renal cell)	Zhang et al. (2014)
	Losartan	β-cell dysfunction	Fu et al. (2011)
	Olmesartan	Autoimmune myocarditis	Younce et al. (2013)
	Telmisartan	Cardiac hypertrophy	Lee et al. (2014)

Fig. 6.7 Chinese philosophy of yin yang (A balance between pro-survival and pro-apoptotic)

6.1.10 Effect of ER Stress Modulators on Hypoxic Survival

Having established the critical role of the altered proteostasis and ER stress response under hypobaric hypoxia, we have evaluated the efficacy of various modulators of this response in improving hypoxic tolerance and survival. Amongst the drugs evaluated we elegantly demonstrated that the chemical chaperone, 4-PBA is most effective in reducing ER stress under acute hypoxia and significantly increased the gasping time, an indicator of hypoxic tolerance. Further we assessed the therapeutic potential of the drug against the high altitude induced cardiac pathology. Clinically approved for use in urea cycle disorders and thalassemia, we find that 4-PBA substantially reduced the cardiac injury and damage induced by chronic hypoxic stress through a downregulation of the ER stress response and activation of cell survival signaling cascades (Jain et al. 2016a).

We have extensively worked on a Himalayan medicinal plant, Seabuckthorn (SBT) (*Hippophae rhamnoides* L.) which has gained a lot of recognition in the recent years for its medicinal and nutritional activity. All parts of this plant are considered to be a good source of large number of bioactive substances such as vitamins, carotenoids, phytosterols, polyunsaturated fatty acids and some essential amino acids with a remarkable activity against common human diseases. Leaves of the plant, rich in flavonoids, have earlier been shown to possess potent antioxidant, anti-inflammatory activity and adaptogenic activity and have shown to protect against hypoxia induced pulmonary vascular leakage and ROS mediated damage (Reviewed comprehensively by us) (Suryakumar and Gupta 2011).

To explore the mechanism behind the protective effects of SBT, various signaling mechanisms were evaluated. Administration of SBT extract led to a marked increase in the expression of hypoxia responsive proteins such as HIF-1α and HO-1. SBT also substantially exacerbated the production of nitric oxide, probably through the higher iNOS and eNOS expression. Interestingly, SBT attenuated the ER stress response as observed by the downregulation of ER stress markers CHOP and

PERK. SBT induced anti-inflammatory and antiapoptotic responses in the hypoxia exposed animals also contributed to the enhanced hypoxic tolerance. Our study draws attention to the modulation of proteostasis signaling mechanisms by SBT which mediate its protective effects against hypoxic injury (Rathor et al. 2015; Jain et al. 2016b).

6.2 Conclusions

Our studies demonstrate the role cytosolic and ER chaperones in mediating protective response to high altitude. We have shown the activation of chaperones in acute and chronic hypoxia and also the role of proper protein folding and proteostasis in better adaptation to hypobaric hypoxia. Further modulation of proteostasis has shown to protect against the deleterious effects of hypoxia. The leads from the present study also potentiate the use of proteostasis modulators either chemical or herbal which can be developed as therapeutic interventions for the treatment of a multitude of hypoxia related pathophysiologies.

Acknowledgements The authors are thankful to Dr. Bhuvnesh Kumar, Director, DIPAS, for his constant support and encouragement.The study was supported by the Defence Research and Development Organisation, Ministry of Defence, Government of India.

References

Adewoye AH, Klings ES, Farber HW, Palaima E, Bausero MA, McMahon L, Odhiambo A, Surinder S, Yoder M, Steinberg MH, Asea A (2005) Sickle cell vaso-occlusive crisis induces the release of circulating serum heat shock protein-70. Am J Hematol 78:240–242

Agrawal A, Rathor R, Suryakumar G (2017) Oxidative protein modification alters proteostasis under acute hypobaric hypoxia in skeletal muscles: a comprehensive in vivo study. Cell Stress Chap 22:429–443

Akerfelt M, Trouillet D, Mezger V, Sistonen L (2007) Heat shock factors at a crossroad between stress and development. Ann N Y Acad Sci 1113:15–27

Alhusaini S, McGee K, Schisano B, Harte A, McTernan P, Kumar S, Tripathi G (2010) Lipopolysaccharide, high glucose and saturated fatty acids induce endoplasmic reticulum stress in cultured primary human adipocytes: salicylate alleviates this stress. Biochem Biophys Res Commun 397:472–478

Anckar J, Sistonen L (2007) Heat shock factor 1 as a coordinator of stress and developmental pathways. Adv Exp Med Biol 594:78–88

Arnold-Schild D, Hanau D, Spehner D, Schmid C, Rammensee HG, de la Salle H, Schild H (1999) Receptor mediated endocytosis of heat shock proteins by professional antigen-presenting cells. J Immunol 162:3757–3760

Asea A (2008) Hsp70: a chaperokine. Novartis Found Symp 291:173–179 discussion 179–83, 221–4

Asea A, Kraeft SK, Kurt-Jones EA, Stevenson MA, Chen LB, Finberg RW, Koo GC, Calderwood SK (2000) HSP70 stimulates cytokine production through a CD14-dependant pathway, demonstrating its dual role as a chaperone and cytokine. Nat Med 6:435–442

Asea A, Rehli M, Kabingu E, Boch JA, Bare O, Auron PE, Stevenson MA, Calderwood SK (2002) Novel signal transduction pathway utilized by extracellular HSP70: role of toll-like receptor (TLR) 2 and TLR4. J Biol Chem 277:15028–15034

Bachar-Wikstrom E, Wikstrom JD, Kaiser N, Cerasi E, Leibowitz G (2013) Improvement of ER stress-induced diabetes by stimulating autophagy. Autophagy 9:626–628

Baeuerle PA, Baltimore D (1988) I kappa B: a specific inhibitor of the NF-kappa B transcription factor. Science 242:540–546

Basseri S, Lhotak S, Sharma AM, Austin RC (2009) The chemical chaperone 4-phenylbutyrate inhibits adipogenesis by modulating the unfolded protein response. J Lip Res 50:2486–2501

Beissinger M, Buchner J (1998) How chaperones fold proteins. Biol Chem 379(3):245–259

Brewer JW, Diehl JA (2000) PERK mediates cell-cycle exit during the mammalian unfolded protein response. Proc Natl Acad Sci 97:12625–12630

Brinkmeier H, Ohlendieck K (2014) Chaperoning heat shock proteins: proteomic analysis and relevance for normal and dystrophin-deficient muscle. Proteom Clin Appl 8:875–895

Brot N, Weissbach H (1982) The biochemistry of methionine sulfoxide residues in proteins. Trends Biochem Sci 7(4):137–139

Butterfield DA, Stadtman ER (1997) Protein oxidation processes in aging brain. Adv Cell Aging Gerontol 2:161–191

Calfon M, Zeng H, Urano F, Till JH, Hubbard SR (2002) IRE1 couples endoplasmic reticulum load to secretory capacity by processing the XBP-1 mRNA. Nature 415:92–96

Campisi J, Leem TH, Fleshner M (2003) Stress-induced extracellular Hsp72 is a functionally significant danger signal to the immune system. Cell Stress Chap 8(3):272–286

Capeillere-Blandin C, Gausson V, Descamps-Latscha B, Witko-Sarsat V (2004) Biochemical and spectrophotometric significance of advanced oxidized protein products. Biochim Biophys Acta 1689:91–102

Chakrabarti A, Chen AW, Varner JD (2011) A review of the mammalian unfolded protein response. Biotechnol Bioeng 108:2777–2793

Cunha DA, Ladriere L, Ortis F, Igoillo-Esteve M, Gurzov EN, Lupi R, Marchetti P, Eizirik DL, Cnop M (2009) Glucagon-like peptide-1 agonists protect pancreatic β-cells from lipotoxic endoplasmic reticulum stress through upregulation of BiP and JunB. Diabetes 58:2851–2862

Dalle-Donne I, Rossi R, Giustarini D, Milzani A, Colombo R (2003) Protein carbonyl groups as biomarkers of oxidative stress. Clin Chim Acta 329:23–38

Davies KJ (1987) Protein damage and degradation by oxygen radicals. I general aspects. J Biol Chem 262(20):9895–9901

Faiss R, Pialoux V, Sartori C, Faes C, Deriaz O, Millet GP (2013) Ventilation, oxidative stress, and nitric oxide in hypobaric versus normobaric hypoxia. Med Sci Sports Exerc 45:253–260

Fu S, Yang L, Li P, Hofmann O, Dicker L, Hide W, Lin X, Watkins SM, Ivanov AR, Hotamisligil GS (2011) Aberrant lipid metabolism disrupts calcium homeostasis causing liver endoplasmic reticulum stress in obesity. Nature 473:528–531

Gelfi C, De Palma S, Ripamonti M, Eberini I, Wait R, Bajracharya A, Marconi C, Schneider A, Hoppeler H, Cerretelli P (2004) New aspects of altitude adaptation in Tibetans: a proteomic approach. FASEB J 18:612–614

Georgopoulos C, Welch WJ (1993) Role of the major heat shock proteins as molecular chaperones. Annu Rev Cell Biol 9:601–634

Gross C, Hansch D, Gastpar R, Multhoff G (2003) Interaction of heat shock protein 70 peptide with NK cells involves the NK receptor CD94. Biol Chem 384:267–279

Gurd JW, Bissoon N, Beesley PW, Nakazawa T, Yamamoto T, Vannucci SJ (2002) Differential effects of hypoxia-ischemia on subunit expression and tyrosine phosphorylation of the NMDA receptor in 7- and 21-day-old rats. J Neurochem 82:848–856

Harding HP, Zhang Y, Ron D (1999) Protein translation and folding are coupled by an endoplasmic reticulum-resident kinase. Nature 397:271–274

Harding HP, Novoa I, Zhang Y, Zeng H, Wek R (2000) Regulated translation initiation controls stress-induced gene expression in mammalian cells. Mol Cell 6:1099–1108

Harding HP, Zeng H, Zhang Y, Jungries R, Chung P, Plesken H, Sabatini DD, Ron D (2001) Diabetes mellitus and exocrine pancreatic dysfunction in perk-/- mice reveals a role for translational control in secretory cell survival. Mol Cell 7:1153–1163

Hartl FU (1996) Molecular chaperones in cellular protein folding. Nature 381:571–580

Haslbeck M (2002) sHSP and their role in the chaperone network. Cell Mol Life Sci 59:1649–1657

Haze K, Yoshida H, Yanagi H, Yura T, Mori K (1999) Mammalian transcription factor ATF6 is synthesized as a transmembrane protein and activated by proteolysis in response to endoplasmic reticulum stress. Mol Biol Cell 10:3787–3799

Herbst R, Schafer U, Seckler R (1997) Equilibrium intermediates in the reversible unfolding of firefly (*Photinus pyralis*) luciferase. J Biol Chem 272:7099–7105

Hetz C (2012) The unfolded protein response: controlling cell fate decisions under ER stress and beyond. Nat Rev Mol Cell Biol 13:89–102

Horowitz S, Koldewey P, Stull F, Bardwell JCA (2017) Folding while bound to chaperones. Curr Opin Struct Biol 48:1–5

Hotamisligil GS (2010) Endoplasmic reticulum stress and atherosclerosis. Nat Med 16:396–399

Jain K, Suryakumar G, Prasad R, Ganju L (2013a) Upregulation of cytoprotective defense mechanisms and hypoxia-responsive proteins imparts tolerance to acute hypobaric hypoxia. High Alt Med Biol 14:65–77

Jain K, Suryakumar G, Prasad R, Singh SN, Ganju L (2013b) Myocardial ER chaperone activation and protein degradation occurs due to synergistic, not individual, cold and hypoxic stress. Biochimie 95(10):1897–1908

Jain K, Suryakumar G, Ganju L, Singh SB (2014) Differential hypoxic tolerance is mediated by activation of heat shock response and nitric oxide pathway. Cell Stress Chap 19:801–812

Jain K, Suryakumar G, Ganju L, Singh SB (2016a) Amelioration of ER stress by phenylbutyric acid reduces chronic hypoxia induced cardiac damage and improves hypoxic tolerance through upregulation of HIF-1α. Vasc Pharmacol 83:36–46

Jain K, Suryakumar G, Prasad R, Ganju L, Singh SB (2016b) Enhanced hypoxic tolerance by Seabuckthorn is due to upregulation of HIF-1a and attenuation of ER stress. J Appl Biomed 14(1):71–83

Jee H (2016) Size dependent classification of heat shock proteins: a mini-review. J Exerc Rehabil 12(4):255–259

Jung TW, Lee SY, Hong HC, Choi HY, Yoo HJ, Baik SH, Choi KM (2014) AMPK activator-mediated inhibition of endoplasmic reticulum stress ameliorates carrageenan-induced insulin resistance through the suppression of selenoprotein P in HepG2 hepatocytes. Mol Cell Endocrinol 382:66–73

Kayyali US, Donaldson C, Huang H, Abdelnour R, Hassoun PM (2001) Phosphorylation of xanthine dehydrogenase/oxidase in hypoxia. J Biol Chem 276:14359–14365

Kim H, Moon SY, Kim JS, Baek CH, Kim M, Min JY, Lee SK (2015) Activation of AMP-activated protein kinase inhibits ER stress and renal fibrosis. Am J Physiol Ren Physiol 308:F226–F236

Koumenis C, Naczki C, Koritzinsky M, Rastani S, Diehl A, Sonenberg N, Koromilas A, Wouters BG (2002) Regulation of protein synthesis by hypoxia via activation of the endoplasmic reticulum kinase PERK and phosphorylation of the translation initiation factor eIF2 alpha. Mol Cell Biol 22:7405–7416

Kulisz A, Chen N, Chandel NS, Shao Z, Schumacker PT (2002) Mitochondrial ROS initiate phosphorylation of p38 MAP kinase during hypoxia in cardiomyocytes. Am J Physiol Lung Cell Mol Physiol 282:L1324–L1329

Kumar GK, Klein JB (2004) Analysis of expression and posttranslational modification of proteins during hypoxia. J Appl Physiol 96:1178–1186

Lee J, Hong SW, Park SE, Rhee EJ, Park CY, Oh KW, Park SW, Lee WY (2014) Exendin-4 attenuates endoplasmic reticulum stress through a SIRT1-dependent mechanism. Cell Stress Chap 19:649–656

Lewis NC, Bailey DM, Dumanoir GR, Messinger L, Lucas SJ, Cotter JD, Donnelly J, McEneny J, Young IS, Stembridge M, Burgess KR, Basnet AS, Ainslie PN (2014) Conduit artery structure and function in lowlanders and native highlanders: relationships with oxidative stress and role of sympatho excitation. J Physiol 592:1009–1024

Li J, Wang Y, Wen X, Ma XN, Chen W, Huang F, Kou J, Qi LW, Liu B, Liu K (2015) Pharmacological activation of AMPK prevents Drp1-mediated mitochondrial fission and alleviates endoplasmic reticulum stress-associated endothelial dysfunction. J Mol Cell Cardiol 86:62–74

Marhfour I, Lopez XM, Lefkaditis D, Salmon I, Allagnat F, Richardson SJ, Morgan NG, Eizirik DL (2012) Expression of endoplasmic reticulum stress markers in the islets of patients with type 1 diabetes. Diabetologia 55:2417–2420

McGinnis G, Kliszczewiscz B, Barberio M, Ballmann C, Peters B, Slivka D, Dumke C, Cuddy J, Hailes W, Ruby B, Quindry J (2014) Acute hypoxia and exercise-induced blood oxidative stress. Int J Sport Nutr Exerc Metab 24:684–693

Miller LE, McGinnis GR, Kliszczewicz B, Slivka D, Hailes W, Cuddy J, Dumke C, Ruby B, Quindry JC (2013) Blood oxidative-stress markers during a high-altitude trek. Int J Sport Nutr Exerc Metab 23:65–72

Minet E, Mottet D, Michel G, Roland I, Raes M, Remacle J, Michielsa C (1999) Hypoxia-induced activation of HIF-1: role of HIF-1alpha-Hsp90 interaction. FEBS Lett 460:251–256

Mishra OP, Ashraf QM, Delivoria-Papadopoulos M (2002) Phosphorylation of cAMP response element binding (CREB) protein during hypoxia in cerebral cortex of newborn piglets and the effect of nitric oxide synthase inhibition. Neuroscience 115:985–991

Mohan RM, Golding S, Paterson DJ (2001) Intermittent hypoxia improves atrial tolerance to subsequent anoxia and reduces stress protein expression. Acta Physiol Scand 172:89–95

Mohanraj P, Merola AJ, Wright VP, Clanton TL (1998) Antioxidants protect rat diaphragmatic muscle function under hypoxic conditions. J Appl Physiol 84:1960–1966

Moore LG (2001) Human genetic adaptation to high altitude. High Alt Med Biol 2:257–279

Nedic O, Rattan SI, Grune T, Trougakos IP (2013) Molecular effects of advanced glycation end products on cell signalling pathways, ageing and pathophysiology. Free Rad Res 47(1):28–38

Niforou K, Chemonidou C, Trougakos IP (2014) Molecular chaperones and proteostasis regulation during redox imbalance. Redox Biol 2:323–332

Nishitoh H (2012) CHOP is a multifunctional transcription factor in the ER stress response. J Biochem 151:217–219

Okada T, Yoshida H, Akazawa R, Negishi M, Mori K (2002) Distinct roles of activating transcription factor 6 (ATF6) and double-stranded RNA-activated protein kinase-like endoplasmic reticulum kinase (PERK) in transcription during the mammalian unfolded protein response. Biochem J 366:585–594

Panjwani NN, Popova L, Srivastava PK (2002) Heat shock proteins gp96 and hsp70 activate the release of nitric oxide by APCs. J Immunol 168:2997–3003

Perez VI, Buffenstein R, Masamsetti V, Leonard S, Salmon AB, Mele J, Andziak B, Yang T, Edrey Y, Friguet B, Ward W, Richardson A, Chaudhuri A (2009) Protein stability and resistance to oxidative stress are determinants of longevity in the longest-living rodent, the naked mole-rat. Proc Natl Acad Sci U S A 106:3059–3064

Perreault K, Courchesne-Loyer A, Fortier M, Maltai M, Barsalani R, Riesco E, Dionne IJ (2016) Sixteen weeks of resistance training decrease plasma heat shock protein 72 (eHSP72) and increase muscle mass without affecting high sensitivity inflammatory markers' levels in sarcopenic men. Aging Clin Exp Res 28:207–214

Pockley AG, De Faire U, Kiessling R, Lemne C, Thulin T, Frostegard J (2002) Circulating heat shock protein and heat shock protein antibody levels in established hypertension. J Hypertens 20:1815–1820

Pockley AG, Georgiades A, Thulin T, de Faire U, Frostegard J (2003) Serum heat shock protein 70 levels predict the development of atherosclerosis in subjects with established hypertension. Hypertension 42:235–238

Powers ET, Morimoto RI, Dillin A, Kelly JW, Balch WE (2009) Biological and chemical approaches to diseases of proteostasis deficiency. Annu Rev Biochem 78:959–991

Puri P, Mirshahi F, Cheung O, Natarajan R, Maher JW, Kellum JM, Sanyal AJ (2008) Activation and dysregulation of the unfolded protein response in nonalcoholic fatty liver disease. Gastroenterology 134:568–576

Qin L, Wang Z, Tao L, Wang Y (2010) ER stress negatively regulates AKT/TSC/mTOR pathway to enhance autophagy. Autophagy 6:239–247

Radak Z, Lee K, Choi W, Sunoo S, Kizaki T, Ohishi S, Suzuki K, Taniguchi N, Ohno H, Asano K (1994) Oxidative stress induced by intermittent exposure at a simulated altitude of 4000 m decreases mitochondrial superoxide dismutase content in soleus muscle of rats. Eu J Appl Physiol 69:392–395

Rathor R, Sharma P, Suryakumar S, Ganju L (2015) A pharmacological investigation of *Hippophae salicifolia* (HS) and *Hippophae rhamnoides turkestanica* (HRT) against multiple stress (C-H-R): an experimental study using rat model. Cell Stress Chap 20:821–831

Roth DM, Balch WE (2011) Modeling general proteostasis: proteome balance in health and disease. Curr Opin Cell Biol 23:126–134

Row BW, Goldbart A, Gozal E, Gozal D (2003) Spatial pre-training attenuates hippocampal impairments in rats exposed to intermittent hypoxia. Neurosci Lett 339:67–71

Rutkowski DT, Hegde RS (2010) Regulation of basal cellular physiology by the homeostatic unfolded protein response. J Cell Biol 189:783–794

Seimon TA, Nadolski MJ, Liao X, Magallon J, Nguyen M, Feric NT, Koschinsky ML, Harkewicz R, Witztum JL, Tsimikas S, Golenbock D, Moore KJ, Tabas I (2010) Atherogenic lipids and lipoproteins trigger CD36-TLR2-dependent apoptosis in macrophages undergoing endoplasmic reticulum stress. Cell Metab 12:467–482

Sen S, Kundu BK, Wu HC, Hashmi SS, Guthrie P, Locke LW, Roy RJ, Matherne GP, Berr SS, Terwelp M, Scott B, Carranza S, Frazier OH, Glover DK, Dillmann WH, Gambello MJ, Entman ML, Taegtmeyer H (2013) Glucose regulation of load-induced mTOR signaling and ER stress in mammalian heart. J Am Heart Assoc 2:e004796

Serrano J, Encinas JM, Salas E, Fernandez AP, Castro-Blanco S, Fernandez-Vizarra P, Bentura ML, Rodrigo J (2003) Hypobaric hypoxia modifies constitutive nitric oxide synthase activity and protein nitration in the rat cerebellum. Brain Res 976:109–119

Sharma P, Jha AB, Dubey RS, Pessarakli M (2012) Reactive oxygen species, oxidative damage, and antioxidative defense mechanism in plants under stressful conditions. J Bot 2012:1–26

Shi Y, Vattem KM, Sood R, An J, Liang J (1998) Identification and characterization of pancreatic eukaryotic initiation factor 2 alpha-subunit kinase, PEK, involved in translational control. Mol Cell Biol 18:7499–7509

Shi Y, Baker JE, Zhang C, Tweddell JS, Su J, Pritchard KA Jr (2002) Chronic hypoxia increases endothelial nitric oxide synthase generation of nitric oxide by increasing heat shock protein 90 association and serine phosphorylation. Circ Res 91:300–306

Sinha S, Ray US, Saha M, Singh SN, Tomar OS (2009a) Antioxidant and redox status after maximal aerobic exercise at high altitude in acclimatized lowlanders and native highlanders. Eur J Appl Physiol 106:807–814

Sinha S, Ray US, Tomar OS, Singh SN (2009b) Different adaptation patterns of antioxidant system in natives and sojourners at high altitude. Resp Physiol Neurobiol 167:255–260

Sondermann H, Becker T, Mayhew M, Wieland F, Hartl FU (2000) Characterization of a receptor for heat shock protein 70 on macrophages and monocytes. Biol Chem 381:1165–1174

Srivastava P (2002) Interaction of heat shock proteins with peptides and antigen presenting cells: chaperoning of the innate and adaptive immune responses. Annu Rev Immunol 20:395–425

Stadtman ER (1992) Protein oxidation and aging. Science 257:1220–1224

Suryakumar G, Gupta A (2011) Medicinal and therapeutic potential of Sea buckthorn (*Hippophae rhamnoides* L.). J Ethnopharmacol 138(2):268–278

Suzen E, Karademir B, Ozer NK (2015) Basic mechanisms in endoplasmic reticulum stress and relation to cardiovascular diseases. Free Rad Biol Med 78:30–41

Tang T, Arbiser JL, Brandt SJ (2002) Phosphorylation by mitogen activated protein kinase mediates the hypoxia-induced turnover of the TAL1/SCL transcription factor in endothelial cells. J Biol Chem 277:18365–18372

Terry DF, McCormick M, Andersen S, Pennington J, Schoenhofen E, Palaima E, Bausero M, Ogawa K, Perls TT, Asea A (2004) Cardiovascular disease delay in centenarian offspring: role of heat shock proteins. Ann N Y Acad Sci 1019:502–505

Tersey SA, Nishiki Y, Templin AT, Cabrera SM, Stull ND, Colvin SC, Evans-Molina C, Rickus JL, Maier B, Mirmira RG (2012) Islet β-cell endoplasmic reticulum stress precedes the onset of type 1 diabetes in the nonobese diabetic mouse model. Diabetes 61:818–827

Trougakos IP, Sesti F, Tsakiri E, Gorgoulis VG (2013) Non-enzymatic posttranslational protein modifications and proteostasis network deregulation in carcinogenesis. J Proteome 92:274–298

Usui M, Yamaguchi S, Tanji Y, Tominaga R, Ishigaki Y, Fukumoto M, Katagiri H, Mori K, Oka Y, Ishihara H (2012) Atf6α-null mice are glucose intolerant due to pancreatic β-cell failure on a high-fat diet but partially resistant to diet-induced insulin resistance. Metabolism 61:1118–1128

Vabulas RM, Ahmad-Nejad P, Ghose S, Kirschning CJ, Issels RD, Wagner H (2002) HSP70 as endogenous stimulus of the Toll/interleukin-1 receptor signal pathway. J Biol Chem 277:15107–15112

Verma G, Datta M (2012) The critical role of JNK in the ER-mitochondrial crosstalk during apoptotic cell death. J Cell Physiol 227:1791–1795

Weinberg JM, Venkatachalam MA, Roeser NF, Senter RA, Nissim I (2001) Energetic determinants of tyrosine phosphorylation of focal adhesion proteins during hypoxia/reoxygenation of kidney proximal tubules. Am J Pathol 158:2153–2164

Witko-Sarsat V, Frielander M, Capeille're-Blandin C, Nguyen-Khoa T, Nguyen AT, Zingraff J, Jungers P, Descamps-Latscha B (1996) Advanced oxidation protein products as a novel marker of oxidative stress in uremia. Kidney Int 49:1304–1313

Witko-Sarsat V, Frielander M, Nguyen-Khoa T, Capeillere-Blandin C, Nguyen AT, Canteloup S, Dayer JM, Jungers P, Drüeke T, Descamps-Latscha B (1998) Advanced oxidation protein products as novel mediators of inflammation and monocyte activation in chronic renal failure. J Immunol 161:2524–2532

Wright BH, Corton JM, El-Nahas AM, Wood RF, Pockley AG (2000) Elevated levels of circulating heat shock protein 70 (Hsp70) in peripheral and renal vascular disease. Heart Vessel 15:18–22

Xin X, Dang H, Zhao X, Wang H (2017) Effects of hypobaric hypoxia on rat retina and protective response of resveratrol to the stress. Int J Med Sci 14(10):943–950

Yamane S, Hamamoto Y, Harashima S, Harada N, Hamasaki A, Toyoda K, Fujita K, Joo E, Seino Y, Inagaki N (2011) GLP-1 receptor agonist attenuates endoplasmic reticulum stress-mediated β-cell damage in Akita mice. J Diabet Investig 2:104–110

Yin J, Gu L, Wang Y, Fan N, Ma Y, Peng Y (2015) Rapamycin improves palmitate-induced ER stress/NF κB pathways associated with stimulating autophagy in adipocytes. Mediat Inflamm 2015:1–12

Yorimitsu T, Klionsky DJ (2007) Endoplasmic reticulum stress: a new pathway to induce autophagy. Autophagy 3:160–162

Yoshida H, Matsui T, Yamamoto A, Okada T, Mori K (2001) XBP1 mRNA is induced by ATF6 and spliced by IRE1 in response to ER stress to produce a highly active transcription factor. Cell 107:881–891

Younce CW, Burmeister MA, Ayala JE (2013) Exendin-4 attenuates high glucose-induced cardiomyocyte apoptosis via inhibition of endoplasmic reticulum stress and activation of SERCA2a. Am J Physiol Cell Physiol 304:C508–C518

Yusta B, Baggio LL, Estall JL, Koehler JA, Holland DP, Li H, Pipeleers D, Ling Z, Drucker DJ (2006) GLP-1 receptor activation improves β cell function and survival following induction of endoplasmic reticulum stress. Cell Metab 4:391–406

Zhang K, Kaufman RJ (2008) From endoplasmic-reticulum stress to the inflammatory response. Nature 454:455–462

Zhang K, Wang S, Malhotra J, Hassler JR, Back SH, Wang G, Chang L, Xu W, Miao H, Leonardi R, Chen YE, Jackowski S, Kaufman RJ (2011) The unfolded protein response transducer IRE1α prevents ER stress-induced hepatic steatosis. EMBO J 30:1357–1375

Zhang J, Li Y, Jiang S, Yu H, An W (2014) Enhanced endoplasmic reticulum SERCA activity by overexpression of hepatic stimulator substance gene prevents hepatic cells from ER stress-induced apoptosis. Am J Physiol Cell Physiol 306:C279–C290

Zinszner H, Kuroda M, Wang X, Batchvarova N, Lightfoot RT (1998) CHOP is implicated in programmed cell death in response to impaired function of the endoplasmic reticulum. Genes Dev 12:982–995

Chapter 7
Role of Chaperone Mediated Autophagy in Viral Infections

Hoorieh Soleimanjahi and Asghar Abdoli

Abstract Chaperone-mediated autophagy (CMA) is a selective mechanism for degradation of soluble cytosolic proteins is responsible for the timed degradation of 30% of cytosolic proteins under conditions of prolonged nutrient deprivation and stress. Molecular chaperones in the lysosomal lumen and in the cytosol ~~and~~ induce this proteolytic pathway. A central molecule for CMA is a receptor in the lysosomal membrane is known as the lysosome-associated membrane protein (LAMP) type 2A. The decrease in CMA leads to cells to be more prone to oxidative stresses and pathogenes. Furthermore, the decreased CMA in aging is stem from reduced LAMP-2A in the lysosomal membrane. Here, we describe the evidence in support of the contribution of chaperokines and CMA in viral infections.

Keywords Autophagy · Chaperone · Immunity · Lysosome · Misflolded proteins · Viral infections

Abbreviations

CHIP	Hsc70-interacting protein
CMA	Chaperone-mediated autophagy
HCV	Hepatitis C virus
HSP	Heat shock proteins
IFNLR1	Interferon α receptor 1
LAMP-2A	Lysosome-Associated Membrane Protein type 2A

H. Soleimanjahi (✉)
Department of Virology, Faculty of Medical Sciences, Tarbiat Modares University, Tehran, Iran
e-mail: soleim_h@modares.ac.ir

A. Abdoli
Department of Hepatitis and AIDS, Pasteur Institute of Iran, Tehran, Iran

A. A. A. Asea, P. Kaur (eds.), *Chaperokine Activity of Heat Shock Proteins*, Heat Shock Proteins 16, https://doi.org/10.1007/978-3-030-02254-9_7

PERK	PKR-like ER kinase
UPR	Unfolded protein response
VSV	Vesicular stomatitis virus

7.1 Introduction

Eukaryotic cells principally use two different mechanisms for protein degradation: the ubiquitin proteasome system and the autophagy machinery. Autophagy is an evolutionarily cytoplasmic conserved process which involves the delivery of cellular content, damaged organelles or misflolded proteins to the lysosomes for degradation and recycling (Cohen-Kaplan et al. 2016; Yan et al. 2015). Christian de Duve coined the term 'autophagy' to describe the presence of single- or double- membrane vesicles that contain parts of the cytoplasm and organelles in 1963, at the Ciba Foundation symposium on lysosomes (de Duve 1963). Based on the cargo delivery mechanism, three types of autophagy can be recognized, including (1) Microautophagy involves engulfing damaged cellular organelles directly by the lysosome (2) macroautophagy, usually named autophagy, is characterized with formation of double-membraned vesicles termed autophagosome to encircle the damaged organelles, misflolded proteins or infectious agents (Taji et al. 2017; Feizi et al. 2017). Subsequently, the autophagosome fuses with the lysosome to degrade its contents(Khateri et al. 2018) and (3) chaperone-mediated autophagy (CMA), in which chaperones targeted and marked selective cargos are translocated to the lysosome(Dice 2007).

7.2 Chaperokine Activity of Heat Shock Proteins

Molecular chaperones family is highly conserved ubiquitously expressed heat shock proteins and contain several highly conserved stress response proteins which protect cells from many stressful conditions (Hartl 1996). Hsp70 proteins are fundamental and most ubiquitous components of the cellular network of molecular chaperones aids proteins adopt native conformation or recover function after misfolding (Wondrak 2015). The stress inducible isoforms of HSP70 (as a molecular chaperone) is overexpressed in the majority of distinct histological origins in human cancers. Elevated basal expression of HSP70 in cancer cells is supposed to facilitate maintenance of protein homeostasis in the high stress condition of cancer microenvironment. Main metabolic controller mTOR as a regulator of cancer microenvironment frequently activated in cancer by overexpression of HSP70, leading survival of tumors in the presence of adverse cellular condition (Ciocca and Calderwood 2005). The impacts of molecular chaperones in biology, medicine are important aspects of chaperone activity. Attention is paid to the functions and applications of molecular chaperones from virome to eukaryotic and prokayotic cells, focusing on the heat shock proteins 90 (Hsp90), 70 (Hsp70) and 40 (Hsp40) classes of chaperones. The role of these classes of chaperones in human diseases is discussed in several studies.

Recent advances have seen the application of chaperones in mediating of autophagy in viral infections. The prospects for this subject are also discussed recently in numerous studies (Slavotinek and Biesecker 2001; Willis and Patterson 2010). Chaperones also play central roles in the translocation of proteins across intracellular membranes into organelles, and facilitate the proteolytic degradation of unstable proteins. Stressful conditions induce the synthesis of more than 20 heat shock proteins that enable cells to adapt to environmental and metabolic changes and to survive stress conditions. Recent evidence indicates that over expression of chaperones have facilitate to develop growing list of human protein-folding-related diseases (Brehme and Voisine 2016).

7.3 Chaperone-Mediated Autophagy (CMA)

CMA is a form of selective and specific autophagy that does not involvevesicular transport. It was described in the late 1990s by AM Cuervo and JDice (Cuervo and Dice 1996) in mammalian cells and is not found in yeast. CMA-mediated protein turnover take place in cytoplasm constitutively; though, it is more activated under stress conditions, such as oxidative and pathologic as well as elongated starvation (Dice 2007; Cuervo 2010). It allows translocation of cytosol protein substrates in the lysosome and may be responsible for the degradation of up to 30% of poorly folded cytosolic proteins in conditions of elongated nutritional stress (Chiang and Dice 1988). During early starvation periods, predominantly macroautophagy is stimulated and reaches its maximum level after 4–6 h, followed by a prompt reduction (Finn and Dice 2005). CMA will be enhanced gradually after longer time periods of starvation, with peak levels after 6–12 h, and will remains activated during the whole starvation period regardless of its length (Cuervo et al. 1995). This is an important crosstalk between these two autophagy pathways as CMA is a selective degradation system and as such it is a more delicate source to retrieve essential amino acids for protein synthesis as compared to macroautophagy. CMA also plays a role in immunity by promoting the presentation of antigens on MHC class I and II molecules (Kettern et al. 2011; Zhou et al. 2005; Crotzer et al. 2010). The CMA divided in three steps: (A) recognition of the proteins to be degraded by the chaperone proteins, (B) the transport of the protein to the lysosome membrane and the unfolding of this protein (C) translocation of the protein in the lumen of the lysosome for degradation and recycling.

(A) *Proteins Recognition by the Chaperone Proteins*

The proteins degraded by CMA all contain a consensus motif of 5 acids amines: KFERQ. Historically, ribonuclease A was the first substrate of the CMA discovered (Dice et al. 1986). The mutation of this motif prevents the degradation of proteins by CMA and inversely, the addition of this motif to the sequence of a protein allows its degradation by CMA (Dice et al. 1986). The KFERQ motif is recognized by a chaperone complex containing heat shock cognate 70 protein (Chiang et al. 1989). In fact, other chaperone/cochaperone proteins such as hsc70-interacting protein(CHIP), HSP90 or HSP40, would participate in targeting the substrate(Agarraberes and Dice 2001).

(B) *Translocation of the Protein to the Lysosome Membrane*

To start translocation, the complex formed by the association of the chaperone protein hsc70 with the protein to be degraded binds to the C-terminal of the Lysosomal membrane protein LAMP-2A (Lysosome-Associated Membrane Protein type 2A) present at the membrane lysosomale (Cuervo and Dice 1996). This protein exist to one of the three splicing variants of the lamp2 gene (LAMP2-A, B and C). These three variants differ in their transmembrane part and cytosolic, and only LAMP2-A is involved in CMA (GOUGH et al. 1995). Substrates are able to bind LAMP-2A in their native conformation, however unfolding is mandatory prior to transport (Salvador et al. 2000).The chaperone/cochaperones linked with the substrate are likely to support in the unfolding process (Agarraberes and Dice 2001). The chaperone lysosomal protein hsp90 contributes to the stabilization of LAMP-2A during its multimerization (Bandyopadhyay et al. 2008). The hsp90 chaperone localizes on both sides of the lysosomal membrane and the domain of hsp90 at the cytosolic side of the lysosomal membrane is suggested to contribute in substrate protein unfolding. Complete translocation of the substrate needs the presence of a resident lysosomal chaperone (lys-hsc70) (Agarraberes et al. 1997).

(C) *Lysosomal Translocation Complex*

Binding of the target-hsc70 protein complex to LAMP-2A results in the multimerization of the LAMP-2 monomers. Organization of LAMP-2A into multimeric complexes required to translocate substrate into lysosomes(Clough and Bent 1998). This leads to the formation of a macromolecular complex of 700 kDa to the lysosome membrane. Once the translocation is complete, the LAMP-2A multimer quickly disassembles and this process is dependent on HSC70. This dissociation is an active process mediated by hsc70 and the regeneration of the LAMP-2 monomers leads to the maintenance of cycles of binding/translocation in order to degrade other target proteins (Bandyopadhyay et al. 2008).

7.4 Virus-Induced Chaperone-Mediated Autophagy

A necessity for the vigorous synthesis of viral proteins poses other obstacle for the virus as it challenges the capacity of the infected cell to appropriately fold and activate proteins. Gathering of sub-optimally folded proteins in the ER of the infected cell triggers a cascade of signaling pathway recognized as the ER stress or the unfolded protein response (UPR) (WELIHINDA et al. 1999). Although the ER protein chaperone BiP is fundamental to starting all divisions of the response, consequent signaling continues through a number of distinct mechanisms that consist of other transmembrane sensors such as PKR-like ER kinase (PERK), ATF6 and IRE1. The stimulation of PERK and following phosphorylation of eIF2α limits translation to reduce the accumulation of unfolded proteins (Malhotra and Kaufman 2007).

Viruses are known to both trigger UPR and able to inhibit these responses via different approach. The second is vital to keep the host cells from ER stress-mediated death, to permit viral proteins synthesis and to produce virus progeny (Waris et al. 2002). Chaperone-mediated autophagy (CMA) plays key role to degrade cytosolic proteins induced under oxidative stress conditions via the PERK and IRE1 branches of the UPR (Boya et al. 2013; Wang et al. 2015).

Chandra P.K and et al. reported that HCV inactivated autophagy response impairs HCV clearance using exogenous IFN-α but not through IFN λ (Chandra et al. 2014). The reason is that HCV replication lead to autophagy induction and reduce the expression of IFNAR1 chain of the type I IFN-receptor, resulting to defective JAK/STAT signaling. The HCV-triggered autophagy response did not influence the expression of type III IFN receptors, which is why IL29 cause HCV clearance. To uncover why HCV induced autophagy response specifically target IFNAR1 but not the Type III IFN-receptors, the mechanisms of IFNAR1 degradation in HCV cell culture model via CMA was evaluated. It was found a CMA consensus amino acid motif (QKVEV) exist at residue of 34 in the human IFNAR1 protein and this motif is absent in the IFNλ receptor (Dash et al. 2016). As previously mentioned CMA is a type of selective autophagy for degradation of cytosolic protein containing certain consensus amino acid motif (KFERQ). The targeting motif is found to be glutamine (Q) flanked at both end by a hydrophobic (F, I, L, V) an acidic (E, D), a basic (R, K) and the second basic amino acid or hydrophobic (Cuervo and Dice 1996).

Vesicular stomatitis virus (VSV) facilitate the degradation of IFNAR1 by induction of PERK-dependent IFNAR1 phosphorylation VSV infected 2fTGH human fibrosarcoma cells showed enhanced expression level of UPR markers. Interestingly, this infection also leads to induce IFNAR1 phosphorylation on Ser535 and declined total levels of IFNAR1. Parallel results and an rise in the extent of IFNAR1 ubiquitination were detected in the KR derivative cell line signifying that VSV infection stimulates IFNAR1 phosphorylation, degradation and ubiquitination in a Tyk2-independent manner (Kurt et al. 2015; Liu et al. 2009). On the other hand, Knockdown of PERK in human 2fTGH cells decreased phosphorylation and downregulation of IFNAR1 in cells infected with VSV(Liu et al. 2009).

Chaperone-mediated autophagy Activators such as 6-aminonicotinamide and nutrient starvation, declined IFNAR1 levels in Huh-7.5 cells. Co-immunoprecipitation, colocalization and siRNA knockdown experiments discovered that IFNAR1 but not IFNλ receptor 1 (IFNLR1) interacts with HSC70 and LAMP2A, which are central elements of chaperone-mediated autophagy (CMA) (Kurt et al. 2015). Silencing CMA pathways enhanced innate antiviral response and restored degradation of interferon-alpha receptor-1 expression (Dash et al. 2016). Kurt and colleagues have described that HCV able to induce Chaperon-mediated autophagy (CMA) by inducing the expression of LAMP2A, a CMA receptor, and HSC70, a cytosolic chaperone essential for CMA. Silencing of CMA machinery boosted innate antiviral response and returned degradation of interferon-alpha receptor-1 expression (Kurt et al. 2015). Taken together, these data suggest that viral infection and expression of structural viral proteins stimulate downregulation and degradation of IFNAR1 through a PERK-dependent signaling and chaperon mediated autophagy.

7.5 Conclusions

Chaperone-mediated autophagy (CMA) is a selective autophagy type with a specific targeting of soluble cytosolic misfolded proteins containing the consensus KFERQ motif by a chaperone complex containing heat shock cognate 70 protein. Next, this complex binds to LAMP2A on the lysosome membrane to internalize and degrade, subsequently. The selectivity of CMA permits the maintenance of protein homeostasis and quality control of proteins by the timed degradation of damaged proteins or malfunctioning proteins. CMA works in a harmonized manner with other autophagic pathways, which able to compensate for each other. A need for the dynamic synthesis of viral proteins poses other problem for the virus to productive progeny production as it challenges the capacity of the infected cell to correctly fold and activate proteins. Here, we review some of the recent findings regarding the role of CMA in viral infections.

Acknowledgments We would like to thank all members of virology department of Tarbiat Modares University and Hepatitis and AIDS department of Pasteur Institute for kind assistant and support.

References

Agarraberes FA, Dice JF (2001) A molecular chaperone complex at the lysosomal membrane is required for protein translocation. J Cell Sci 114(13):2491–2499

Agarraberes FA, Terlecky SR, Dice JF (1997) An intralysosomal hsp70 is required for a selective pathway of lysosomal protein degradation. J Cell Biol 137(4):825–834

Bandyopadhyay U, Kaushik S, Varticovski L, Cuervo AM (2008) The chaperone-mediated autophagy receptor organizes in dynamic protein complexes at the lysosomal membrane. Mol Cell Biol 28(18):5747–5763

Boya P, Reggiori F, Codogno P (2013) Emerging regulation and functions of autophagy. Nat Cell Biol 15(7):713

Brehme M, Voisine C (2016) Model systems of protein-misfolding diseases reveal chaperone modifiers of proteotoxicity. Dis Model Mech 9(8):823–838

Chandra PK, Bao L, Song K, Aboulnasr FM, Baker DP, Shores N, Wimley WC, Liu S, Hagedorn CH, Fuchs SY (2014) HCV infection selectively impairs type I but not type III IFN signaling. Am J Pathol 184(1):214–229

Chiang H, Dice JF (1988) Peptide sequences that target proteins for enhanced degradation during serum withdrawal. J Biol Chem 263(14):6797–6805

Chiang H-L, Terlecky SR, Plant CP, Dice JF (1989) A role for a 70-kilodaton heat shock protein in lysosomal degradation of intracellular proteins. Science 246(4928):382

Ciocca DR, Calderwood SK (2005) Heat shock proteins in cancer: diagnostic, prognostic, predictive, and treatment implications. Cell Stress Chaperones 10(2):86–103

Clough SJ, Bent AF (1998) Floral dip: a simplified method forAgrobacterium-mediated transformation of Arabidopsis thaliana. Plant J 16(6):735–743

Cohen-Kaplan V, Livneh I, Avni N, Cohen-Rosenzweig C, Ciechanover A (2016) The ubiquitin-proteasome system and autophagy: coordinated and independent activities. Int J Biochem Cell Biol 79:403–418

Crotzer VL, Glosson N, Zhou D, Nishino I, Blum JS (2010) LAMP-2-deficient human B cells exhibit altered MHC class II presentation of exogenous antigens. Immunology 131(3):318–330
Cuervo AM (2010) Chaperone-mediated autophagy: selectivity pays off. Trends Endocrinol Metab 21(3):142–150
Cuervo AM, Dice JF (1996) A receptor for the selective uptake and degradation of proteins by lysosomes. Science 273(5274):501
Cuervo A, Knecht E, Terlecky SR, Dice JF (1995) Activation of a selective pathway of lysosomal proteolysis in rat liver by prolonged starvation. Am J Phys Cell Phys 269(5):C1200–C1208
Dash S, Chava S, Aydin Y, Chandra PK, Ferraris P, Chen W, Balart LA, Wu T, Garry RF (2016) Hepatitis C virus infection induces autophagy as a prosurvival mechanism to alleviate hepatic ER-stress response. Viruses 8(5):150
de Duve C (1963) Ciba foundation symposium on lysosomes (de Reuck AVS, Cameron MP, eds). JA Churchill Ltd., London
Dice JF (2007) Chaperone-mediated autophagy. Autophagy 3(4):295–299
Dice J, Chiang H, Spencer E, Backer J (1986) Regulation of catabolism of microinjected ribonuclease A. Identification of residues 7–11 as the essential pentapeptide. J Biol Chem 261(15):6853–6859
Feizi N, Mehrbod P, Romani B, Soleimanjahi H, Bamdad T, Feizi A, Jazaeri EO, Targhi HS, Saleh M, Jamali A (2017) Autophagy induction regulates influenza virus replication in a time-dependent manner. J Med Microbiol 66(4):536–541
Finn PF, Dice JF (2005) Ketone bodies stimulate chaperone-mediated autophagy. J Biol Chem 280(27):25864–25870
Gough NR, Hatem CL, Fambrough DM (1995) The family of LAMP-2 proteins arises by alternative splicing from a single gene: characterization of the avian LAMP-2 gene and identification of mammalian homologs of LAMP-2b and LAMP-2c. DNA Cell Biol 14(10):863–867
Hartl FU (1996) Molecular chaperones in cellular protein folding. Nature 381(6583):571
Kettern N, Rogon C, Limmer A, Schild H, Höhfeld J (2011) The Hsc/Hsp70 co-chaperone network controls antigen aggregation and presentation during maturation of professional antigen presenting cells. PLoS One 6(1):e16398
Khateri M, Abdoli A, Motevalli F, Fotouhi F, Bolhassani A, Arashkia A, Jazaeri EO, Shahbazi S, Mehrbod P, Naziri H (2018) Evaluation of autophagy induction on HEV 239 vaccine immune response in a mouse model. IUBMB Life 70:207–214
Kurt R, Chandra PK, Aboulnasr F, Panigrahi R, Ferraris P, Aydin Y, Reiss K, Wu T, Balart LA, Dash S (2015) Chaperone-mediated autophagy targets IFNAR1 for lysosomal degradation in free fatty acid treated HCV cell culture. PLoS One 10(5):e0125962
Liu J, HuangFu W-C, Kumar KS, Qian J, Casey JP, Hamanaka RB, Grigoriadou C, Aldabe R, Diehl JA, Fuchs SY (2009) Virus-induced unfolded protein response attenuates antiviral defenses via phosphorylation-dependent degradation of the type I interferon receptor. Cell Host Microbe 5(1):72–83
Malhotra JD, Kaufman RJ (2007) The endoplasmic reticulum and the unfolded protein response. Semin Cell Dev Biol 18:716–731
Salvador N, Aguado C, Horst M, Knecht E (2000) Import of a cytosolic protein into lysosomes by chaperone-mediated autophagy depends on its folding state. J Biol Chem 275(35):27447–27456
Slavotinek AM, Biesecker LG (2001) Unfolding the role of chaperones and chaperonins in human disease. Trends Genet 17(9):528–535
Taji F, Kouchesfahani HM, Sheikholeslami F, Romani B, Baesi K, Vahabpour R, Edalati M, Teimoori-Toolabi L, Jazaeri EO, Abdoli A (2017) Autophagy induction reduces telomerase activity in HeLa cells. Mech Ageing Dev 163:40–45
Wang D-w, Peng Z-j, Ren G-f, Wang G-x (2015) The different roles of selective autophagic protein degradation in mammalian cells. Oncotarget 6(35):37098
Waris G, Tardif KD, Siddiqui A (2002) Endoplasmic reticulum (ER) stress: hepatitis C virus induces an ER-nucleus signal transduction pathway and activates NF-κB and STAT-3. Biochem Pharmacol 64(10):1425–1430

Welihinda AA, Tirasophon W, Kaufman RJ (1999) The cellular response to protein misfolding in the endoplasmic reticulum. Gene Expr 7(4–5):293–300

Willis MS, Patterson C (2010) Hold me tight: role of the heat shock protein family of chaperones in cardiac disease. Circulation 122(17):1740–1751

Wondrak GT (2015) Stress response pathways in cancer. Springer, Netherlands

Yan MM, Ni JD, Song D, Ding M, Huang J (2015) Interplay between unfolded protein response and autophagy promotes tumor drug resistance. Oncol Lett 10(4):1959–1969

Zhou D, Li P, Lin Y, Lott JM, Hislop AD, Canaday DH, Brutkiewicz RR, Blum JS (2005) Lamp-2a facilitates MHC class II presentation of cytoplasmic antigens. Immunity 22(5):571–581

Chapter 8
Chaperones in Sterile Inflammation and Injury

Asmita Choudhury and Pranoti Mandrekar

Abstract Heat shock proteins (Hsp) are highly conserved proteins and their expression increases in response to stress stimuli for the maintenance of cellular homeostasis. Cellular stress conditions like hypoxia, change in pH, and metabolic changes are typically associated with upregulation of HSP. In addition to their traditional role of protein refolding or degradation during maintenance of homeostasis, Hsp also acts as key modulators of innate and adaptive immune pathways and play an important role in inflammation and host defense mechanisms. Several pathological conditions such as chronic inflammatory diseases and cancer are typically marked with an elevated expression of HSP. Increasing evidence demonstrates that Hsp can be secreted in the extracellular milieu during disease conditions. These proteins are released into the extracellular spaces either by necrotic cell death or by active secretion via exosomes. Exosomes containing Hsp can communicate and release exosomal contents to elicit immune cell activation. Thus, increased levels of Hsp can be considered as a danger signal both intracellularly, for its ability to chaperone inflammatory signaling molecules as well as extracellularly, for eliciting the activation of immune cells via exosomal communications. This chapter summarizes the chaperone role of HSP in serving as regulators of the immune response in chronic disease conditions.

Keywords Chaperones · DAMPs · Exosome · Heat shock factor 1 · Heat shock proteins · Inflammation

---

A. Choudhury · P. Mandrekar (✉)
Department of Medicine, University of Massachusetts Medical School, Worcester, MA, USA
e-mail: Pranoti.Mandrekar@umassmed.edu

A. A. A. Asea, P. Kaur (eds.), *Chaperokine Activity of Heat Shock Proteins*, Heat Shock Proteins 16, https://doi.org/10.1007/978-3-030-02254-9_8

Abbreviations

17-AAG	17-allylamino-17-dimethoxy-geldanamycin
17-DMAG	17-Dimethylaminoethylamino-17-dimethoxygeldenamycin
AP-1	Activator protein-1
APC	Antigen presenting cells
CDC37	Cell division cycle-37
CpG ODN	CpG-containing oligodeoxynuceotide
CTL	Cytotoxic T lymphocyte
CXCL10	C-X-C motif ligand-10
DAMP	Damage associated molecular pattern
DC	Dendritic cell
DMA	5,5-(N-N-Dimethyl)-amiloride hydrochloride
ERK	Extracellular signal-regulated kinases
HDGF	Hepatoma derived growth factor
HCV	Hepatitis C virus
HMGB1	High mobility group box 1
HSE	Heat shock element
HSF1	Heat shock factor 1
HSP	Heat shock protein
HSR	Heat shock response
I/R	Ischemia/reperfusion
IFN	Interferon
IKK	IκB kinase
IL	Interleukin
IRAK	IL-1 receptor-associated kinase
IRF3	IFN regulatory factor 3
JAK	Janus kinase
JNK	c-Jun amino-terminal kinases
LPS	Lipopolysaccharide
MAPK	Mitogen-activated protein kinases
MCP-1	Monocyte chemoattractant protein-1
MHC	Major histocompatibility complex
MKK	MAP kinase kinase
MMP-2	Matrix metalloproteinases-2
NEMO	NF-κB essential modulator
NF-κB	Nuclear factor-kappa B
NK	Natural killer
NLR	Nod-like receptor
PAMP	Pathogen associated molecular pattern
PBMC	Peripheral blood mononuclear cells
PRR	Pattern recognition receptor
RA	Rheumatoid arthritis
RAGE	Receptor for advanced glycation end products

RANTES	Regulated on activation, normal T cell expressed and secreted
RIP	Receptor interacting kinase
RLR	RIG-1-like receptor
RNS	Reactive nitrogen species
ROS	Reactive oxygen species
TAB	TAK-1 binding proteins
TAK1	TGF-β activated kinase 1
TANK	TRAF-associated NF-κB activator
TBK1	TANK-binding kinase 1
TGF-β	Tumor growth factor-beta
TIR	Toll/IL-1 receptor
TLR	Toll-like receptor
TNFR1	Tumor necrosis factor receptor1
TNFα	Tumor necrosis factor-alpha
TRAF6	Tumor necrosis factor receptor (TNF-R)-associated factor-6
TRAM	TRIF related adaptor molecule
TRIF	Toll/IL- receptor domain containing adaptor inducing IFN-β

8.1 Introduction

Multicellular organisms have a well-developed defense and repair mechanism to combat infection or trauma through the activation of innate immune signaling pathways. Inflammation is a host defense mechanism that involves recruitment of inflammatory cells, phagocytosis of the pathogens and production of cytokines to activate the adaptive immune response. In the absence of infection, inflammation induced by tissue injury, metabolic or chemical injury, ischemia, or transplantation is termed as "sterile inflammation". Tissue damage caused by necrotic cell death recruits immune cells that secrete inflammatory cytokines. Necrosis leads to the loss of integrity of plasma membrane and releases the intracellular components into the extracellular milieu (Zheng et al. 2011). These intracellular components can act as adjuvants to activate the antigen presenting cells (APC) (Kono and Rock 2008). Persistence of sterile inflammation increases tissue destruction and fibrosis and has been observed in various pathological conditions (Land 2015). Immune signaling pathways activated during sterile inflammation involve crucial molecular adapters and kinases. The stability of these intermediate signaling molecules is paramount to an efficient immune response and is facilitated by molecular chaperones. Understanding the role of chaperones in immune signaling pathways during chronic inflammatory diseases will uncover crucial mechanisms related to organ injury.

The function of Hsp as molecular chaperones and its role in chronic inflammatory diseases is being extensively investigated. Studies show that Hsp play an important part in mediating sterile inflammation under pathological conditions (Gao et al. 2009; Zheng et al. 2001). Studying regulation of immune signaling in the context of their role as cellular chaperones, either intracellular or via communication of extracellular vesicles containing HSP, would unravel the multiple layers of regulation of inflammatory pathways. These studies will also further extend our knowledge of their potential use as possible therapeutic targets for inflammatory diseases.

8.2 Heat Shock Proteins Serve as Chaperones

The 3D structure and conformation of the proteins are encoded within the polypeptide chains which govern the characteristic folding pattern essential for its functioning. Cellular stressors can hinder the protein folding and render them inactive. In addition, the introduction of random errors during DNA replication, mRNA translation, protein folding, and assembly can lead to the formation of protein aggregates. All cells are equipped with highly efficient and conserved protein homeostasis (proteostasis) pathways including the heat shock response (HSR) that detects metastable proteins, assists them to either refold or mediates their removal preventing accumulation of protein aggregates (Gidalevitz et al. 2011). A balance between protein folding and clearance of misfolded and aggregated proteins is crucial for cellular homeostasis (Bar-Lavan et al. 2012). The fluctuations in the cellular milieu are sensed by the heat shock factor 1 (HSF1), the hierarchical regulator of HSR which regulates the expression of HSP (Lis and Wu 1993). Under non-stress conditions activity of Hsf1 is repressed and it exists as an inert monomer in the nucleus and cytosol. Molecular chaperones such as Hsp40, Hsp70, and Hsp90 bind to HSF1 and maintain it in an inert state (Gomez-Pastor et al. 2017). Exposure to stress or elevated levels of misfolded proteins activates Hsf1. Subsequently, extensive post-translational modifications such as hyperphosphorylation, sumoylation, and acetylation at multiple residues of Hsf1 occurs. Activated and transcriptionally competent Hsf1 trimers bind to the heat shock elements (HSE) present in the promoter region of target Hsp genes and mediate its transcription (Kline and Morimoto 1997).

Hsp, known for their function as molecular chaperones, were first discovered by Ritossa in 1962. These are highly conserved and ubiquitously expressed proteins ranging from 10 to 100 kDa molecular weight. Based on their molecular weight, stress proteins are classified as (1) high molecular weight Hsp which includes Hsp110, Hsp90, Hsp70, Hsp60, and Hsp40 and (2) small molecular Hsp such as Hsp27, αA- and αB-crystallins, Hsp20, HspB2, HspB3, HspB7, HspB8, and HspB9 (Jee 2016). Each family of HSP is composed of members which are either constitutively expressed or induced upon exposure to cellular stressors. They are further categorized based on the different subcellular compartmentalization. The function and physiological role of Hsp vary depending on their size and specific locations. Exposure to stress such as oxygen deprivation, pH, and nutrient deprivation, induces the expression of Hsp. In addition, inflammation, infection and anti-cancer agents also act as potent inducers of Hsp (Lindquist and Craig 1988; Morimoto 1998). Hsp are ubiquitously and constitutively expressed proteins and have various housekeeping functions essential for normal growth and development. They facilitate the transport of proteins across the membranes and interact with key components of the various signal transduction pathways (Nollen and Morimoto 2002). Hsp play a crucial role in the immune signaling pathways and regulation of host defense (Muralidharan and Mandrekar 2013). In addition, Hsf1 also regulates the expression of various inflammatory genes. Role of Hsp in the functioning of innate and adaptive immune system, T cell proliferation, cross-presentation of antigens, and proliferation and suppressive function of regulatory T cells is widely investigated (Kolinski et al. 2016).

8.3 Sterile Inflammation: DAMPs as Mediators

In an attempt to explain why T cell immunity is generated to transplants, tumors, and in autoimmune diseases, Matzinger proposed the danger hypothesis, according to which abnormal cell death caused either by infection or any other pathological conditions, is a sign of danger (Matzinger 1994). These danger signals are recognized by a set of receptors leading to the initiation of the inflammatory response. Based on the source of cellular damage, the danger signals have been classified to (1) pathogen associated molecular patterns (PAMP) or (2) damage associated molecular patterns (DAMP). PAMPs are exogenous microbial molecules such as lipopolysaccharide (LPS), peptidoglycan, bacterial DNA, viral DNA/RNA, and mannans in yeast cell walls. DAMPs, also known as "alarmins" are the endogenous equivalent of PAMPs that signal tissue and cell damage. Tissue damage caused by trauma such as wounded tissues by mechanical force, excessive heat, cold, chemicals, radiations, hypoxia, withdrawal or excess of nutrients, and chemotherapeutics leads to secretion of DAMPs. DAMPs are secreted by the necrotic cells but are not released by the apoptotic cells, and act in an autocrine or paracrine manner alerting the host for the presence of tissue injury (Giuliano et al. 2011). Upon its release to the external milieu, DAMPs trigger inflammation under a sterile condition. A diverse range of structurally and chemically heterogeneous molecule serves as DAMPs. Some of the common DAMPs are high mobility group box 1 (HMGB1), hepatoma derived growth factor (HDGF), extracellular Hsp, IL-1α, and uric acid (Bianchi 2007; Brenner et al. 2013). Kono and Rock enlisted four criteria to establish a molecule as DAMP. These include: (1) DAMPs should be active as a highly purified molecule, (2) its biological activity should not be attributed to the contamination of PAMPs that may purify along with the host molecules, (3) DAMPs should be active at concentrations in pathophysiological conditions, and (4) inactivation of DAMPs should inhibit biological activity of dead cells both in vitro and in vivo (Kono and Rock 2008). Recognition of DAMPs occurs via receptors such as toll-like receptors (TLR), NOD-like receptors (NLR), RIG-I-like receptors (RLR), AIM2-like receptors, and receptor for advanced glycation end products (RAGE) (Tang et al. 2012). Activation of pattern recognition receptors (PRR) by DAMPs leads to nuclear factor-kappa B (NF-κB) mediated secretion of pro-inflammatory cytokines and interferon (IFN) regulatory factor (IRF) mediated secretion of interferons (Park et al. 2006). Various disease conditions such as atherosclerosis (Galkina and Ley 2009), cardiovascular diseases (Zheng et al. 2011), cardiac ischemia/reperfusion (I/R) injury (Toldo et al. 2016), non-alcoholic steatohepatitis (Wree et al. 2016), type 2 diabetes mellitus, and Alzheimer's disease (Land 2015) demonstrate role of DAMPs as a mediator of sterile inflammation.

8.4 Signaling Pathways in Sterile Inflammation

The first line of host defense involves innate immune signaling mediated by the monocytes, macrophages, natural killer (NK) cells, dendritic cells (DC), mast cells, and neutrophils. These cells mount responses for induction of inflammation and clearance of infection through various mechanisms. Neutrophils recruited to the site of infection or inflammation eliminate pathogens by the generation of reactive oxygen species (ROS) and reactive nitrogen spices (RNS), which demonstrate a potent antimicrobial activity as well as facilitate inflammatory responses. ROS also mediates TLR activation of pro-inflammatory signaling in the neutrophils. NK cells are non-phagocytic cells that kill intracellular pathogens and infected cells by release of pore-forming proteins, perforins and proteolytic enzymes, granzymes. Monocytes, macrophages, and DCs are primary cells that recognize PAMPs and DAMPs through PRRs and activate signaling pathways that induce pro-inflammatory cytokines and chemokines, initiating the subsequent adaptive immune response. These cells also directly engulf the pathogens and degrade them intracellularly by lysosomal degradation and increase the production of ROS (Nathan and Shiloh 2000). Macrophages and DCs also mediate antigen presentation of intracellularly degraded immunogenic proteins to adaptive immune cells like B and T cells initiating an immune response resulting in clearance of pathogens (Medzhitov and Janeway 1997).

TLRs are membrane spanning receptors crucial for activation of pro-inflammatory signaling cascade. Till date, 11 TLRs have been identified in humans, TLR1 to TLR11. Of these TLR1, 2, 4, 5, and 6 recognize the bacterial components whereas viral DNA and RNA components are recognized by TLR3, 7, 8, and 9. Activation of TLRs by the PAMPs and DAMPs lead to receptor dimerization and initiation of downstream signaling pathways. TLR signaling (Fig. 8.1) occurs either via the MyD88 dependent pathway, used by all TLRs except TLR3 or MyD88 independent pathways, used by TLR3 and 4 (Akira and Takeda 2004). In the MyD88 dependent pathway, the cytoplasmic domain of TLR known as Toll/IL-1 receptor (TIR) domain binds to MyD88 and recruits IL-1 receptor-associated kinase (IRAK-1 and -4). IRAKs phosphorylate tumor necrosis factor receptor (TNF-R)-associated factor-6 (TRAF6), that activates tumor growth factor-beta (TGF-β) activated kinase 1 (TAK1)/TAK-1 binding proteins (TAB) complex. This complex activates the mitogen-activated protein kinases (MAPK) or NF-κB pathway. TAK1 complex phosphorylates the IκB kinase (IKK) complex that consists of IKKα, IKKβ, and NF-κB essential modulator (NEMO) (Wang et al. 2001). Phosphorylation of IKK releases NF-κB which is otherwise sequestered by IKKβ (Woronicz et al. 1997). TAK1 also activates the MAPK signaling pathway which

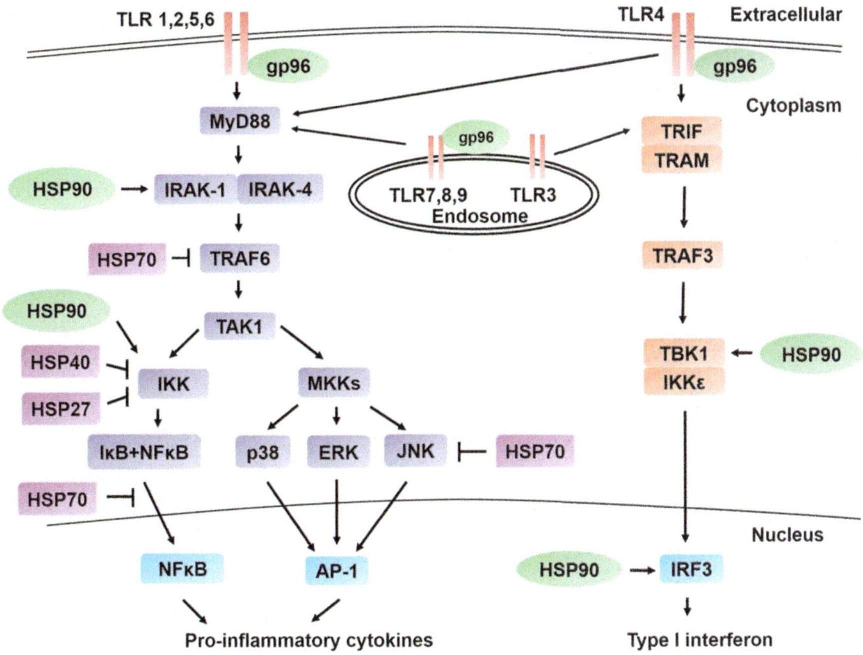

Fig. 8.1 Graphical representation of TLR pathways depicts the interaction of HSP with the pro-inflammatory signaling molecules. Interaction of PAMPs and DAMPs with TLRs initiates the MyD88 dependent or independent pathway that leads to activation of downstream signaling cascade. This induces expression of pro-inflammatory cytokines such as TNFα, IL-6, and IL-1β or type I interferons. Chaperone proteins interact with the adapter and kinase molecules at the various step of signaling cascade and regulate their activity

phosphorylates the downstream molecules p38, c-Jun amino-terminal kinases (JNK), and extracellular signal-regulated kinases (ERK) that activate the transcription factor activator protein-1 (AP1) (Dong et al. 2002). NF-κB and AP1 promote the expression of various pro-inflammatory chemokines and cytokines (Hayden et al. 2006). The other arm of TLR signaling is the MyD88 independent pathway, mediated through recruitment of an adaptor molecule Toll/1L-1 receptor domain containing adaptor inducing IFN-β (TRIF) to the TIR domain (Takeda and Akira 2007). TRIF requires additional adaptor molecule TRIF-related adaptor molecule (TRAM). This complex interacts with TRAF3 that acts as a bridge to TRAF-associated NF-κB activator (TANK)-binding kinase 1 (TBK1) and activates IKKε kinase. IKKε kinase further activates IRF3. IRF3 is a transcription factor that regulates expression of type I interferons and regulated on activation normal T cell expressed and secreted (RANTES) which are responsible for viral clearance (Meylan et al. 2004; Takeda and Akira 2007). Cellular chaperones play a crucial role in various steps in the TLR signaling pathways as well as modulate various aspects of immune activation and immune signaling pathways (Muralidharan and Mandrekar 2013).

8.5 Intracellular Chaperones in Sterile Inflammation: A New Perspective

Direct cross-talk between immune activation and stress response pathways is an important area of investigation. Inflammation is a stress condition and is often marked by elevated HSP. Increased production of ROS during inflammation induces Hsf1 and increases the level of HSP (Lee et al. 2015). Early studies demonstrate a beneficial role of Hsp since they ameliorate mortality associated with inflammation during sepsis (Yang et al. 1998). However, several studies demonstrate that HSP possesses an immunomodulatory role and can either activate the inflammatory immune signaling pathway for defense against pathogenic infection or inflammation or they may also possess an anti-inflammatory signal to prevent excessive inflammation (Muralidharan and Mandrekar 2013). By virtue of its role in mediating pro-inflammatory signaling through stabilizing adapters and kinases, intracellular chaperones may promote inflammation. Increase in the level of intracellular HSP is observed in various pathologic conditions like cancer, alcoholic liver disease, and cardiovascular disease (Adhuna et al. 1997; Ambade et al. 2014; Goldstein and Li 2009). Use of HSP inhibitors has been shown to significantly ameliorate organ injury supporting the notion that intracellular chaperones promote inflammation. For instance, blockade of Hsp90 leads to inhibition of T cell proliferation and decreases secretion of IFN-γ, TNFα, and IL-17 (Tukaj et al. 2014). Inhibition of Hsp90 in vivo reduces pro-inflammatory cytokine secretion and attenuates LPS mediated liver injury (Ambade et al. 2012). Hsp90aa1 is increased in human alcoholic cirrhosis patients and in mouse models of alcoholic liver injury. Pharmacological inhibition of Hsp90, using 17-Dimethylaminoethylamino-17-dimethoxygeldenamycin (17-DMAG), in these pre-clinical models alleviates chronic alcohol induced liver injury and decreases pro-inflammatory cytokine production (Ambade et al. 2014). In addition, Hsp90 inhibitors are also shown to be beneficial in models of chronic inflammatory diseases such as rheumatoid arthritis (RA) and uveitis (Poulaki et al. 2007; Rice et al. 2008). These findings demonstrate the key role of HSP in promoting pro-inflammatory cytokine production.

Hsp are known to interact with several adapters and kinases of pro-inflammatory signaling pathways (Fig. 8.1). Among HSP, Hsp90 and Hsp70 are most extensively studied as molecular chaperones in immune signaling. Hsp90, both cytoplasmic and its paralogue in the endoplasmic reticulum (gp96/Hsp90b1) has a wide range of clientele that includes the proteins with immunomodulatory roles. For instance, gp96 is involved in maturation, folding, assembly, and export of TLRs except for TLR3, and in absence of gp96, TLRs are retained intracellularly (Liu et al. 2010; Randow and Seed 2001; Yang et al. 2007). Following LPS stimulation, constitutive isoforms Hsc70 and Hsp90ab1 associate with TLR4 in lipid microdomains and are essential for LPS recognition as well as assisting its trafficking to the Golgi apparatus (Triantafilou and Triantafilou 2004). Hsp90 plays an important role in activation of both MyD88 dependent and independent pathway. Hsp90 along with cell division cycle-37 (CDC37) associates with the IKK complex by binding to the IKKα and IKKβ kinase domains (Chen et al. 2002). The two important components of MyD88 independent pathway, TBK1 and IRF3 are client proteins of Hsp90. Hsp90 stabilizes TBK1 and inhibition of Hsp90 leads to proteasome mediated degradation of TBK1. Also, knock-down of Hsp90 has been demonstrated to attenuate IRF3 activation and its target gene expression. In addition, Hsp90 forms a complex between TBK1 and IRF3 facilitating their close proximity and promoting signal transduction from TBK1 to IRF3 (Yang et al. 2006). Receptor interacting kinase (RIP) is one of the components of TNFR1 complex and plays an important role in TNF mediated NF-κB activation. Hsp90 interacts and stabilizes RIP and is recruited to the TNFR1 upon stimulation with TNF to activate NF-κB (Lewis et al. 2000). Janus kinase (JAK)1/2 are client proteins of Hsp90 and play an important role in type I and II interferon pathways (Shang and Tomasi 2006). Hspa5 elicits the TLR3 mediated IRF3 dependent innate immune response in Hepatitis C virus (HCV) infected hepatocytes. Also, hepatic expression of Hspa5 correlates with the level of chemokines such as RANTES and (C-X-C motif) ligand-10 (CXCL10), which are observed to be elevated in HCV infected livers (Wei et al. 2016). Hsp27 mediates the interaction between p38 and NF-κB for the induction of pro-inflammatory mediators and activates TAK1, MAP kinase kinase (MKK), and IKK signaling pathways downstream of cytokine stimulation in fibroblasts, illustrating its pro-inflammatory role (Rajaiya et al. 2012). Hsp27 is also required for IL-1 mediated expression of pro-inflammatory mediators IL-6 and IL-8 (Alford et al. 2007). These studies highlight the key chaperone functions of intracellular Hsp that contribute to pro-inflammatory cytokine production and suggest their role in sterile inflammation.

8.6 Inflammation and Extracellular Exosomal Chaperones

Plasma membrane provides a limiting boundary to the cell constituents and prevents the extracellular exposure of intracellular contents. Hsp are released to the extracellular milieu by two mechanisms (1) passive (necrosis) and (2) active (secretion). Necrosis but not apoptosis leads to the release of the intracellular constituents to the surrounding. Necrotic cell death releases molecular chaperones Hsp70, Hsp90, and gp96 (Basu et al. 2000; Berwin et al. 2001). Exposure of DCs to epithelial cells undergoing necrosis results in activation of the NF-κB and their maturation (Basu et al. 2000; Sauter et al. 2000). Given that injured cells have elevated expression of intracellular Hsp and necrotic cells release extracellular Hsp that can activate the immune signaling pathways, Hsp can act as a potent danger signal (Campisi et al. 2003). This hypothesis was validated by experiments conducted by Daniels et al. in which they induced the expression of Hsp70 in the tumor cells in vivo and observed that release of Hsp70 from these cells undergoing necrosis mediate antitumor immunity (Daniels et al. 2004). In addition to tissue necrosis, Hsp are also released by the cell through active secretion independent of cell death. This concept was first demonstrated by Hightower in 1989, as they observed the release of stress inducible

Hsp110 and Hsp70 after heat shock of cultured rat embryo cells (Hightower and Guidon Jr. 1989). In response to cellular stress, Hsp are released by various cell types such as human intestinal cells (Broquet et al. 2003), glial cells (Guzhova et al. 2001), peripheral blood mononuclear cells (PBMC) (Hunter-Lavin et al. 2004), cancer cells (McCready et al. 2010), B lymphocytes (Raposo et al. 1996) T lymphocytes (Blanchard et al. 2002), DCs (Zitvogel et al. 1998), mast cells (Skokos et al. 2001), and cardiac myocytes (Gupta and Knowlton 2007). Hsp do not contain secretory signal sequence and its secretion is resistant to brefeldin A, an inhibitor to classical ER- Golgi pathway (Broquet et al. 2003; Hightower and Guidon Jr. 1989; Hunter-Lavin et al. 2004). Hsp incorporate into the lipid bilayer, oligomerize within the membrane, and can be released through the formation of exosomes (Armijo et al. 2014). Exosomes are nanometer sized (40–90 nm in diameter) membrane vesicles, that contain signal proteins/peptides, micro RNA, mRNAs, and lipid (Thery et al. 2001; Valadi et al. 2007). Various cellular Hsp such as Hsp27 (Graner et al. 2009), Hsp40 (Takeuchi et al. 2015), Hsp60 (Merendino et al. 2010), Hsc70 and Hsp70 (Hegmans et al. 2004), and Hsp90 are known to be secreted in exosomes (Clayton et al. 2005). Use of an exosomal inhibitor 5,5-(N-N-Dimethyl)-amiloride hydrochloride (DMA) on tumor cell lines that demonstrate active release of HSP, results in significant reduction of Hsp secretion (Merendino et al. 2010).

In response to heat stress, increased expression of selected Hsp in the exosomal compartment is observed (Clayton et al. 2005). Molecular chaperone Hsp90aa1 in exosome is secreted by invasive cancer cells (McCready et al. 2010), DCs sensitized to tumor antigenic peptides (Thery et al. 2001), and human skin cells in response to TGFα (Cheng et al. 2008). Exosomes secreted from mouse and human mast cell lines are demonstrated to have functional mRNA. Upon in vitro translation of these mRNA to proteins, Hspa8 was observed to be present which demonstrates that exosome carries functional and intact HSP mRNAs that can be translated to proteins in presence of functional protein machinery (Valadi et al. 2007). In vivo studies demonstrate that exposure of rats to acute stressors leads to increase in the surface expression of Hsp70 in the circulating plasma exosomes (Beninson et al. 2014). PBMCs release exosomes in both basal and stress induced states. However, the content of Hsp70 in the exosomes is higher under stress induced conditions (Lancaster and Febbraio 2005). It is observed that Hsp70 secreted from the human prostate carcinoma cell lines are released through exosomes (Mambula and Calderwood 2006). There are two ways by which the extracellular Hsp70 levels can be increased in the exosomes. Either stress stimulus increases the number of exosomes containing Hsp70 (Clayton et al. 2005) or concentration of Hsp70 increases within the exosome (Lancaster and Febbraio 2005).

Several properties of HSP make them a highly plausible candidate to serve as DAMPs. The role of HSP as DAMPs is evident by the fact that necrotic cell death leads to release of Hsp70, Hsp90, and gp96 (Basu et al. 2000). Extracellular Hsp are known to be involved in various pathological conditions and demonstrate immunomodulatory roles. Targeting gp96 onto the cell surface primes tumor-specific T cell response in vivo and induces maturation of DCs leading to secretion of pro-inflammatory cytokines and chemokines such as IL-1β and monocyte chemoattractant protein-1 (MCP1). It also upregulates major histocompatibility complex (MHC) I and II, CD80, CD86, and CD40 (Zheng et al. 2001). A similar approach in prostate tumor cells expressing the secretory form of ER-resident chaperone grp170 also demonstrates induction of systemic T cell responses. These tumor cells demonstrate increased production of pro-inflammatory cytokine TNFα by DCs hence establishing extracellular Hsp as potential danger molecules (Gao et al. 2009). Various properties of extracellular Hsp contribute to their potential as DAMPs. HSP are highly abundant and share high sequence homology between human and microbes and can thus serve as a predominant autoimmune targets in various pathological conditions (Choudhury and Khole 2013; Gaston et al. 1989). Chaperone proteins are powerful modulators of immune response as Hsp bound peptides elicit potent T cell responses (Srivastava 2002; Tobian et al. 2004). The complex of Hsp and antigenic peptides are taken up by distinct receptors CD91 and presented to MHCI molecules (Basu et al. 2001). Antigenic peptides derived from various sources like viral, bacterial, and tumor antigens bind to the peptide binding sites identified on Hsp70, Hsp90, and gp96 (Binder 2014). Free antigenic peptides or serum albumin bound peptides demonstrate a lower efficiency of presentation to MHCI molecules compared to the peptides chaperoned to Hsp90 or Hsp70. Disruption of peptide binding to Hsp70 and Hsp90 by use of deoxyspergualin or tanespimycin or by use of small interfering RNA to Hsp90aa1, significantly decrease peptide presentation to MHCI (Binder et al. 2001; Callahan et al. 2008; Kunisawa and Shastri 2006). Uptake of Hsp chaperoned peptides by APC leads to priming of CD8+ T cells (Tobian et al. 2004). Studies have demonstrated that tumor derived Hsp70 chaperoned peptides mediates its transfer to immature DCs for MHC class I restricted T cell maturation, thus demonstrating the immunogenic potential to instruct DC to cross-present endogenously expressed tumor peptides. The signaling of DC maturation does not require any external signal such as exogenous TNFα, suggesting it as an intrinsic property of Hsp70 peptide complex (Noessner et al. 2002).

Exosomes are known to demonstrate immunological activities and modulate antigen presentation, immune activation, immune suppression, immune surveillance, and intercellular communication (Greening et al. 2015). Exosomal Hsp have been shown to modulate the immunological response and demonstrate pro-immunogenic properties. Extracellular vesicles isolated from the ascites of mice with aggressive T cell lymphoma contain Hsp70 and demonstrate unusual expression of Hsp90 on their surface. Injecting these vesicles to the syngenic mice induces both humoral and cellular immune responses (Menay et al. 2017). Using a colorectal cancer cell line, Gastpar et al. demonstrated that Hsp70 present on the surface of the exosome can directly activate NK cells, whereas exosomes lacking surface Hsp70 fail to do so (Gastpar et al. 2005). HSP containing exosomes are known to modulate the tumor immune response directly or indirectly. Heat-stressed carcinoembryonic antigen positive tumor cells contain more Hsp70 in their exosomes compared to the non-heat shock counterpart. In addition, these exosomes can induce a potent cytotoxic T lymphocyte (CTL) response in vitro and in vivo and induce more efficient DC maturation (Dai et al. 2005). Hsp70 positive exosome derived from the Hsp70 overexpressing tumor cells activate mouse NK cells in vitro to kill the tumor cells.

Further, stimulation of splenocytes in culture by Hsp70 positive exosome enhances the capability of NK cells to kill the target. This study highlights the importance of Hsp70 in immune surveillance (Elsner et al. 2007). Using human cancer xenograft models in mice, it is demonstrated that radiation causes the release of exosomes containing Hsp70 which act as the potent stimulators of pro-inflammatory cytokines TNFα and IL-6 and increase CTL and NK activity (Hurwitz et al. 2010). Hsp70-enriched exosomal extracts from heat-treated tumors induce strong Th1 immune responses, resulting in elimination of cancer cells in allogeneic hosts in vivo characterized by the elevated production of IFN-γ and IgG2a in murine models (Cho et al. 2009). DCs could activate CD8+ T cells following uptake of human myeloma exosomes rich in Hsp70 and full-length tumor antigens. These exosomes also mediate tumor rejection in vivo (Wolfers et al. 2001). Exposure of hepatocellular carcinoma cells to anticancer drug leads to the active release of membrane Hsp bearing exosomes containing Hsp60, Hsp70, and Hsp90. These Hsp bearing exosomes stimulate NK cells leading to granzyme B production as marked by increased expression of CD94 and down-regulation of CD69, NKG2D, and Nkp44 that initiate apoptosis of cancer cells (Lv et al. 2012). Exosomes containing Hsp70 on their surface act as potent inducers of macrophage activation that leads to a very high increase in TNFα production compared to that of recombinant Hsp70 (Vega et al. 2008). Bausero et al. demonstrated that breast adenocarcinoma cell and erythroleukemic cells treated with IFN-γ induces Hsp70 expression within the exosomes and upregulates CD83 expression and IL-12 release from naive DCs (Bausero et al. 2005). B lymphoblast derived exosomes are capable of activating human and mouse antigen-specific T cell clones, demonstrating a possible role of the exosome in the modulation of T cell immunity (Raposo et al. 1996). Mast cells activate the B cells and T lymphocytes through the release of mast cell derived extracellular vesicles that contain a high amount of Hsp60 and Hsp70 and promote DCs maturation (Skokos et al. 2003). Hsp90 is exported from cancer cells in the form of exosomes and promotes breast cancer cell migration and invasiveness through matrix metalloproteinases-2 (MMP-2) activation (Sims et al. 2011). Given that Hsp are frequently found in the exosomes and immunomodulatory signals are associated with the exosomes (Fleshner and Crane 2017), it can be concluded that the Hsp associated with exosomes can act as DAMP and evoke sterile inflammation by acting as a chaperokine.

8.7 Extracellular Chaperones in Pathological Conditions

Extracellular Hsp are observed in various clinical conditions (De Maio and Vazquez 2013). Hsp90aa1 is secreted in exosomes by invasive cancer cell lines (McCready et al. 2010). Brain tumor demonstrates higher release of Hsp by exosomes containing Hsp90, Hsp70, Hsp60, Hsp40, and Hsp27 (Chalmin et al. 2010; Graner et al. 2007). Hsp70 is observed in lung tumor exosomes (Li et al. 2016). Cardiac myocytes release Hsp60 to the extracellular spaces via exosomes and induction of stress condition like hypoxia triples the release of exosome (Gupta and Knowlton 2007).

Patients with sarcoidosis demonstrate significantly higher expression of Hsp70 in exosomes isolated from lungs (Qazi et al. 2010). Drosophila models of polyglutamine disease demonstrate the presence of exosomes containing Hsp40, Hsp70, and Hsp90 (Takeuchi et al. 2015). Hsp70 is released from the prostate carcinoma cells through the endolysosomal compartment (Mambula and Calderwood 2006). Aging heart subjected to global I/R demonstrate higher levels of secreted Hsp27 through exosomes and augmented Hsp27 is associated with greater production of TNFα and IL-1β (Ao et al. 2016).

Hsp are present inside the exosomes and alteration in vesicle integrity increases the leakiness of exosomes leading to release of Hsp into the extracellular milieu. Evidence suggest that pathological stress conditions such as hypoxia, pH imbalance, ROS, gamma-irradiation, tissue injury cytokines like TGFα, and altered metabolic states can trigger the release of Hsp from exosomes (Li et al. 2012). ROS production mediated by pathological conditions like fever, acidosis, and moderate and heavy alcohol consumption increases the leakiness of exosomes that activates the innate immune system via TLR4 leading to cardiac myocyte apoptosis (Malik et al. 2013). Along similar lines, in several clinical conditions, expression of tissue as well as serum Hsp, are elevated suggesting a stress-mediated response and possibility of release of Hsp to the extracellular milieu. Hsp27 is highly induced in lungs upon exposure to a variety of stressors in vitro and in vivo (Hastie et al. 1997; Merendino et al. 2002). Serum level of Hsp60 and Hsp70 correlates with the development of acute liver injury (Ganter et al. 2006; Pespeni et al. 2005). The bronchial epithelium of patients with asthma exhibit higher levels of Hsp60, Hsp70, and Hsp90 (Fajac et al. 1997). In liver, hepatocytes challenged with acetaminophen demonstrate an increase of Hsp70 in the culture supernatants (Martin-Murphy et al. 2010). Hsp60 is elevated in the synovial membrane of patients with juvenile chronic arthritis (Boog et al. 1992). Expression of gp96 is increased in the synovial tissues as well as in the synovial fluid of RA patients compared to arthritis free healthy controls (Huang et al. 2009). Transient I/R associated with cardiac surgery is marked by an increase in expression of Hsp27, Hsp60, and Hsp70 in the heart as well as in the serum (Dybdahl et al. 2002; Zou et al. 2008). Human primary breast tumor cells express elevated levels of Hsp27 and release Hsp27 into circulation elevating its level in the serum (Banerjee et al. 2011). Hsp70 and Hsp90 are frequently detected on the surface various tumor cells (Ferrarini et al. 1992; Hance et al. 2014; Multhoff et al. 1997) as well as on the surface of primary melanoma cells and advanced malignant melanomas (Becker et al. 2004).

8.8 Targeting HSP in Inflammation and Injury

Though HSP are highly abundant in normal cells and demonstrate housekeeping activities, it might seem futile to develop a drug against this target. However, the use of HSP inhibitors as a drug target has demonstrated a great deal of success in various pathological conditions (Shrestha et al. 2016). HSP bind to a plethora of client proteins, and it has been recognized as a potential target in several diseases.

Pharmacological targeting of Hsp90 as a treatment strategy was primarily developed in the context of cancer (Neckers 2007). Some of the commonly used inhibitors of Hsp90 are geldanamycin, its less toxic analog, 17-allylamino-17-dimethoxy-geldanamycin (17AAG), 17-DMAG, and radicicol (Soti et al. 2005). These drugs were primarily generated to destabilize oncogenes. Inhibition of Hsp90 as treatment strategy has been effective in the in vivo models of various disease conditions like RA (Rice et al. 2008), renal I/R injury (O'Neill et al. 2015), uveitis (Poulaki et al. 2007), and alcohol-mediated liver injury (Ambade et al. 2014). Success of Hsp90 inhibitors has been attributed to higher binding affinity and selective inhibition of stress-inducible isoforms of Hsp90 that exhibit distinct post-translational modifications such deacetylation and are different from constitutive pools in normal cells (Scroggins et al. 2007).

Hsp90 tightly regulates NF-κB signaling that plays a central role in the inflammatory signaling pathways. Small molecule inhibition of Hsp90 using SNX-7081 blocks the NF-κB nuclear translocation potently inhibiting cytokine production in vivo and reduces growth factor activation of ERK and JNK signaling in rat arthritis models (Rice et al. 2008). IRAK-1 is one of the client proteins of Hsp90 and pharmacological inhibition of Hsp90 leads to degradation of IRAK-1 hence affecting the TLR signaling pathway (De Nardo et al. 2005). The cell surface expression and activity of CD14 is regulated by Hsp90. Treatment with Hsp90 inhibitor significantly decreases the membrane CD14 levels (Poulaki et al. 2007). Using 17-AAG, Bae et al. demonstrated that inhibiting Hsp90 downregulates surface expression of antigens and costimulatory molecules on T cells and simultaneously decreases activating receptors on NK cells (Bae et al. 2013). Inhibition of Hsp90 by radicicol, geldanamycin, or 17-AAG inhibits cross-priming and prevents loading of MHCI with peptides (Callahan et al. 2008). Geldanamycin has also been demonstrated to inhibit alcohol-induced elevation in LPS induced NF-κB and TNFα production (Mandrekar et al. 2008). Along similar lines, inhibition of Hsp90 in macrophages activated by LPS and CpG-containing oligodeoxynucleotide (CpG ODN) also significantly inhibits production of TNFα, IL-6, and IL-1β (Wax et al. 2003; Zhu and Pisetsky 2001). Inhibition of Hsp90 decrease LPS mediated hepatic injury and induction of pro-inflammatory cytokines like TNFα and IL-6 in the liver through Hsf1 activation (Ambade et al. 2012). In vivo use of 17-DMAG significantly ameliorates steatosis and macrophage activation in chronic alcohol-mediated liver injury in mice. In addition to NF-κB inhibition and decrease in pro-inflammatory cytokines, Hsp90 inhibition also decreases fatty acid synthesis and favors fatty acid oxidation (Ambade et al. 2014). 17-DMAG has been demonstrated to have beneficial effects and increases survival rate in sepsis models by its anti-inflammatory and anti-oxidant effects (Wang et al. 2016). Inhibiting Hsp90 leads to loss of pro-inflammatory client proteins and induces the expression of anti-inflammatory genes, Hsp70 and Hsf1. Hsf1 negatively regulates the expression of various cytokine genes such as TNFα, IL-1β, IL-6, and IL-8 (Ambade et al. 2014; De Maio 2011; Singh et al. 2008; Xie et al. 2002). In accordance, Hsp70 has been shown to dampen the effect of the inflammatory response. Exposure of macrophages to heat shock reduces TNFα expression which correlates with increased expression of Hsp70 (Snyder et al. 1992). In similar lines, overexpression of Hsp70 in rat has been demonstrated to prevent LPS induced increase in TNFα and IL-6 which is associated with the inhibition of IκB degradation (Dokladny et al. 2010).

8.9 Conclusions

In this chapter, we have made an effort to draw the attention of the reader to recent advances in understanding the cross-talk between chaperones and immune signaling pathways. Hsp are multifaceted proteins that have a wide range of cellular functions. Apart from maintaining cellular homeostasis, HSP regulate the pro-inflammatory signaling pathways via their chaperone function. Inflammation is the first line of defense that activates signaling cascades to eliminate injury. Apart from playing a crucial role in NF-κB mediated secretion of chemokines and cytokines, Hsp also modulate the innate and adaptive immune response through antigen presentation and molecular mimicry. Hsp are released into the extracellular milieu either by necrosis or via exosomes. Increased secretion of exosomal Hsp has been demonstrated in various pathological conditions. In addition, exosomal Hsp activate the immune cells. Taken together, these facts point to the role of Hsp as DAMPs (1) intracellularly by stabilizing the pro-inflammatory signaling pathway, and (2) extracellularly by activating the immune cells via the exosomal Hsp (Fig. 8.2). Though this field is widely explored some intriguing questions remain unanswered. (1) In

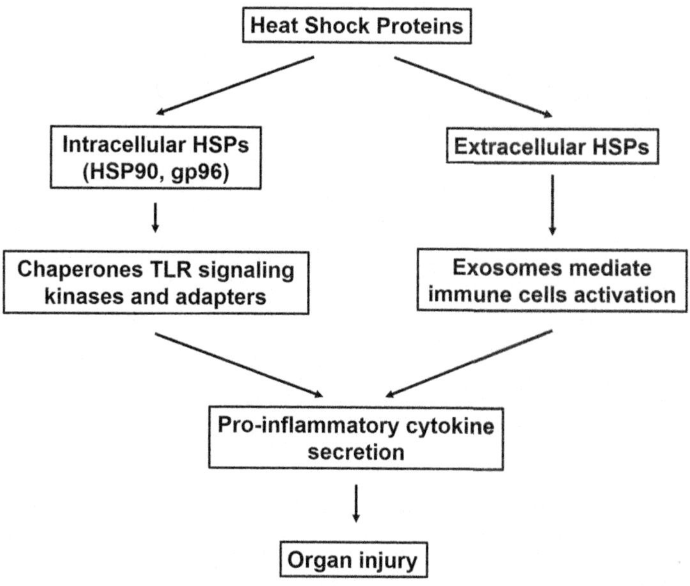

Fig. 8.2 Schematic representation illustrates the role of chaperones in mediating sterile inflammation leading to organ injury

various pathological conditions, HSP have been reported to be overexpressed in the tissue and the presence of extracellular Hsp in the serum has been demonstrated. However, exosomal Hsp as a source of extracellular Hsp has been demonstrated in very few pathological conditions. Understanding the contribution of exosomal Hsp to the extracellular milieu in disease conditions could help in better understanding of disease pathophysiology. (2) Exosomes can mediate an autocrine and paracrine role. Identifying the cross talk between exosomes and cells in the immediate vicinity as well as distant organs via circulation could help in better characterization of the role of chaperones/Hsp in organ injury. (3) Multiple properties of Hsp make them a plausible candidate for DAMPs. One such property includes the ability of Hsp to act as adjuvants and present antigenic peptides to MHC complexes to activate the adaptive immune response. Future research to understand whether exosomal Hsp demonstrate the peptide binding ability and antigen presentation will unravel the novel role of Hsp in inflammation. (4) So far, the release of Hsp through exosomal vesicles has been studied using very few drugs. Screening the levels of secreted Hsp via exosomes using other drugs could be useful in establishing the efficacy of the drugs and establish their potential as biomarkers. (5) In addition, the efficacy of Hsp inhibitors to target exosomal Hsp could be evaluated as a treatment strategy. Overall the importance of Hsp as chaperones in inflammation and organ injury, either intracellularly or extracellularly, justifies their extensive investigation to uncover new mechanisms and potential drug targets.

Acknowledgements This work was supported by grant # RO1 AA17986-01 and RO1 AA25289-01 (to PM) from the National Institute of Alcohol Abuse and Alcoholism, U.S. National Institutes of Health, Bethesda, MD, USA.

References

Adhuna P, Mukhopadhyay B, Bhatnagar R (1997) Modulation of macrophage heat shock proteins (HSPs) expression in response to intracellular infection by virulent and avirulent strains of Leishmania donovani. Biochem Mol Biol Int 43:1265–1275

Akira S, Takeda K (2004) Toll-like receptor signalling. Nat Rev Immunol 4:499–511

Alford KA, Glennie S, Turrell BR, Rawlinson L, Saklatvala J, Dean JL (2007) Heat shock protein 27 functions in inflammatory gene expression and transforming growth factor-beta-activated kinase-1 (TAK1)-mediated signaling. J Biol Chem 282:6232–6241

Ambade A, Catalano D, Lim A, Mandrekar P (2012) Inhibition of heat shock protein (molecular weight 90 kDa) attenuates proinflammatory cytokines and prevents lipopolysaccharide-induced liver injury in mice. Hepatology 55:1585–1595

Ambade A, Catalano D, Lim A, Kopoyan A, Shaffer SA, Mandrekar P (2014) Inhibition of heat shock protein 90 alleviates steatosis and macrophage activation in murine alcoholic liver injury. J Hepatol 61:903–911

Ao L, Zhai Y, Jin C, Cleveland JC, Fullerton DA, Meng X (2016) Attenuated recovery of contractile function in aging hearts following global ischemia/reperfusion: role of extracellular HSP27 and TLR4. Mol Med 23:863

Armijo G, Okerblom J, Cauvi DM, Lopez V, Schlamadinger DE, Kim J, Arispe N, De Maio A (2014) Interaction of heat shock protein 70 with membranes depends on the lipid environment. Cell Stress Chaperones 19:877–886

Bae J, Munshi A, Li C, Samur M, Prabhala R, Mitsiades C, Anderson KC, Munshi NC (2013) Heat shock protein 90 is critical for regulation of phenotype and functional activity of human T lymphocytes and NK cells. J Immunol 190:1360–1371

Banerjee S, Lin CF, Skinner KA, Schiffhauer LM, Peacock J, Hicks DG, Redmond EM, Morrow D, Huston A, Shayne M et al (2011) Heat shock protein 27 differentiates tolerogenic macrophages that may support human breast cancer progression. Cancer Res 71:318–327

Bar-Lavan Y, Kosolapov L, Frumkin A, Ben-Zvi A (2012) Regulation of cellular protein quality control networks in a multicellular organism. FEBS J 279:526–531

Basu S, Binder RJ, Suto R, Anderson KM, Srivastava PK (2000) Necrotic but not apoptotic cell death releases heat shock proteins, which deliver a partial maturation signal to dendritic cells and activate the NF-kappa B pathway. Int Immunol 12:1539–1546

Basu S, Binder RJ, Ramalingam T, Srivastava PK (2001) CD91 is a common receptor for heat shock proteins gp96, hsp90, hsp70, and calreticulin. Immunity 14:303–313

Bausero MA, Gastpar R, Multhoff G, Asea A (2005) Alternative mechanism by which IFN-gamma enhances tumor recognition: active release of heat shock protein 72. J Immunol 175:2900–2912

Becker B, Multhoff G, Farkas B, Wild PJ, Landthaler M, Stolz W, Vogt T (2004) Induction of Hsp90 protein expression in malignant melanomas and melanoma metastases. Exp Dermatol 13:27–32

Beninson LA, Brown PN, Loughridge AB, Saludes JP, Maslanik T, Hills AK, Woodworth T, Craig W, Yin H, Fleshner M (2014) Acute stressor exposure modifies plasma exosome-associated heat shock protein 72 (Hsp72) and microRNA (miR-142-5p and miR-203). PLoS One 9:e108748

Berwin B, Reed RC, Nicchitta CV (2001) Virally induced lytic cell death elicits the release of immunogenic GRP94/gp96. J Biol Chem 276:21083–21088

Bianchi ME (2007) DAMPs, PAMPs and alarmins: all we need to know about danger. J Leukoc Biol 81:1–5

Binder RJ (2014) Functions of heat shock proteins in pathways of the innate and adaptive immune system. J Immunol 193:5765–5771

Binder RJ, Blachere NE, Srivastava PK (2001) Heat shock protein-chaperoned peptides but not free peptides introduced into the cytosol are presented efficiently by major histocompatibility complex I molecules. J Biol Chem 276:17163–17171

Blanchard N, Lankar D, Faure F, Regnault A, Dumont C, Raposo G, Hivroz C (2002) TCR activation of human T cells induces the production of exosomes bearing the TCR/CD3/zeta complex. J Immunol 168:3235–3241

Boog CJ, de Graeff-Meeder ER, Lucassen MA, van der Zee R, Voorhorst-Ogink MM, van Kooten PJ, Geuze HJ, van Eden W (1992) Two monoclonal antibodies generated against human hsp60 show reactivity with synovial membranes of patients with juvenile chronic arthritis. J Exp Med 175:1805–1810

Brenner C, Galluzzi L, Kepp O, Kroemer G (2013) Decoding cell death signals in liver inflammation. J Hepatol 59:583–594

Broquet AH, Thomas G, Masliah J, Trugnan G, Bachelet M (2003) Expression of the molecular chaperone Hsp70 in detergent-resistant microdomains correlates with its membrane delivery and release. J Biol Chem 278:21601–21606

Callahan MK, Garg M, Srivastava PK (2008) Heat-shock protein 90 associates with N-terminal extended peptides and is required for direct and indirect antigen presentation. Proc Natl Acad Sci U S A 105:1662–1667

Campisi J, Leem TH, Fleshner M (2003) Stress-induced extracellular Hsp72 is a functionally significant danger signal to the immune system. Cell Stress Chaperones 8:272–286

Chalmin F, Ladoire S, Mignot G, Vincent J, Bruchard M, Remy-Martin JP, Boireau W, Rouleau A, Simon B, Lanneau D et al (2010) Membrane-associated Hsp72 from tumor-derived exosomes mediates STAT3-dependent immunosuppressive function of mouse and human myeloid-derived suppressor cells. J Clin Invest 120:457–471

Chen G, Cao P, Goeddel DV (2002) TNF-induced recruitment and activation of the IKK complex require Cdc37 and Hsp90. Mol Cell 9:401–410

Cheng CF, Fan J, Fedesco M, Guan S, Li Y, Bandyopadhyay B, Bright AM, Yerushalmi D, Liang M, Chen M et al (2008) Transforming growth factor alpha (TGFalpha)-stimulated secretion of HSP90alpha: using the receptor LRP-1/CD91 to promote human skin cell migration against a TGFbeta-rich environment during wound healing. Mol Cell Biol 28:3344–3358

Cho JA, Lee YS, Kim SH, Ko JK, Kim CW (2009) MHC independent anti-tumor immune responses induced by Hsp70-enriched exosomes generate tumor regression in murine models. Cancer Lett 275:256–265

Choudhury A, Khole VV (2013) HSP90 antibodies: a detrimental factor responsible for ovarian dysfunction. Am J Reprod Immunol 70:372–385

Clayton A, Turkes A, Navabi H, Mason MD, Tabi Z (2005) Induction of heat shock proteins in B-cell exosomes. J Cell Sci 118:3631–3638

Dai S, Wan T, Wang B, Zhou X, Xiu F, Chen T, Wu Y, Cao X (2005) More efficient induction of HLA-A*0201-restricted and carcinoembryonic antigen (CEA)-specific CTL response by immunization with exosomes prepared from heat-stressed CEA-positive tumor cells. Clin Cancer Res 11:7554–7563

Daniels GA, Sanchez-Perez L, Diaz RM, Kottke T, Thompson J, Lai M, Gough M, Karim M, Bushell A, Chong H et al (2004) A simple method to cure established tumors by inflammatory killing of normal cells. Nat Biotechnol 22:1125–1132

De Maio A (2011) Extracellular heat shock proteins, cellular export vesicles, and the stress observation system: a form of communication during injury, infection, and cell damage. It is never known how far a controversial finding will go! Dedicated to Ferruccio Ritossa. Cell Stress Chaperones 16:235–249

De Maio A, Vazquez D (2013) Extracellular heat shock proteins: a new location, a new function. Shock 40:239–246

De Nardo D, Masendycz P, Ho S, Cross M, Fleetwood AJ, Reynolds EC, Hamilton JA, Scholz GM (2005) A central role for the Hsp90.Cdc37 molecular chaperone module in interleukin-1 receptor-associated-kinase-dependent signaling by toll-like receptors. J Biol Chem 280:9813–9822

Dokladny K, Lobb R, Wharton W, Ma TY, Moseley PL (2010) LPS-induced cytokine levels are repressed by elevated expression of HSP70 in rats: possible role of NF-kappaB. Cell Stress Chaperones 15:153–163

Dong C, Davis RJ, Flavell RA (2002) MAP kinases in the immune response. Annu Rev Immunol 20:55–72

Dybdahl B, Wahba A, Lien E, Flo TH, Waage A, Qureshi N, Sellevold OF, Espevik T, Sundan A (2002) Inflammatory response after open heart surgery: release of heat-shock protein 70 and signaling through toll-like receptor-4. Circulation 105:685–690

Elsner L, Muppala V, Gehrmann M, Lozano J, Malzahn D, Bickeboller H, Brunner E, Zientkowska M, Herrmann T, Walter L et al (2007) The heat shock protein HSP70 promotes mouse NK cell activity against tumors that express inducible NKG2D ligands. J Immunol 179:5523–5533

Fajac I, Roisman GL, Lacronique J, Polla BS, Dusser DJ (1997) Bronchial gamma delta T-lymphocytes and expression of heat shock proteins in mild asthma. Eur Respir J 10:633–638

Ferrarini M, Heltai S, Zocchi MR, Rugarli C (1992) Unusual expression and localization of heat-shock proteins in human tumor cells. Int J Cancer 51:613–619

Fleshner M, Crane CR (2017) Exosomes, DAMPs and miRNA: features of stress physiology and immune homeostasis. Trends Immunol 38:768–776

Galkina E, Ley K (2009) Immune and inflammatory mechanisms of atherosclerosis (*). Annu Rev Immunol 27:165–197

Ganter MT, Ware LB, Howard M, Roux J, Gartland B, Matthay MA, Fleshner M, Pittet JF (2006) Extracellular heat shock protein 72 is a marker of the stress protein response in acute lung injury. Am J Phys Lung Cell Mol Phys 291:L354–L361

Gao P, Sun X, Chen X, Subjeck J, Wang XY (2009) Secretion of stress protein grp170 promotes immune-mediated inhibition of murine prostate tumor. Cancer Immunol Immunother 58:1319–1328

Gaston JS, Life PF, Bailey LC, Bacon PA (1989) In vitro responses to a 65-kilodalton mycobacterial protein by synovial T cells from inflammatory arthritis patients. J Immunol 143:2494–2500
Gastpar R, Gehrmann M, Bausero MA, Asea A, Gross C, Schroeder JA, Multhoff G (2005) Heat shock protein 70 surface-positive tumor exosomes stimulate migratory and cytolytic activity of natural killer cells. Cancer Res 65:5238–5247
Gidalevitz T, Prahlad V, Morimoto RI (2011) The stress of protein misfolding: from single cells to multicellular organisms. Cold Spring Harb Perspect Biol 3:a009704
Giuliano JS Jr, Lahni PM, Wong HR, Wheeler DS (2011) Pediatric sepsis – part V: extracellular heat shock proteins: alarmins for the host immune system. Open Inflamm J 4:49–60
Goldstein MG, Li Z (2009) Heat-shock proteins in infection-mediated inflammation-induced tumorigenesis. J Hematol Oncol 2:5
Gomez-Pastor R, Burchfiel ET, Thiele DJ (2017) Regulation of heat shock transcription factors and their roles in physiology and disease. Nat Rev Mol Cell Biol 19:4–19
Graner MW, Cumming RI, Bigner DD (2007) The heat shock response and chaperones/heat shock proteins in brain tumors: surface expression, release, and possible immune consequences. J Neurosci 27:11214–11227
Graner MW, Alzate O, Dechkovskaia AM, Keene JD, Sampson JH, Mitchell DA, Bigner DD (2009) Proteomic and immunologic analyses of brain tumor exosomes. FASEB J 23:1541–1557
Greening DW, Gopal SK, Xu R, Simpson RJ, Chen W (2015) Exosomes and their roles in immune regulation and cancer. Semin Cell Dev Biol 40:72–81
Gupta S, Knowlton AA (2007) HSP60 trafficking in adult cardiac myocytes: role of the exosomal pathway. Am J Physiol Heart Circ Physiol 292:H3052–H3056
Guzhova I, Kislyakova K, Moskaliova O, Fridlanskaya I, Tytell M, Cheetham M, Margulis B (2001) In vitro studies show that Hsp70 can be released by glia and that exogenous Hsp70 can enhance neuronal stress tolerance. Brain Res 914:66–73
Hance MW, Nolan KD, Isaacs JS (2014) The double-edged sword: conserved functions of extracellular hsp90 in wound healing and cancer. Cancers (Basel) 6:1065–1097
Hastie AT, Everts KB, Zangrilli J, Shaver JR, Pollice MB, Fish JE, Peters SP (1997) HSP27 elevated in mild allergic inflammation protects airway epithelium from H2SO4 effects. Am J Phys 273:L401–L409
Hayden MS, West AP, Ghosh S (2006) NF-kappaB and the immune response. Oncogene 25:6758–6780
Hegmans JP, Bard MP, Hemmes A, Luider TM, Kleijmeer MJ, Prins JB, Zitvogel L, Burgers SA, Hoogsteden HC, Lambrecht BN (2004) Proteomic analysis of exosomes secreted by human mesothelioma cells. Am J Pathol 164:1807–1815
Hightower LE, Guidon PT Jr (1989) Selective release from cultured mammalian cells of heat-shock (stress) proteins that resemble glia-axon transfer proteins. J Cell Physiol 138:257–266
Huang QQ, Sobkoviak R, Jockheck-Clark AR, Shi B, Mandelin AM 2nd, Tak PP, Haines GK 3rd, Nicchitta CV, Pope RM (2009) Heat shock protein 96 is elevated in rheumatoid arthritis and activates macrophages primarily via TLR2 signaling. J Immunol 182:4965–4973
Hunter-Lavin C, Davies EL, Bacelar MM, Marshall MJ, Andrew SM, Williams JH (2004) Hsp70 release from peripheral blood mononuclear cells. Biochem Biophys Res Commun 324:511–517
Hurwitz MD, Kaur P, Nagaraja GM, Bausero MA, Manola J, Asea A (2010) Radiation therapy induces circulating serum Hsp72 in patients with prostate cancer. Radiother Oncol 95:350–358
Jee H (2016) Size dependent classification of heat shock proteins: a mini-review. J Exerc Rehabil 12:255–259
Kline MP, Morimoto RI (1997) Repression of the heat shock factor 1 transcriptional activation domain is modulated by constitutive phosphorylation. Mol Cell Biol 17:2107–2115
Kolinski T, Marek-Trzonkowska N, Trzonkowski P, Siebert J (2016) Heat shock proteins (HSPs) in the homeostasis of regulatory T cells (Tregs). Cent Eur J Immunol 41:317–323
Kono H, Rock KL (2008) How dying cells alert the immune system to danger. Nat Rev Immunol 8:279–289

Kunisawa J, Shastri N (2006) Hsp90alpha chaperones large C-terminally extended proteolytic intermediates in the MHC class I antigen processing pathway. Immunity 24:523–534
Lancaster GI, Febbraio MA (2005) Exosome-dependent trafficking of HSP70: a novel secretory pathway for cellular stress proteins. J Biol Chem 280:23349–23355
Land WG (2015) The role of damage-associated molecular patterns (DAMPs) in human diseases: part II: DAMPs as diagnostics, prognostics and therapeutics in clinical medicine. Sultan Qaboos Univ Med J 15:e157–e170
Lee JH, Lee YK, Lim JJ, Byun HO, Park I, Kim GH, Xu WG, Wang HJ, Yoon G (2015) Mitochondrial respiratory dysfunction induces claudin-1 expression via reactive oxygen species-mediated heat shock factor 1 activation, leading to hepatoma cell invasiveness. J Biol Chem 290:21421–21431
Lewis J, Devin A, Miller A, Lin Y, Rodriguez Y, Neckers L, Liu ZG (2000) Disruption of hsp90 function results in degradation of the death domain kinase, receptor-interacting protein (RIP), and blockage of tumor necrosis factor-induced nuclear factor-kappaB activation. J Biol Chem 275:10519–10526
Li W, Sahu D, Tsen F (2012) Secreted heat shock protein-90 (Hsp90) in wound healing and cancer. Biochim Biophys Acta 1823:730–741
Li X, Wang S, Zhu R, Li H, Han Q, Zhao RC (2016) Lung tumor exosomes induce a pro-inflammatory phenotype in mesenchymal stem cells via NFkappaB-TLR signaling pathway. J Hematol Oncol 9:42
Lindquist S, Craig EA (1988) The heat-shock proteins. Annu Rev Genet 22:631–677
Lis J, Wu C (1993) Protein traffic on the heat shock promoter: parking, stalling, and trucking along. Cell 74:1–4
Liu B, Yang Y, Qiu Z, Staron M, Hong F, Li Y, Wu S, Li Y, Hao B, Bona R et al (2010) Folding of toll-like receptors by the HSP90 paralogue gp96 requires a substrate-specific cochaperone. Nat Commun 1:79
Lv LH, Wan YL, Lin Y, Zhang W, Yang M, Li GL, Lin HM, Shang CZ, Chen YJ, Min J (2012) Anticancer drugs cause release of exosomes with heat shock proteins from human hepatocellular carcinoma cells that elicit effective natural killer cell antitumor responses in vitro. J Biol Chem 287:15874–15885
Malik ZA, Kott KS, Poe AJ, Kuo T, Chen L, Ferrara KW, Knowlton AA (2013) Cardiac myocyte exosomes: stability, HSP60, and proteomics. Am J Physiol Heart Circ Physiol 304:H954–H965
Mambula SS, Calderwood SK (2006) Heat shock protein 70 is secreted from tumor cells by a non-classical pathway involving lysosomal endosomes. J Immunol 177:7849–7857
Mandrekar P, Catalano D, Jeliazkova V, Kodys K (2008) Alcohol exposure regulates heat shock transcription factor binding and heat shock proteins 70 and 90 in monocytes and macrophages: implication for TNF-alpha regulation. J Leukoc Biol 84:1335–1345
Martin-Murphy BV, Holt MP, Ju C (2010) The role of damage associated molecular pattern molecules in acetaminophen-induced liver injury in mice. Toxicol Lett 192:387–394
Matzinger P (1994) Tolerance, danger, and the extended family. Annu Rev Immunol 12:991–1045
McCready J, Sims JD, Chan D, Jay DG (2010) Secretion of extracellular hsp90alpha via exosomes increases cancer cell motility: a role for plasminogen activation. BMC Cancer 10:294
Medzhitov R, Janeway CA Jr (1997) Innate immunity: impact on the adaptive immune response. Curr Opin Immunol 9:4–9
Menay F, Herschlik L, De Toro J, Cocozza F, Tsacalian R, Gravisaco MJ, Di Sciullo MP, Vendrell A, Waldner CI, Mongini C (2017) Exosomes isolated from ascites of T-cell lymphoma-bearing mice expressing surface CD24 and HSP-90 induce a tumor-specific immune response. Front Immunol 8:286
Merendino AM, Paul C, Vignola AM, Costa MA, Melis M, Chiappara G, Izzo V, Bousquet J, Arrigo AP (2002) Heat shock protein-27 protects human bronchial epithelial cells against oxidative stress-mediated apoptosis: possible implication in asthma. Cell Stress Chaperones 7:269–280
Merendino AM, Bucchieri F, Campanella C, Marciano V, Ribbene A, David S, Zummo G, Burgio G, Corona DF, Conway de Macario E et al (2010) Hsp60 is actively secreted by human tumor cells. PLoS One e9247:5

Meylan E, Burns K, Hofmann K, Blancheteau V, Martinon F, Kelliher M, Tschopp J (2004) RIP1 is an essential mediator of Toll-like receptor 3-induced NF-kappa B activation. Nat Immunol 5:503–507

Morimoto RI (1998) Regulation of the heat shock transcriptional response: cross talk between a family of heat shock factors, molecular chaperones, and negative regulators. Genes Dev 12:3788–3796

Multhoff G, Botzler C, Jennen L, Schmidt J, Ellwart J, Issels R (1997) Heat shock protein 72 on tumor cells: a recognition structure for natural killer cells. J Immunol 158:4341–4350

Muralidharan S, Mandrekar P (2013) Cellular stress response and innate immune signaling: integrating pathways in host defense and inflammation. J Leukoc Biol 94:1167–1184

Nathan C, Shiloh MU (2000) Reactive oxygen and nitrogen intermediates in the relationship between mammalian hosts and microbial pathogens. Proc Natl Acad Sci U S A 97:8841–8848

Neckers L (2007) Heat shock protein 90: the cancer chaperone. J Biosci 32:517–530

Noessner E, Gastpar R, Milani V, Brandl A, Hutzler PJ, Kuppner MC, Roos M, Kremmer E, Asea A, Calderwood SK et al (2002) Tumor-derived heat shock protein 70 peptide complexes are cross-presented by human dendritic cells. J Immunol 169:5424–5432

Nollen EA, Morimoto RI (2002) Chaperoning signaling pathways: molecular chaperones as stress-sensing 'heat shock' proteins. J Cell Sci 115:2809–2816

O'Neill S, Humphries D, Tse G, Marson LP, Dhaliwal K, Hughes J, Ross JA, Wigmore SJ, Harrison EM (2015) Heat shock protein 90 inhibition abrogates TLR4-mediated NF-kappaB activity and reduces renal ischemia-reperfusion injury. Sci Rep 5:12958

Park JS, Gamboni-Robertson F, He Q, Svetkauskaite D, Kim JY, Strassheim D, Sohn JW, Yamada S, Maruyama I, Banerjee A et al (2006) High mobility group box 1 protein interacts with multiple Toll-like receptors. Am J Phys Cell Phys 290:C917–C924

Pespeni M, Mackersie RC, Lee H, Morabito D, Hodnett M, Howard M, Pittet JF (2005) Serum levels of Hsp60 correlate with the development of acute lung injury after trauma. J Surg Res 126:41–47

Poulaki V, Iliaki E, Mitsiades N, Mitsiades CS, Paulus YN, Bula DV, Gragoudas ES, Miller JW (2007) Inhibition of Hsp90 attenuates inflammation in endotoxin-induced uveitis. FASEB J 21:2113–2123

Qazi KR, Torregrosa Paredes P, Dahlberg B, Grunewald J, Eklund A, Gabrielsson S (2010) Proinflammatory exosomes in bronchoalveolar lavage fluid of patients with sarcoidosis. Thorax 65:1016–1024

Rajaiya J, Yousuf MA, Singh G, Stanish H, Chodosh J (2012) Heat shock protein 27 mediated signaling in viral infection. Biochemistry 51:5695–5702

Randow F, Seed B (2001) Endoplasmic reticulum chaperone gp96 is required for innate immunity but not cell viability. Nat Cell Biol 3:891–896

Raposo G, Nijman HW, Stoorvogel W, Liejendekker R, Harding CV, Melief CJ, Geuze HJ (1996) B lymphocytes secrete antigen-presenting vesicles. J Exp Med 183:1161–1172

Rice JW, Veal JM, Fadden RP, Barabasz AF, Partridge JM, Barta TE, Dubois LG, Huang KH, Mabbett SR, Silinski MA et al (2008) Small molecule inhibitors of Hsp90 potently affect inflammatory disease pathways and exhibit activity in models of rheumatoid arthritis. Arthritis Rheum 58:3765–3775

Sauter B, Albert ML, Francisco L, Larsson M, Somersan S, Bhardwaj N (2000) Consequences of cell death: exposure to necrotic tumor cells, but not primary tissue cells or apoptotic cells, induces the maturation of immunostimulatory dendritic cells. J Exp Med 191:423–434

Scroggins BT, Robzyk K, Wang D, Marcu MG, Tsutsumi S, Beebe K, Cotter RJ, Felts S, Toft D, Karnitz L et al (2007) An acetylation site in the middle domain of Hsp90 regulates chaperone function. Mol Cell 25:151–159

Shang L, Tomasi TB (2006) The heat shock protein 90-CDC37 chaperone complex is required for signaling by types I and II interferons. J Biol Chem 281:1876–1884

Shrestha L, Bolaender A, Patel HJ, Taldone T (2016) Heat shock protein (HSP) drug discovery and development: targeting heat shock proteins in disease. Curr Top Med Chem 16:2753–2764

Sims JD, McCready J, Jay DG (2011) Extracellular heat shock protein (Hsp)70 and Hsp90alpha assist in matrix metalloproteinase-2 activation and breast cancer cell migration and invasion. PLoS One 6:e18848

Singh IS, Gupta A, Nagarsekar A, Cooper Z, Manka C, Hester L, Benjamin IJ, He JR, Hasday JD (2008) Heat shock co-activates interleukin-8 transcription. Am J Respir Cell Mol Biol 39:235–242

Skokos D, Le Panse S, Villa I, Rousselle JC, Peronet R, Namane A, David B, Mecheri S (2001) Nonspecific B and T cell-stimulatory activity mediated by mast cells is associated with exosomes. Int Arch Allergy Immunol 124:133–136

Skokos D, Botros HG, Demeure C, Morin J, Peronet R, Birkenmeier G, Boudaly S, Mecheri S (2003) Mast cell-derived exosomes induce phenotypic and functional maturation of dendritic cells and elicit specific immune responses in vivo. J Immunol 170:3037–3045

Snyder YM, Guthrie L, Evans GF, Zuckerman SH (1992) Transcriptional inhibition of endotoxin-induced monokine synthesis following heat shock in murine peritoneal macrophages. J Leukoc Biol 51:181–187

Soti C, Nagy E, Giricz Z, Vigh L, Csermely P, Ferdinandy P (2005) Heat shock proteins as emerging therapeutic targets. Br J Pharmacol 146:769–780

Srivastava P (2002) Roles of heat-shock proteins in innate and adaptive immunity. Nat Rev Immunol 2:185–194

Takeda K, Akira S (2007) Toll-like receptors. Curr Protoc Immunol Chapter 14:Unit 14.12

Takeuchi T, Suzuki M, Fujikake N, Popiel HA, Kikuchi H, Futaki S, Wada K, Nagai Y (2015) Intercellular chaperone transmission via exosomes contributes to maintenance of protein homeostasis at the organismal level. Proc Natl Acad Sci U S A 112:E2497–E2506

Tang D, Kang R, Coyne CB, Zeh HJ, Lotze MT (2012) PAMPs and DAMPs: signal 0s that spur autophagy and immunity. Immunol Rev 249:158–175

Thery C, Boussac M, Veron P, Ricciardi-Castagnoli P, Raposo G, Garin J, Amigorena S (2001) Proteomic analysis of dendritic cell-derived exosomes: a secreted subcellular compartment distinct from apoptotic vesicles. J Immunol 166:7309–7318

Tobian AA, Canaday DH, Boom WH, Harding CV (2004) Bacterial heat shock proteins promote CD91-dependent class I MHC cross-presentation of chaperoned peptide to CD8+ T cells by cytosolic mechanisms in dendritic cells versus vacuolar mechanisms in macrophages. J Immunol 172:5277–5286

Toldo S, Quader M, Salloum FN, Mezzaroma E, Abbate A (2016) Targeting the innate immune response to improve cardiac graft recovery after heart transplantation: implications for the donation after cardiac death. Int J Mol Sci 17:E958

Triantafilou M, Triantafilou K (2004) Heat-shock protein 70 and heat-shock protein 90 associate with Toll-like receptor 4 in response to bacterial lipopolysaccharide. Biochem Soc Trans 32:636–639

Tukaj S, Zillikens D, Kasperkiewicz M (2014) Inhibitory effects of heat shock protein 90 blockade on proinflammatory human Th1 and Th17 cell subpopulations. J Inflamm (Lond) 11:10

Valadi H, Ekstrom K, Bossios A, Sjostrand M, Lee JJ, Lotvall JO (2007) Exosome-mediated transfer of mRNAs and microRNAs is a novel mechanism of genetic exchange between cells. Nat Cell Biol 9:654–659

Vega VL, Rodriguez-Silva M, Frey T, Gehrmann M, Diaz JC, Steinem C, Multhoff G, Arispe N, De Maio A (2008) Hsp70 translocates into the plasma membrane after stress and is released into the extracellular environment in a membrane-associated form that activates macrophages. J Immunol 180:4299–4307

Wang C, Deng L, Hong M, Akkaraju GR, Inoue J, Chen ZJ (2001) TAK1 is a ubiquitin-dependent kinase of MKK and IKK. Nature 412:346–351

Wang YL, Shen HH, Cheng PY, Chu YJ, Hwang HR, Lam KK, Lee YM (2016) 17-DMAG, an HSP90 inhibitor, ameliorates multiple organ dysfunction syndrome via induction of HSP70 in Endotoxemic rats. PLoS One 11:e0155583

Wax S, Piecyk M, Maritim B, Anderson P (2003) Geldanamycin inhibits the production of inflammatory cytokines in activated macrophages by reducing the stability and translation of cytokine transcripts. Arthritis Rheum 48:541–550

Wei D, Li NL, Zeng Y, Liu B, Kumthip K, Wang TT, Huo D, Ingels JF, Lu L, Shang J et al (2016) The molecular chaperone GRP78 contributes to toll-like receptor 3-mediated innate immune response to hepatitis C virus in hepatocytes. J Biol Chem 291:12294–12309

Wolfers J, Lozier A, Raposo G, Regnault A, Thery C, Masurier C, Flament C, Pouzieux S, Faure F, Tursz T et al (2001) Tumor-derived exosomes are a source of shared tumor rejection antigens for CTL cross-priming. Nat Med 7:297–303

Woronicz JD, Gao X, Cao Z, Rothe M, Goeddel DV (1997) IkappaB kinase-beta: NF-kappaB activation and complex formation with IkappaB kinase-alpha and NIK. Science 278:866–869

Wree A, Mehal WZ, Feldstein AE (2016) Targeting cell death and sterile inflammation loop for the treatment of nonalcoholic steatohepatitis. Semin Liver Dis 36:27–36

Xie Y, Chen C, Stevenson MA, Auron PE, Calderwood SK (2002) Heat shock factor 1 represses transcription of the IL-1beta gene through physical interaction with the nuclear factor of interleukin 6. J Biol Chem 277:11802–11810

Yang RC, Wang CI, Chen HW, Chou FP, Lue SI, Hwang KP (1998) Heat shock treatment decreases the mortality of sepsis in rats. Kaohsiung J Med Sci 14:664–672

Yang K, Shi H, Qi R, Sun S, Tang Y, Zhang B, Wang C (2006) Hsp90 regulates activation of interferon regulatory factor 3 and TBK-1 stabilization in Sendai virus-infected cells. Mol Biol Cell 17:1461–1471

Yang Y, Liu B, Dai J, Srivastava PK, Zammit DJ, Lefrancois L, Li Z (2007) Heat shock protein gp96 is a master chaperone for toll-like receptors and is important in the innate function of macrophages. Immunity 26:215–226

Zheng H, Dai J, Stoilova D, Li Z (2001) Cell surface targeting of heat shock protein gp96 induces dendritic cell maturation and antitumor immunity. J Immunol 167:6731–6735

Zheng Y, Gardner SE, Clarke MC (2011) Cell death, damage-associated molecular patterns, and sterile inflammation in cardiovascular disease. Arterioscler Thromb Vasc Biol 31:2781–2786

Zhu FG, Pisetsky DS (2001) Role of the heat shock protein 90 in immune response stimulation by bacterial DNA and synthetic oligonucleotides. Infect Immun 69:5546–5552

Zitvogel L, Regnault A, Lozier A, Wolfers J, Flament C, Tenza D, Ricciardi-Castagnoli P, Raposo G, Amigorena S (1998) Eradication of established murine tumors using a novel cell-free vaccine: dendritic cell-derived exosomes. Nat Med 4:594–600

Zou N, Ao L, Cleveland JC Jr, Yang X, Su X, Cai GY, Banerjee A, Fullerton DA, Meng X (2008) Critical role of extracellular heat shock cognate protein 70 in the myocardial inflammatory response and cardiac dysfunction after global ischemia-reperfusion. Am J Physiol Heart Circ Physiol 294:H2805–H2813

Chapter 9
Chaperone-Like Proteins in Inflammation and Immunomodulation: Examples of Resistin and PPIases

Saurabh Pandey, Javeed Ahmad, and Nasreen Zafar Ehtesham

Abstract Hsp and other small proteins that function as an accessory chaperones interact with the cellular signaling network. They come up as an immediate response to stress when cells face the challenge of its own programmed death response. Chaperokines, with their inflammation and immune modulatory potential, try to strike the balance between recovery to normalcy and programmed death. Various disease pathways, environmental stresses, toxins and infections are the challenges that mount fatal stress. Resistin, an adipokine in murine model, small macrophage secreted protein in human and PPIases of *Mycobacterium tuberculosis* are the potential chaperokines. Though according to classical definition they do not fall under the category of HSP having α-crystalline domain, they function as small chaperone and are part of biological diversity where evolutionary conserved and adapted survival mechanisms work in parallel to protect and sustain life.

Keywords Chaperokine · Heat shock protein · Immunomodulation · Inflammation · PPIase · Resistin

Abbreviations

ATP	Adenosine triphosphate
CAP-1	Adenylyl cyclase-associated protein 1
Cyp	Cyclophilin

S. Pandey
Inflammation Biology and Cell Signaling Laboratory, National Institute of Pathology, New Delhi, India

Perelman School of Medicine, University of Pennsylvania, Philadelphia, PA, USA

J. Ahmad · N. Z. Ehtesham (✉)
Inflammation Biology and Cell Signaling Laboratory, National Institute of Pathology, New Delhi, India

A. A. A. Asea, P. Kaur (eds.), *Chaperokine Activity of Heat Shock Proteins*, Heat Shock Proteins 16, https://doi.org/10.1007/978-3-030-02254-9_9

E. coli	*Escherichia coli*
ER	Endoplasmic reticulum
FKBP	FK506 binding protein
FoxP3	Forkhead Box P3
hRes	Human Resistin
HSP	Heat shock protein family
Hsp	Heat shock proteins
IL	Interleukin
IRF-1	Interferon regulatory factor 1
M.tb	*Mycobacterium tuberculosis*
MIP	Macrophage infectivity potentiator
NFAT	Nuclear factor of activated T cells
NFκB	Nuclear factor kappa B
PIN1	Peptidyl-prolyl cis-trans isomerase NIMA-interacting 1
PPIases	Peptidyl-prolyl cis/trans isomerases
SDS	Sodium dodecyl sulfate
SH3	Src homology 3
TLR	Toll-like receptors
TNF	Tumor necrosis factor
Treg	Regulatory T cells
UPR	Unfolded protein response
WD40	WD (Trp-Asp) or beta-transducin repeat motif

9.1 Introduction

In the course of evolution, nature equips the living organisms to survive in stress-induced challenges. In the molecular world, within the cells, ubiquitously expressed protein modifiers and chaperones take up these challenges. Heat shock proteins (Hsp) are one among these stress responders. Heat shock protein family (HSP) was discovered about half a century ago from *Drosophila melanogaster* where temperature induced puffing pattern was observed on the polytene chromosome (Ritossa 1962). Lindquist and Craigled discovered a group of proteins that are heat induced (Lindquist and Craig 1988). Now, a wide variety of heat shock proteins family (HSP) including Hsp90, Hsp70 and other small heat shock protein perform the function of homeostasis of proteins, controlling their quality thus supporting the cells to overcome the stresses induced by nutrient deprivation, low oxygen, exposure to toxin, infectious agents, inflammation, radiation, water or salt imbalance etc. Cells mount the survival response to avert the stresses leading to cell death or rather fine-tunes the cellular fate by selecting survival over death.

Chronic environmental stresses induce unfolded protein response (UPR), halt protein translation and may trigger programmed cell death as observed in case of degenerative diseases and cancer (Hishiya and Takayama 2008). In response to UPR, the levels of molecular chaperones are elevated in the endoplasmic reticulum (ER) through higher expression and localization to prevent molecular crowding. Paradoxically, sensing the irreversible damage, some of the molecular chaperones signal for terminal-fate of programmed death (Hishiya and Takayama 2008).

With the discovery of moonlighting function, several proteins were identified that do not fit directly as a chaperone or accessory chaperone based on sequence homology alone, such as MoxR1 and peptidyl-prolylcis/trans isomerases (PPIase) from *Mycobacterium tuberculosis* (*M.tb*), but work as a molecular chaperone (Bhuwan et al. 2016; Pandey et al. 2016). Inversely, molecular chaperones also participate in other cellular functions such as Chaperonin 10 and Chaperonin 60 in immunomodulation (Piatigorsky 2007). Moreover, Hsp60 and Hsp90 are involved in the capacitation of mouse sperm to make them physiologically competent to fertilization (Asquith 2004). Resistin and PPIases are two such examples of protein moonlighting where chaperone-like function and immunomodulation have been observed. This makes them fall under the group of proteins working as chaperokine, a new term indicating chaperone with a function of folding assistance and cytokine with immune modulatory activity.

Resistin was first reported for their association with type II diabetes mellitus in murine model where it resist the insulin action (Steppan et al. 2001a, b; Kim et al. 2001). It was secreted from adipose tissues of the mouse and serves as a connecting link between obesity and diabetes, as a modulator of glucose homeostasis (Steppan et al. 2001a, b; Kim et al. 2001). Human Resistin (hRes) is predominantly secreted from macrophages (Silswal et al. 2005) and has remarkable biophysical similarity with chaperones and show chaperone-like activity (Suragani et al. 2013). Being secreted from macrophages, instead of adipose tissues, experimental observation indicated the role of hRes in inflammation, stress adaptation and chaperoning (Suragani et al. 2013).

Peptidyl-prolyl cis/trans isomerases (PPIases) are proteins that were discovered for their immunomodulatory properties. But soon their biophysical properties and function as a foldase facilitating the isomerization of peptidyl-prolyl bond have been discovered. PPIases are moonlighting proteins involved in diverse functions. Role of PPIases has also been observed in cell-cycle regulatory functions (Lin et al. 2015), inflammatory pathways (Tan et al. 2014; Liu et al. 1991), and chromatin remodeling as a regulator of nuclear events (Wu et al. 2000). PPIases of pathogenic organisms are involved in virulence, inflammation and immune modulation (Arora et al. 2005; Unal and Steinert 2014; Pandey et al. 2017). In this review, we will focus on the chaperoning, inflammatory and immune modulatory functions of resistin along with the genomic and biophysical characteristics. Functions of PPIases with references to pathogenic organisms such as *M.tb* and relevant host-pathogen interactions will be described. Some applications of Resistin and PPIases in basic biology and medicine will also be cited.

9.2 Genomic and Biophysical Features of Resistin

Resistin along with other resistin-like molecules has a specific pattern of the cysteine that imparts the unusual stability and ability to form chaperone-like higher order oligomerization (Patel 2004; Suragani et al. 2013). This cysteine signature sequence is C-X11-C-X8-C-X-C-X3-C-X10-C-X-C-X-C-X9-CC-X3–6-END where X denotes any amino acid (Steppan et al. 2001b). Moreover, mouse has two other genes coding for resistin-like molecules retn1α, retn1β along with retn coding for resistin. Human counterpart has homolog for only two of them retn and retn1β (Steppan et al. 2001b). Synteny mapping indicates the presence of human resistin on chromosome 19 while mouse resistin is on chromosome 8 (Ghosh et al. 2003). Further, a notable feature of mouse resistin encoding gene, unlike the hRes, is the availability of additional intron region close to 3′ end that has several regulatory transcription factor interacting motifs, most importantly PPARγ binding region, regulating insulin resistance mechanism in mouse adipose tissues (Ghosh et al. 2003).

X-ray crystallography revealed the formation of stable tri- and hexameric structures of mouse resistin where disulfide bridges between cysteines are crucial for structural stability. The C-terminal cysteine rich region forms the aggregation of β-sheets whereas α-helical region forms the tail region in the tri- and hexameric structures in in-vitro crystallization (Patel 2004). Similarly, oligomers are also reported in the mouse circulatory system (Patel 2004). The biophysical study shows the concentration dependent variation in the secondary structural distribution in hRes that can be explained by the structural oligomerization (Aruna et al. 2008). Further, the trimeric form of the hRes is stable to urea denaturation. However, hRes is susceptible to sodium halides (NaCl and NaF). Higher-order oligomers formed by trimeric building blocks are destabilized by urea denaturation. hRes shows the SDS insensitivity but is sensitive to β-mercaptoethanol denaturation (Aruna et al. 2003). These observations directly indicate that ionic interactions are important for trimer formation whereas hydrophobic interactions form the higher-order structures (Aruna et al. 2008). This exposed hydrophobic region potentiates any chaperone-like molecule to interact with the denatured or partially denatured proteins and this binding as accessory chaperone in energy independent manner facilitate the subsequent action of ATP dependent chaperone complex for refolding by shifting the thermodynamic energy barriers (McHaourab et al. 2009; Basha et al. 2012).

9.3 Resistin in Chaperoning, Inflammation and Immune Modulation

Human resistin, unlike murine resistin, is not associated with either obesity or resistance to insulin (Lin et al. 2015). hRes, a small protein of 10–12 kDa, as shown to have increased expression under cellular stress including infection (Ehtesham et al. 2011). hRes protects citrate synthase, a heat labile protein, from thermal aggregation, protects and refolds back Nde1restrictionenzyme to regain activity and displays an affinity for denatured proteins – features that point to it being a protein with chaperone-like activity (Suragani et al. 2013). The association-dissociation kinetics in in-silico molecular dynamics simulation of hRes oligomer indicates it to be a putative chaperone (Suragani et al. 2013). Moreover, *Escherichia coli* (*E. coli*) cells over-expressing hRes show better survival when subjected to temperature shock (Suragani et al. 2013). Also, chemically induced unfolded protein response reduces extra-cellular secretion and enhances ER localization of hRes, and shows protection against UPR driven apoptosis in mammalian system (Suragani et al. 2013). These experimental evidences establish the role of hRes as a small chaperone-like protein.

Stimulation of human macrophage cells by hRes induces the synthesis and secretion of proinflammatory cytokines, TNF-α and IL-12 via activation of NFκB NFκB pathway (Silswal et al. 2005; Pang and Le 2006). Moreover, relation of hRes with inflammation is also reversibly implied where proinflammatory stress enhances the cellular resistin levels (Kaser et al. 2003; Lehrke et al. 2004). Resistin diminishes the synthesis of proinflammatory cytokines signaling and suppresses T-cell proliferation in dendritic cells (Son et al. 2008). The antigen endocytosis by lipoteichoic acid stimulated dendritic cells is impaired upon hRes treatment. These impaired dendritic cells show alteration in the FoxP3 mediated Treg expansion. In PBMCs but not CD4+ T-cells, FoxP3 is expressed upon resistin stimulation (Son et al. 2008). hRes impairs the efficacy of T-cell mediated immune response via IRF-1 mediated Treg signaling in human monocyte derived dendritic cells (Son et al. 2010). Further, ligands of Toll-like receptors (TLR) 3, 4, 7/8 and 9 suppress the release of proinflammatory cytokine possibly by NFκB pathway (Schaffler and Scholmerich 2010). This attracts the initial speculation that homo or hetero dimer of cell surface TLRs could serve as receptor for macrophage secreted circulatory resistin signaling in mammalian cells, that can impact cellular inflammation and immunological efficacy of the cells. Recently, adenylyl cyclase-associated protein 1 (CAP1) was reported as receptor for human resistin where CAP1 binds with resistin via central proline-rich Src homology 3 (SH3) domain (Lee et al. 2003). These findings highlight (Fig. 9.1) the immune modulatory potential of human resistin, a chaperokine.

9.4 Structure and Folding Action of PPIases

Proline is an unusual amino acid where instead of amino it has imino moiety that reduces the difference in the thermodynamic energetic preferences in the folded protein space. This lowered difference populates the functionally folded protein space with both cis and trans forms peptidyl-prolyl bonds. PPIases-assisted isomerization of peptidyl-prolyl bond is essential in both nascent and misfolded/unfolded proteins to reach the native form. Folding bottlenecks otherwise will arise in absence of active PPIases. However, all peptidyl-prolyl bonds are not folded and both cis and trans appear in native protein structures. Earlier studies showed that PPIases have substrate specificity based on the experiments performed with small synthetic

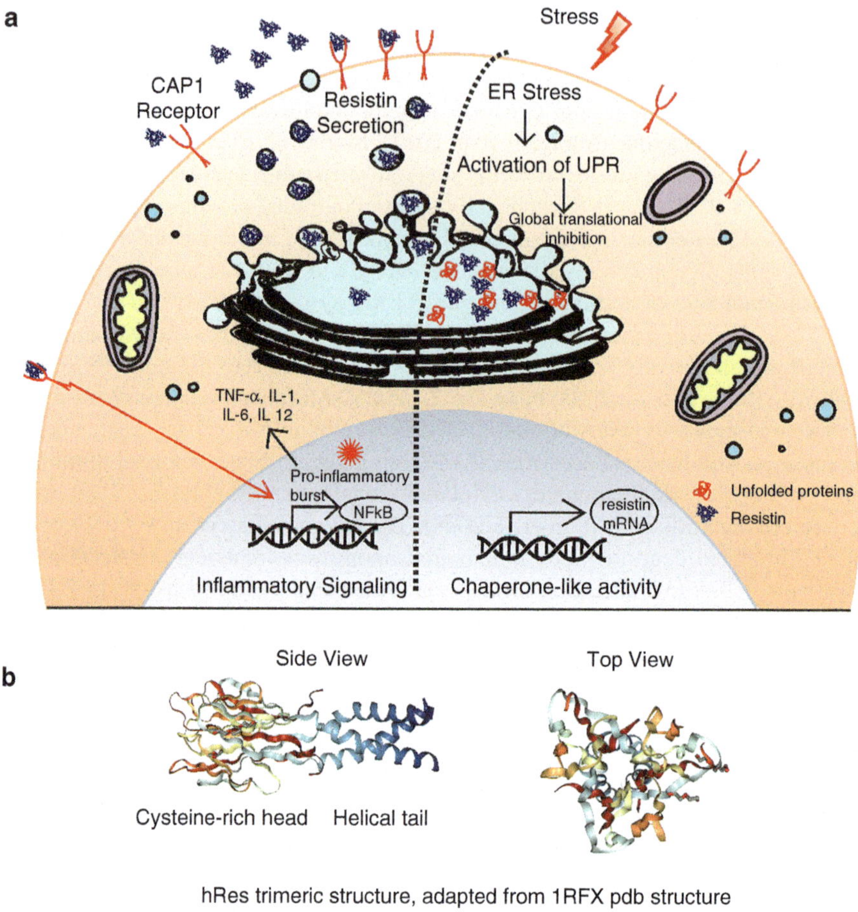

Fig. 9.1 (**a**) Resistin chaperokine-signaling pathway. During normal resistin secretion, CAP1 dependent activation of NFκB, causes the proinflammatory burst and promotes the secretion of TNF-α, IL-1, IL-6 and IL-12. Upon endoplasmic reticulum stress, unfolded protein load increases in the ER. This reduces the resistin secretion from the cells, causes retention of resistin in ER. Though global translation halts, resistin translates possibly through CAP1 independent translation. The resistin, performing a chaperone-like activity, assists in the folding and reducing the protein load. (**b**) Side and top view of structure of Resistin. Cysteine-rich resistin trimer forms the helical head that is composed of β-sheet interconnected with loop region. α-helices compose the tail of the trimer. Top view of resistin structure shows the formation of a gap in the trimeric structural core

peptide substrate (Harrison and Stein 1990). Other reports too mention selection preferences in the isomerase activity of the PPIases, as in the case of FKBP12, PIN1 or Macrophage infectivity potentiator (MIP) proteins (Fishbein et al. 2015; Yaffe 1997; McNaughton et al. 2010). Later, it appeared that shape of the PPIase and accessibility to the substrate site is the reason for selection bias than sequence specificity. Thus steric hindrance inhibits the PPIase enzyme to reach the site of peptidyl-prolyl bond for isomerization and the natively folded proteins remain finally in trans form (Scholz et al. 1998; Gothel and Marahiel 1999) form (Scholz et al. 1998; Gothel and Marahiel 1999). PPIases either exist in only enzyme domain as in the case of human FKBP12, human CypA, M.tb PpiA or with additional domains as tetratrico-peptide, WD40, zinc finger, RNA recognition or calcium binding domain (D'Andrea and Regan 2003; Blackburn and Walkinshaw 2011; Schiene-Fischer et al. 2013). These additional domains are important for protein-protein, protein-nucleic acid interaction, protein processing, chaperoning or cell signaling, emphasizing the functional importance of PPIases (Kumar et al. 2017). The enzyme domain of PPIases is composed of 100–120 amino acids and is significantly conserved in prokaryotes and eukaryotes. However, subfamilies of PPIases including cyclophilins, FKBPs and parvulins show structural variations (Blackburn and Walkinshaw 2011). PPIases are present in more abundance in eukaryotes compared to prokaryotes (Fig. 9.2).

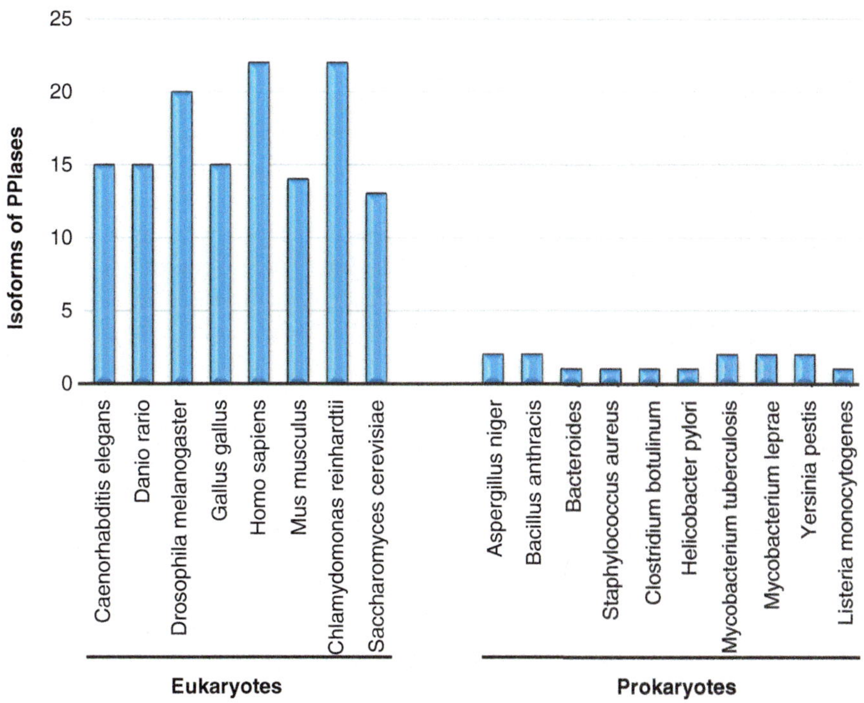

Fig. 9.2 Isoform diversity of PPIases in prokaryotes and eukaryotes. Prokaryotes have a few PPIases but eukaryotic organisms have surprisingly large number of PPIases that are localized to various subcellular locations depending upon signal sequences

9.5 PPIases in Chaperoning, Inflammation and Immune Modulation

Heat shock proteins, stress responders, foldases and chaperones are proteins that directly participate in host-pathogen interaction (Henderson 2010). These interactions stimulate the inflammatory and immunomodulatory responses, as in the case of M.tb PPIases. M.tb has two PPIases, PpiA and PpiB, both are functional isomerases and involved in chaperone-like activities, in-vivo and in-vitro (Pandey et al. 2016). Recombinant M.tb PPIases expressed in *E. coli* expression system prevent the aggregation of heat labile maltodextrin glucosidase protein from thermal aggregation, protect *E. coli* transformants from thermal shock and when overexpressed in the mammalian 293T cells, give protection against oxidative and hypoxic stresses (Pandey et al. 2016). But, chaperone activity of PPIases is not limited to bacterial PPIases, rather PPIases from archaea, yeasts, plants and humans too work as a chaperone. For example, wheat FKBP family PPIase, FKBP73 (Kurek et al. 2002), *E. coli* trigger factor (Kramer et al. 2004), archeal PhFKBP29 (Ideno et al. 2002), yeast PPIases Cpr7 (Mayr et al. 2000) and human cyp40 (Freeman et al. 1996) function as molecular chaperone along with PPIase activity. Functional diversity of PPIases is presented in Fig. 9.3.

In the recent finding from our group, we have observed that *M.tb* PPIases are involved in the cellular inflammation and immune modulation pathways (Pandey et al. 2017). *M.tb* PpiA secrets the pro-inflammatory cytokines TNF-α, IL-6 in macrophage stimulation assay while PpiB suppresses the proinflammatory cytokines and induces the IL-10 secretion (Pandey et al. 2017). Interestingly, other *M.tb* heat shock proteins also show proinflammatory burst for example, human monocytes-derived macrophages, upon stimulation by HSP65, secrete TNF-α and IL-1β (Peetermans et al. 1994). The secretion of cytokines induced by bacterial secretory chaperones tinkers with the host immunology in a contextual manner. Proinflammatory signals serve as a possible inducer of chemo-attractants that causes the infiltration of various immune cells for clearance. Further, dampening the inflammation by anti-inflammatory signals may help the pathogen to hide, mostly within macrophages, that increases the longevity of the infected host cell by stopping the lysosome fusion and phagosome maturation. In an interesting study, human chaperone-like protein of PPIase family, Pin-1 interferes in viral-mediated innate immune response via interferon-regulatory factor 3 (Saitoh et al. 2006).

The drug cyclosporine A, an immunosuppressant used in organ transplant for protection against graft rejection, binds with PPIases e.g. human CypA and FKBP12 to suppress the nuclear factor of activated T cells (NFAT) mediated cytokine response (Liu et al. 1991). Another, human PPIase, Pin1 also participates in T-cell mediated immune response in knockout mouse model (Esnault et al. 2007). These shreds of evidence establish the role of PPIases as a molecule with potential for both immune modulation and chaperone-like activity that qualifies it to be a chaperokine.

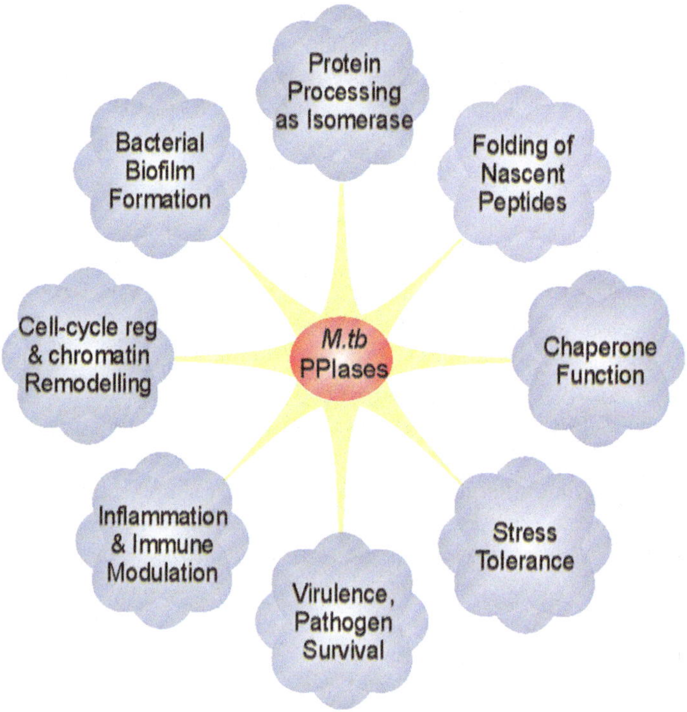

Fig. 9.3 Functional diversity of *M.tb* PPIases. *M.tb* PPIases perform many functions like folding of nascent peptides as part of the ribosomal complex, chaperoning the unfolded proteins, stress adaptation, survival of pathogen, inflammation, immune modulation bacterial biofilm formation as a part of proteins assembled at the cell surface and isomerase function

9.6 Resistin and PPIases in Biology and Medicine: Some Examples

Resistin has been associated with disease conditions in several clinical cross-sectional studies including atherosclerosis and cancers (Reilly et al. 2005; Nakajima et al. 2010a; Sun et al. 2010; Hlavna et al. 2011; Danese et al. 2012; Dalamaga et al. 2013). But among most important applications of the resistin is a biomarker for disease conditions (Wang et al. 2017; Perovic et al. 2017) and treatment endpoint determination for tuberculosis (Ehtesham et al. 2011). In our lab, we have reported that treatment endpoint of tuberculosis can be evaluated efficiently by using resistin as a surrogate biomarker (Ehtesham et al. 2011). This can be helpful to patients as in tuberculosis, a cocktail of medicines is given for 6–9 months for treatment. The stress adaptive potential of PPIases can be harnessed directly. In water and other abiotic stress, tolerance is increased in edible crops like rice and sorghum in presence of PPIase also known as Cyclophilins (Sharma and Singh 2003; Trivedi et al. 2013). PPIases are involved in pathogenesis of several disease processes. Blocking CypA has strong antiviral effect against of many viral diseases including hepatitis C and blocking CypD has direct positive impact on the mitochondrial dysfunction (Naoumov 2014). Thus, PPIases can serve as a potential therapeutic target for the liver disease and other virus born diseases.

9.7 Conclusions

Resistin and PPIases while serving as chaperokine, influence the host immunology on the axis of host-pathogen interaction. They can provide a newer approach for the development of host-directed therapies. As drugs, targeting molecular pathways within pathogens, mount the selection pressure responsible for drug resistance. Further, rebooting the immune system provides surprising disease clearance as in case of multiple sclerosis and skin cancer. Thus, it would be reasonable to focus future attention to research on the chaperokines for the development of the strategies to boost and target host immune system to cope up with infections, cancer and other diseases.

Acknowledgements NZE would like to thank Centre of Excellence research grant (BT/PR12817/COE/34/23/2015) from the Department of Biotechnology, Ministry of Science and Technology (DBT), Government of India and research grant (No.5/9/1019/2011-RHN) from ICMR, Ministry of Health and Family welfare, Government of India for financial support.

References

Arora K, Gwinn WM, Bower MA, Watson A, Okwumabua I, MacDonald HR, Bukrinsky MI, Constant SL (2005) Extracellular cyclophilins contribute to the regulation of inflammatory responses. J Immunol 175:517

Aruna B, Ghosh S, Singh AK, Mande SC, Srinivas V, Chauhan R, Ehtesham NZ (2003) Human recombinant resistin protein displays a tendency to aggregate by forming intermolecular disulfide linkages. Biochemistry 42:10554–10559

Aruna B, Islam A, Ghosh S, Singh AK, Vijayalakshmi M, Ahmad F, Ehtesham NZ (2008) Biophysical analyses of human resistin: oligomer formation suggests novel biological function. Biochemistry 47:12457–12466

Asquith KL, Baleato RM, McLaughlin EA, Nixon B, Aitken RJ (2004) Tyrosine phosphorylation activates surface chaperones facilitating sperm-zona recognition. J Cell Sci 117:3645–3657

Basha E, O'Neill H, Vierling E (2012) Small heat shock proteins and α-crystallins: dynamic proteins with flexible functions. Trends Biochem Sci 37:106–117

Bhuwan M, Arora N, Sharma A, Khubaib M, Pandey S, Chaudhuri TK, Hasnain SE, Ehtesham NZ (2016) Interaction of Mycobacterium tuberculosis virulence factor RipA with chaperone MoxR1 is required for transport through the TAT secretion system. MBio 7:1–12

Blackburn EA, Walkinshaw MD (2011) Targeting FKBP isoforms with small-molecule ligands. Curr Opin Pharmacol 11:365–371

D'Andrea LD, Regan L (2003) TPR proteins: the versatile helix. Trends Biochem Sci 28:655–662

Dalamaga M, Karmaniolas K, Papadavid E, Pelekanos N, Sotiropoulos G, Lekka A (2013) Hyperresistinemia is associated with postmenopausal breast cancer. Menopause 20:845–851
Danese E, Montagnana M, Minicozzi AM, Bonafini S, Ruzzenente O, Gelati M, De Manzoni G, Lippi G, Guidi GC (2012) The role of resistin in colorectal cancer. Clin Chim Acta 413:760–764
Ehtesham NZ, Nasiruddin M, Alvi A, Kumar BK, Ahmed N, Peri S, Murthy KJR, Hasnain SE (2011) Treatment end point determinants for pulmonary tuberculosis: human resistin as a surrogate biomarker. Tuberculosis 91:293–299
Esnault S, Braun RK, Shen ZJ, Xiang Z, Heninger E, Love RB, Sandor M, Malter JS (2007) Pin1 modulates the type 1 immune response. PLoS One 2:e226
Fischer G, Bang H, Ludwig B, Mann K, Hacker J (1992) Mip protein of Legionella pneumophila exhibits peptidyl-prolyl-cis/trans isomerase (PPlase) activity. Mol Microbiol 6:1375–1383
Freeman BC, Toft DO, Morimoto RI (1996) Molecular chaperone machines: chaperone activities of the cyclophilin Cyp-40 and the steroid aporeceptor-associated protein p23. Science 274:1718–1720
Ghosh S, Singh AK, Aruna B, Mukhopadhyay S, Ehtesham NZ (2003) The genomic organization of mouse resistin reveals major differences from the human resistin: functional implications. Gene 305:27–34
Gothel SF, Marahiel MA (1999) Peptidyl-prolylcis-trans isomerases, a superfamily of ubiquitous folding catalysts. Cell Mol Life Sci 55:423–436
Harrison R, Stein RL (1992) Mechanistic studies of enzymatic and nonenzymatic prolyl cis-trans isomerization. J Am Chem Soc 114:3464–3471
Henderson B (2010) Integrating the cell stress response: a new view of molecular chaperones as immunological and physiological homeostatic regulators. Cell Biochem Funct 28:1–14
Hishiya A, Takayama S (2008) Molecular chaperones as regulators of cell death. Oncogene 27:6489–6506
Hlavna M, Kohut L, Lipkova J, Bienertova-Vasku J, Dostalova Z, Chovanec J, Vasku A (2011) Relationship of resistin levels with endometrial cancer risk. Neoplasma 58:124–128
Ideno A, Furutani M, Iba Y, Kurosawa Y, Maruyama T (2002) FK506 binding protein from the hyper-thermophilic archaeon Pyrococcus horikoshii suppresses the aggregation of proteins in *Escherichia coli*. Appl Environ Microbiol 68:464–469
Kaser S, Kaser A, Sandhofer A, Ebenbichler CF, Tilg H, Patsch JR (2003) Resistin messenger-RNA expression is increased by proinflammatory cytokines in vitro. Biochem Biophys Res Commun 309:286–290
Kim KH, Lee K, Moon YS, Sul HS (2001) A cysteine-rich adipose tissue-specific secretory factor inhibits adipocyte differentiation. J Biol Chem 276:11252–11256
Kramer G, Patzelt H, Rauch T, Kurz TA, Vorderwülbecke S, Bukau B, Deuerling E (2004) Trigger factor peptidyl-prolylcis/trans isomerase activity is not essential for the folding of cytosolic proteins in *Escherichia coli*. J Biol Chem 279:14165–14170
Kumar A, Alam A, Rani M, Ehtesham NZ, Hasnain SE (2017) Biofilms: survival and defense strategy for pathogens. Int J Med Microbiol 307:481–489
Kurek I, Pirkl F, Fischer E, Buchner J, Breiman A (2002) Wheat FKBP73 functions in vitro as a molecular chaperone independently of its peptidylprolylcis-trans isomerase activity. Planta 215:119–126
Lee JH, Chan JL, Yiannakouris N, Kontogianni M, Estrada E, Seip R, Orlova C, Mantzoros CS (2003) Circulating resistin levels are not associated with obesity or insulin resistance in humans and are not regulated by fasting or leptin administration: cross-sectional and interventional studies in normal, insulin-resistant, and diabetic subjects. J Clin Endocrinol Metab 88:4848–4856
Lehrke M, Reilly MP, Millington SC, Iqbal N, Rader DJ, Lazar MA (2004) An inflammatory cascade leading to hyperresistinemia in humans. PLoS Med 1:161–168
Lin C-H, Li H-Y, Lee Y-C, Calkins MJ, Lee K-H, Yang C-N, Lu P-J (2015) Landscape of Pin1 in the cell cycle. Exp Biol Med 240:403–408
Lindquist S, Craig EA (1988) The heat-shock proteins. Annu Rev Genet 22:631–677

Liu J, Farmer JD, Lane WS, Friedman J, Weissman I, Schreiber SL (1991) Calcineurin is a common target of cyclophilin-cyclosporin A and FKBP-FK506 complexes. Cell 66:807–815

Mayr C, Richter K, Lilie H, Buchner J (2000) Cpr6 and Cpr7, two closely related Hsp90-associated immunophilins from *Saccharomyces cerevisiae*, differ in their functional properties. J Biol Chem 275:34140–34146

Mchaourab HS, Godar JA, Stewart PL (2009) Structure and mechanism of protein stability sensors: chaperone activity of small heat shock proteins. Biochemistry 48:3828–3837

McNaughton L, Li Z, Van Roey P, Hanes SD, LeMaster DM (2010) Restricted domain mobility in the *Candida albicans* Ess1 prolylisomerase. Biochim Biophys Acta 1804:1537–1541

Nakajima TE, Yamada Y, Hamano T, Furuta K, Gotoda T, Katai H, Kato K, Hamaguchi T, Shimada Y (2009) Adipocytokine levels in gastric cancer patients: resistin and visfatin as biomarkers of gastric cancer. J Gastroenterol 44:685–690

Nakajima TE, Yamada Y, Hamano T, Furuta K, Matsuda T, Fujita S, Kato K, Hamaguchi T, Shimada Y (2010a) Adipocytokines as new promising markers of colorectal tumors: adiponectin for colorectal adenoma, and resistin and visfatin for colorectal cancer. Cancer Sci 101:1286–1291

Naoumov NV (2014) Cyclophilin inhibition as potential therapy for liver diseases. J Hepatol 61:1166–1174

Pandey S, Sharma A, Tripathi D, Kumar A, Khubaib M, Bhuwan M, Chaudhuri TK, Hasnain SE, Ehtesham NZ (2016) Mycobacterium tuberculosis peptidyl-prolyl isomerases also exhibit chaperone like activity in-vitro and in-vivo. PLoS One 11:e0150288

Pandey S, Tripathi D, Khubaib M, Kumar A, Shekh AJ, Gaddam S, Ehtesham NZ, Hasnain SE (2017) Mycobacterium tuberculosis peptidyl-prolyl isomerases are immunogenic, alter cytokine profile and aid in intracellular survival. Front Cell Infect Microbiol 7:1–9

Pang SS, Le YY (2006) Role of resistin in inflammation and inflammation-related diseases. Cell Mol Immunol 3:29–34

Patel SD, Rajala MW, Rossetti L, Scherer PE, Shapiro L (2004) Disulfide-dependent multimeric assembly of resistin family hormones. Science 304:1154–1158

Peetermans WE, Raats CJI, Langermans JAM, Van Furth R (1994) Mycobacterial heat-shock protein 65 induces proinflammatory cytokines but does not activate human mononuclear phagocytes. Scand J Immunol 39:613–617

Perovic E, Mrdjen A, Harapin M, Tesija KA, Simundic AM (2017) Diagnostic and prognostic role of resistin and copeptin in acute ischemic stroke. Top Stroke Rehabil 24(8):614–618

Piatigorsky J (2007) Gene sharing and evolution: the diversity of protein functions. Harvard University Press, Cambridge, MA. isbn:9780674023413

Reilly MP, Lehrke M, Wolfe ML, Rohatgi A, Lazar MA, Rader DJ (2005) Resistin is an inflammatory marker of atherosclerosis in humans. Circulation 111:932–939

Ritossa F (1962) A new puffing pattern induced by temperature shock and DNP in drosophila. Experientia 18:571–573

Saitoh T, Tun-Kyi A, Ryo A, Yamamoto M, Finn G, Fujita T, Akira S, Yamamoto N, Lu KP, Yamaoka S (2006) Negative regulation of interferon-regulatory factor 3–dependent innate antiviral response by the prolylisomerase Pin1. Nat Immunol 7:598–605

Schaffler A, Scholmerich J (2010) Innate immunity and adipose tissue biology. Trends Immunol 31:228–235

Schiene-Fischer C, Aumüller T, Fischer G (2013) Peptide bond cis/trans isomerases: a biocatalysis perspective of conformational dynamics in proteins. Top Curr Chem 328:35–67

Scholz C, Mücke M, Rape M, Pecht A, Pahl A, Bang H, Schmid FX (1998) Recognition of protein substrates by the prolylisomerase trigger factor is independent of proline residues. J Mol Biol 277:723–732

Sharma AD, Singh P (2003) Effect of water stress on expression of a 20 kD cyclophilin-like protein in drought susceptible and tolerant cultivars of sorghum. J Plant Biochem Biotechnol 12:77–80

Silswal N, Singh AK, Aruna B, Mukhopadhyay S, Ghosh S, Ehtesham NZ (2005) Human resistin stimulates the pro-inflammatory cytokines TNF-alpha and IL-12 in macrophages by NFκB-dependent pathway. Biochem Biophys Res Commun 334:1092–1101

Son YM, Ahn SM, Jang MS, Moon YS, Kim SH, Cho KK, Han SH, Yun CH (2008) Immunomodulatory effect of resistin in human dendritic cells stimulated with lipoteichoic acid from *Staphylococcus aureus*. Biochem Biophys Res Commun 376:599–604

Son YM, Ahn SM, Kim GR, Moon YS, Kim SH, Park Y-M, Lee WK, Min TS, Han SH, Yun C-H (2010) Resistin enhances the expansion of regulatory T cells through modulation of dendritic cells. BMC Immunol 11:33

Steppan CM, Bailey ST, Bhat S, Brown EJ, Banerjee RR, Wright CM, Patel HR, Ahima RS, Lazar MA (2001a) The hormone resistin links obesity to diabetes. Nature 409:307–312

Steppan CM, Brown EJ, Wright CM, Bhat S, Banerjee RR, Dai CY, Enders GH, Silberg DG, Wen X, Wu GD, Lazar MA (2001b) A family of tissue-specific resistin-like molecules. Proc Natl Acad Sci U S A 98:502–506

Sun CA, Wu MH, Chu CH, Chou YC, Hsu GC, Yang T, Chou WY, Yu CP, Yu JC (2010) Adipocytokine resistin and breast cancer risk. Breast Cancer Res Treat 123:869–876

Suragani M, Aadinarayana VD, Pinjari AB, Tanneeru K, Guruprasad L, Banerjee S, Pandey S, Chaudhuri TK, Ehtesham NZ (2013) Human resistin, a proinflammatory cytokine, shows chaperone-like activity. Proc Natl Acad Sci U S A 110:20467–20472

Tan J, Huang Z, Guo X, Li T, Deng W (2014) Expression and significance of cyclophilin A in synovial fibroblasts from patients with rheumatoid arthritis. Zhonghua Yi Xue Za Zhi 94:1330–1333

Trivedi DK, Ansari MW, Tuteja N (2013) Multiple abiotic stress responsive rice cyclophilin: (OsCYP-25) mediates a wide range of cellular responses. Commun Integr Biol 6:e25260

Unal CM, Steinert M (2014) Microbial peptidyl-prolylcis/trans isomerases (PPIases): virulence factors and potential alternative drug targets. Microbiol Mol Biol Rev 78:544–571

Wang J-H, Lee C-J, Yang C-F, Chen Y-C, Hsu B-G (2017) Serum resistin as an independent marker of aortic stiffness in patients with coronary artery disease. PLoS One 12:e0183123

Wu X, Wilcox CB, Devasahayam G, Hackett RL, Arévalo-Rodríguez M, Cardenas ME, Heitman J, Hanes SD (2000) The Ess1 prolylisomerase is linked to chromatin remodeling complexes and the general transcription machinery. EMBO J 19:3727–3738

Yaffe MB (1997) Sequence-specific and phosphorylation-dependent proline isomerization: a potential mitotic regulatory mechanism. Science 278:1957–1960

Chapter 10
Exercise-Induced Chaperokine Activity of Hsp70: Possible Role in Chronic Diseases

Mahdieh Molanouri Shamsi, Zuhair Mohammad Hassan, and Reza Gharakhanlou

Abstract In recent years, strong effects of exercise on the immune system have been demonstrated in many studies. Indeed, beneficial effects of regular exercise on immune responses in diseases such as cancer, cardiovascular disease, neuromuscular disorders and diabetes have been observed. Growing evidence indicates that the stress-inducible form of Hsp70 (Hsp70; 72 kDa) is found in the extracellular milieu and can exert chaperoning and regulatory effects on various immunocompetent cells. In this regard, extracellular Hsp72 can stimulate immune responses such as macrophages, T lymphocytes and NK cells. We propose that extracellular Hsp72 is released following acute exercise and acts to stimulate the immune system. The chaperokine activity of eHsp72 would constitute one mechanism for cross talk between tissues after exercise and the adaptation to physical exercise. Here, we discuss the immunostimulatory and immunosuppressive activities of eHsp72 as a possible mechanism for the protective effects of regular physical exercise in chronic disease.

Keywords Chaperokine · Chronic diseases · Exercise · Extracellular Hsp70 · Hsp70 · Immune response

---

M. Molanouri Shamsi (✉) · R. Gharakhanlou
Department of Physical Education & Sport Sciences, Faculty of Humanities, Tarbiat Modares University, Tehran, Iran
e-mail: molanouri@modares.ac.ir

Z. M. Hassan
Department of Immunology, School of Medical Sciences, Tarbiat Modares University, Tehran, Iran

A. A. A. Asea, P. Kaur (eds.), *Chaperokine Activity of Heat Shock Proteins*, Heat Shock Proteins 16, https://doi.org/10.1007/978-3-030-02254-9_10

Abbreviations

APC	Antigen-presenting cells
CD	Cluster of differentiation
CTLs	Cytotoxic T lymphocytes
DC	Dendritic cell
HSP	Heat shock protein family
Hsp	Heat shock proteins
IL-12	Interleukin-12
IL-1β	Interleukin-1β
IL-6	Interleukin-6
LOX-1	Lectin-like oxidized low-density lipoprotein (LDL) receptor-1
NEFA	Non-esterified fatty acids
NF-kβ	Nuclear factor kappa β
NK	Natural killer
SR-A	Macrophage scavenger receptor type A
Th1	T helper 1
TLR	Toll like receptors
TNF-α	Tumor necrosis factor-alpha

10.1 Introduction

Mammalian cells have developed a range of adaptations to survive against acute and prolonged stresses (Beckmann et al. 1992). The heat shock or stress response is one of the most highly conserved adaptive responses in eukaryotes (Locke and Noble 1995). In single cell organisms, the stress response confers tolerance to a variety of stresses including hyperthermia, hyperoxia, hypoxia, and other perturbations, which alter protein synthesis. In response to a wide variety of stressful stimuli, there is a marked increase in total heat shock protein (Hsp) synthesis (Haworth et al. 1997), known as the cellular stress response. In addition to critical role of Hsp within the intracellular environment; a growing literature is providing evidence that some Hsp can be actively released from cells (Lancaster and Febbraio 2005). Recent data indicates that these released stress proteins have the ability to modulate the immune responses (Moseley 2000).

Of all heat shock proteins, the HSP70 family constitutes the most conserved and best-studied class (Bukau and Horwich 1998; Lehner et al. 2004; Wang et al. 2005; Asea 2005). The HSP70 family ranges in size from 70 to 78 kDa (Noble et al. 2008) and it has been demonstrated to be the most inducible and abundant of all Hsp (Madden et al. 2008). Growing evidence proposed that the stress inducible form of Hsp70 (Hsp70; 72 kDa) is found in the extracellular milieu and could exert regulatory effects on various immunocompetent cells (Clayton et al. 2005). Asea (2005) suggested that exogenously added Hsp72 possesses potent chaperone and cytokine activity, a term now referred to as the chaperokine ability of Hsp72 and this extracellular Hsp72 binds with high affinity and avidity to specific cells of the immune system and activates specific signal transduction cascades depending on which receptor it bind (Asea 2005). In this regard extracellular Hsp72 could stimulate immune responses of macrophages, T lymphocytes and NK cells (Moseley 2000).

Possible relations between extracellular Hsp70 levels and chronic disease proposed in some studies (Krause and Rodrigues-Krause Jda 2011; Genth-Zotz et al. 2004; van Eden et al. 2005). Although in some diseases such as diabetes chronic increasing in extracellular levels of Hsp70 is associated with protein damage, oxidative stress and insulin resistance (Rodrigues-Krause et al. 2012; Molanouri Shamsi et al. 2016). The absence and/or the inhibition of Hsp70 expression in neurodegenerative disorders has been shown to result in increased vulnerability of the cells to stress (Rubio et al. 2002; Nakata et al. 1993). Moreover, it is proposed that pro-immune functions of extracellular Hsp especially Hsp70 had rolled in cancer prevention and treatment (Weng et al. 2013; Yu et al. 2013; Hashemi et al. 2008).

Numerous studies showed that circulating levels of Hsp72 was significantly increased following physical exercise as an effective physiological stress (Walsh et al. 2001; Febbraio et al. 2002). As we know regular exercise could be an important strategy against some chronic disease such as neurodegenerative diseases, type 2 diabetes (Tuomilehto et al. 2001; Krause and Rodrigues-Krause Jda 2011), cardiovascular diseases (Nocon et al. 2008), and postmenopausal breast cancer (Monninkhof et al. 2007; Molanouri Shamsi et al. 2017a, b). Recent evidence suggest that Hsp such as Hsp72 may implicated in exercise-induced adaptation against chronic disease (Henstridge et al. 2016). Chaperokine activities of Hsp during exercise especially in diseases did not obvious until now. In this chapter we assessed chaperokine activities of HSP70 family especially Hsp72. Also, possible effects of extracellular Hsp72 in relate to exercise and chronic disease assessed in current chapter.

10.1.1 Chaperokine Activity of Hsp70

Important role of stress proteins in modulation of immune responses demonstrated based on addressed model by Matzinger and Fuchs (1996). This model that is named as Danger Theory of Immunity, ability of organism for detecting and addressing danger signals against murdered situation by infectious agents or dies due to necrosis or ischemia is depended on stress protein-peptide complexes in to the extracellular environment (Moseley 2000). Based on this theory it is shown that in addition to important role of stress proteins in survival of environmental stresses as a chaperokine, the stress proteins such as Hsp72 can activate cellular immune responses. In fact the extracellular Hsp72 (eHsp72) can act as both chaperone and cytokine (Asea 2003).

Potential roles of extracellular Hsp72 on immune responses have been shown in some studies. Extracellular Hsp72 play a functional role in antigen-presenting cells

(APC) and stimulation of an immune response (Asea 2003, 2005). Some studies showed that tumor derived exogenous Hsp72 bind to APC and stimulate the tumor specific cytotoxic T cell response (Multhoff et al. 1995; Asea 2006; Moehler et al. 2005). In addition, detection by immune system has improved when extracellular Hsp bind to microbial peptides (Triantafilou and Triantafilou 2004; Asea et al. 2002). Moreover, it is suggested that innate immune cells are activated by danger signals that are derived from stressed or damaged self-proteins such as eHsp72 (Campisi et al. 2003; Asea et al. 2000; Asea 2006). Numerous studies showed that some leukocytes such as natural killer (NK) cells (Gross et al. 2003), monocytes (Asea et al. 2000), macrophages (Sondermann et al. 2000) and dendritic cells (Asea et al. 2002) and some toll like receptors (TLR) 2, 4 and 7, CD14, CD40, CD36, CD91, Lectin-like oxidized low-density lipoprotein (LDL) receptor-1 (LOX-1) and macrophage scavenger receptor type A (SR-A) bind eHsp72 induce innate immune activation (Asea 2006; Whitham and Fortes 2008).

Also, a key potential role of eHsp72 in the interphase between innate and adaptive immunity by activation of CD40 on APC and CD40L on T cells had proposed in some studies (Caux et al. 1994; Whitham and Fortes 2008). Asea et al. (2000) showed that binding eHsp72 to monocytes activate nuclear factor kappa β (NF-kβ) and lead to up-regulation of pro-inflammatory cytokines. Significant increase in some cytokines such as Tumor necrosis factor-alpha (TNF-α), Interleukin-1β (IL-1β), Interleukin-6 (IL-6) and Interleukin-12 (IL-12) observed following exposure of APC to exogenous eHsp72 (Asea et al. 2000, 2002). In sum, the role of eHsp72 in activation of immune responses and also act as alert for immune system during times of stress has been approving chaperokine activity of eHsp72.

10.1.2 Exercise-Induced Extracellular Hsp70

Many genes and/or proteins are activated following acute physical exercise or after a prolonged exercise training period. It is shown that cellular changes that occurs after acute bout of exercise will finally induce adaptive response following exercise training. Numerous studies showed that many stressors associate with exercise such as increased body temperature (Mizzen and Welch 1988), oxidative stress, ischemia (Smolka et al. 2000), protein degradation (Chiang et al. 1989), hypoxia (Guttman et al. 1980), acidosis (Peart et al. 2011), and glucose deprivation (Febbraio et al. 2004) can induce intracellular Hsp expression in many tissues. Acute exercise increased intracellular expression of Hsp72 in variety of tissues such as heart (Salo et al. 1991), liver (Salo et al. 1991), brain (Walters et al. 1998), and skeletal muscle (Salo et al. 1991; Febbraio and Koukoulas 2000; Febbraio et al. 2002).

In addition, some studies have indicated that acute exercise increases Hsp72 in the peripheral circulation (Lancaster et al. 2004a, b; Marshall et al. 2006; Suzuki et al. 2006). Walsh et al. (2001) for the first time demonstrated that acute moderate exercise (60 min of aerobic exercise; 60% VO2 peak) can increase Hsp72 in the systemic circulation. Also, increasing in eHsp72 had observed following acute high or moderate intensity exercise in some other studies (Suzuki et al. 2006; Febbraio et al. 2004; Ogura et al. 2008). It seems that exercise in different types, intensity and duration affects the released Hsp72 following exercise (Lancaster et al. 2004a, b).

Additionally, some situations such as aging, infectious and inflammatory diseases impressed releasing Hsp72 following physical exercise. It is showed that following moderate exercise training eHsp72 activation in obese animals closed to those of the healthy lean ones (García et al. 2013). Also, activation of eHsp72 decreased in aging situation following exercise (Park et al. 2009). While eHsp70 levels increased about 2000% following an ironman triathlon and 100 km run (Suzuki et al. 2006; Gomez-Merino et al. 2006).

The source and effects of the released Hsp72 following physical exercise is not clear until now. While it is hypothesized that since skeletal muscle expresses Hsp72 following physical exercise, it was assumed that this rise was due to secretion from muscle. However, Walsh et al. (2001) observed increased levels of eHsp72 did not relate to muscle damage. Additionally, it is demonstrated that appearance of extracellular Hsp72 overtake the increase in skeletal muscle Hsp72 levels (Walsh et al. 2001). Current data have shown that main possible sources of eHsp72 during the physical exercise are lymphocytes (Heck et al. 2011), liver (Febbraio et al. 2002) and brain (Lancaster et al. 2004a, b) and skeletal muscles has minor role in the concentration of serum Hsp72 following exercise. Johnson et al. (2005) proposed that alpha-1-adrenoreceptors on the liver tissue have critical role in increased systemic Hsp72 levels. Additionally, Lancaster and Febbraio (2005) suggested that sympathoadrenergically innervated tissues that activated during physical exercise may release eHsp72.

Numerous studies proposed that body temperature is not the important reason in exercise-induced of eHsp72 (Whitham and Fortes 2008; Ruell et al. 2006). Many evidences proposed that some physiological changes induced by physical exercise and also environmental factors have rolled in increased levels of eHsp70 following exercise. Ogura et al. (2008) suggested that effects of body temperature on elevation of extracellular Hsp72 is induced when the other exercise stressor such as reactive oxygen species, hormones and metabolic disturbances combined with it. Whitham and Fortes (2008) suggested that about 86% of eHsp72 concentration was explained by changes in blood glucose. As we know decreased levels of blood glucose during physical exercise can stimulate glycogenolysis in the liver and this probably associated with some released factors from this tissue to maintain its fuel sources (Coggan 1991). During endurance physical exercise other fuel sources such as fat storage in skeletal muscle and adipose tissue can supply the energy sources (Horowitz and Klein 2000). Released eHsp72 from liver tissue can be a signal to other tissues such as skeletal muscle and adipose tissue to supply fat as a potent fuel during exercise (van Hall 2015). TLR2 and 4 as toll like receptors that binds with eHsp72 was found in adipocytes and myocytes (Lin et al. 2000; Lang et al. 2003). Also, Febbraio et al. (2004) demonstrated that glucose ingestion during exercise attenuates the circulating Hsp72. Additionally, Civitarese et al. 2005 showed that glucose ingestion during exercise decreases non-esterified fatty acids (NEFA) levels and lipid metabolism, and significant positive correlations were found between eHsp72 and NEFA. Périard et al. (2012)) proposed that increased metabolic demands and needing for energy conversion possibly stimulate eHsp72 during exercise in heat stress. However, because of less studies in this field further studies is required to determination the metabolic effects of eHsp72 during physical exercise.

10.1.2.1 Extracellular Hsp70 and Immune Function During Exercise

In recent years, strong effects of exercise on the immune system have been approving in many studies. Numerous studies showed that various parts of immune responses including, presenting antigens, neutrophils, natural killer cells, Toll-like receptors, major histocompatibility complex class II and anti and pro-inflammatory cytokines have changed during physical exercise (Gleeson 2006, 2007; Brolinson and Elliott 2007; Peake et al. 2017; Molanouri Shamsi et al. 2015). Also, it is demonstrated that these mentioned changes in immune cells dependent on intensity and duration of the activity performed. Some studies proposed that acute moderate physical exercise induced cell-mediated immunity and lead to a pro-inflammatory response with Th1 immune activation (Lancaster et al. 2004a, b; Gleeson 2007).

Although the exact role of eHsp72 in moderate or intensive exercise on the immune responses is not clear until now. It is proposed that extracellular Hsp72 as a stress protein released following acute exercise associated to the activation of the immune system (Heck et al. 2011; Krause and Rodrigues-Krause Jda 2011; Whitham and Fortes 2008; Asea 2003). Some studies suggested that eHsp72 especially may involve in modulation of innate/inflammatory immune responses during moderate exercise (Giraldo et al. 2010; Fleshner and Johnson 2005; Ortega et al. 2006). A single session of moderate exercise increased microbicide capacity of neutrophils (Ortega et al. 2006), and eHsp72 as a possible "danger signal" stimulates neutrophils' chemotaxis (Campisi et al. 2003). Activation of neutrophils by Hsp may act as a danger signal to induce a more effective response in macrophages (Zhen et al. 2004). Stimulatory effects of moderate exercise on macrophage phagocytosis (Kovalchin et al. 2006; Panjwani et al. 2002), and activation of macrophages phagocytosis by eHsp70 was observed in some studies (Panjwani et al. 2002).

Moreover, Horn et al. (2007) showed that following endurance exercise eHsp70 induced more recruitment of NK cells in heat situation. Additionally, with considering to effects of eHsp72 in cytokine production in immune cells (Johnson and Fleshner 2006; Asea et al. 2000). It is believed that eHsp72 have role in exercise-related inflammatory reaction (Asea 2003). Effects of eHsp72 on NK cell activation and cytokine production following acute moderate exercise can propose the role of eHsp72 in cellular arm of innate immunity.

On the other hand, many studies have shown that various immune cell function including neutrophil respiratory burst, lymphocyte proliferation, monocyte TLR, and major histocompatibility complex class II protein expression are temporarily impaired following acute bouts of continuous, prolonged (1.5 h), of moderate to high intensity exercise that lasts ~3 to 24 h after exercise (Nieman 1994; Pedersen and Bruunsgaard 1995; Ronsen et al. 2001; Gleeson 2007). Depression of immune function linked with prolonged exercise resulted increasing of susceptibility to infection by intracellular microorganisms (Gleeson 2007). Substantial increase in eHsp72 levels following acute bouts of prolonged, continuous heavy exercise especially in heat environment were observed. Extracellular Hsp72 concentration significantly increased immediately after an ironman triathlon and 100 km run (Suzuki et al. 2006; Gomez-Merino et al. 2006). Also, elevation of eHsp70 accrued immediately after high-intensity exercise (Bittencourt and Porto 2017; Murase et al. 2016). It is proposed that these changes in eHsp70 associated with depression of immune function for elite athletes (Heck et al. 2011).

Some studies proposed that unlike to moderate acute exercise that activate pro-inflammatory response, long term moderate or high intensity sessions of exercise in order to control skeletal muscle damage activate anti-inflammatory immune response (Lancaster et al. 2004a, b; LaVoy et al. 2017). In addition to pro-inflammatory and immune activation effects of eHsp70, anti-inflammatory effects and suppression of immunity with this factor with some changes in immune system such as inhibitory effects on maturation of DC and inducing interleukin-10 production and inhibits T-cell proliferation were observed in some studies (Motta et al. 2007; Theriault et al. 2006). It is proposed that using the Hsp vaccines at lower doses appeared to favor induction of Cytotoxic T lymphocytes (CTLs) over the immunoregulatory response (Udono and Srivastava 1993; Liu et al. 2009; Molanouri Shamsi et al. 2017b). It is probable the released higher doses of eHsp70 following prolonged, of moderate to high intensity exercise associated with immune suppression and induced anti-inflammatory effects. More studies in this field will be necessary to determine the relationships between doses of extracellular Hsp70 and immune responses following physical exercise in different duration and intensity.

In addition, skeletal muscle resident immune cell population that induce pro or anti-inflammatory phenotypes play a fundamental role in exercise training adaptation and repair and pathogen clearance (Pillon et al. 2013; Furrer et al. 2017; Molanouri Shamsi et al. 2014). In the case of exercise, muscle contraction induces IL-6 release to the circulation, which subsequently impacts on hepatic glucose production and adipose tissue lipolysis (Pedersen et al. 2001). It is proposed that regulated cross-talk between myofibers and immune cells induce paracrine/autocrine and contact interactions especially during exercise (Pillon et al. 2013; Furrer et al. 2017). Welc and Clanton (2013) proposed possible relations between extracellular Hsp and skeletal muscle released cytokines especially IL-6. Also, Lee et al. (2017) was observed similar changes in extracellular Hsp72 and IL-6 with whole body precooling after acute exercise. While, it is demonstrated that changes in skeletal muscle released cytokines such as IL-6 that introduced as myokine following exercise have rolled in exercise training adaptation. Double-edged sword effects of released IL-6 following exercise was proposed by Bente Pedersen's group in Copenhagen (Starkie et al. 2003; Gleeson 2007). Although inhibition of IL-6 response decreased lipolysis and expression of a number of metabolic genes in the exercised muscle (Starkie et al. 2003; Pilegaard et al. 2002). But, it is showed that decreased levels of release of IL-6 from contracting muscles with antioxidant supplementation and glucose consumption lead to limiting in exercise induced depression of immune function (Peters et al. 1993; Gleeson et al. 2004). Decreased levels of eHsp72 with using antioxidant and glucose ingestion following long-term exercise were observed in some studies (Febbraio et al. 2004; Lancaster and Febbraio 2005). Heck et al. (2011) proposed that moderate loads of exercise do increase intracellular Hsp70s in lymphocytes, whereas higher loads increase eHsp70 release from these cells, as stated above, suggest that eHsp70 may also function as an actual immune signal to target organs and this may be a physiological universal during exercise efforts. It is possible eHsp70 as a danger signal from immune system can modulate releasing IL-6 from skeletal muscle and indirectly play an important role in exercise training adaptation.

10.1.3 Possible Role of Exercise-Induced Extracellular Hsp70 in Chronic Diseases

Physical inactivity has been identified an independent risk factor for the occurrence and development of numerous diseases including obesity, type 2 diabetes, neurodegenerative complications, cancer and cardiovascular diseases (Apostolopoulos et al. 2014; Handschin and Spiegelman 2008; Pedersen 2011). Regular exercise is effective in increasing physical health and correcting some risk factors for the development of disease complications (Kujala 2004). Also, it is clear that, physical inactivity has been associated with immune system disorders and chronic inflammation (Booth et al. 2012; Knight 2012). It is showed that moderate physical exercise has direct effects on immune cells and these responses on immune system have beneficial outcomes on disease (Apostolopoulos et al. 2014). Immune regulatory effects of exercise through some released factors by exercise were proposed as one of important mechanism against occurrence and development of chronic disease (Ortega 2016). Increasing in extracellular Hsp70 and immunostimulatory and immunosuppressive activities of this factor were proposed in some clinical situation such as cancer, cardiovascular disease, diabetes, and neurodegenerative disorders (De Maio 2011; De Maio and Vazquez 2013). De Maio and Vazquez (2013) proposed that multiple role of extracellular Hsp in disease depended on the cellular milieu or type of target cell (De Maio and Vazquez 2013). Also, Campisi et al. (2003) proposed that elevated levels of Hsp72 stimulate immune function during times of increased risk of pathogenic challenge and tissue damage.

It is suggested that extracellular Hsp70 with acting via TLR2 induced cardiomyocyte inflammation and decreased contractility (Mathur et al. 2011). On the other hand, TLR2 deficiency may increase cardiac hypertrophy and ventricular remodeling after transverse aortic constriction (Bualeong et al. 2016). Lin and Knowlton (2014) demonstrated that a mild innate immune activation may control the cardiac injury and induce tissue repair. Numerous studies showed that moderate acute exercise result myocardial dysfunction and cardiac injury and activation of inflammatory cytokines act as possible mechanism in cardiomyocyte damage following exercise (Shave et al. 2010; Bishopric et al. 2001; Scharhag et al. 2008).

Arisi et al. (2017) proposed that mesenchymal stem cell activation may induce with cardiomyocyte damage in exercise. Protective effects of moderate exercise on heart diseases was demonstrated, but effects of released Hsp70 in relate to cardiac tissue is no obvious until now. As a hypothesis exercise induced eHsp70 with activation of inflammatory cytokines can roll in cardio protection by providing protection against myocardial dysfunction and infarction following exercise training (Hamilton et al. 2003; Radford et al. 1996).

Moreover, low-grade chronic inflammation is associated with cardiovascular, type-2 diabetes and some other chronic pathological diseases. It is showed that one of main mechanism in positive effects of regular physical activity is anti-inflammatory effects of exercise (Gleeson et al. 2011). Changed inflammatory milieu and activation of cytotoxic immune cells following acute physical exercise introduced as a mechanism for anti-inflammatory effects of regular exercise training (Gleeson et al. 2011; Hojman 2017). Asea (2003) suggested that exercise-related inflammatory reaction may regulate by eHsp72. This regulatory effect can result lower incidence and progression of some chronic disease such as cancer, cardiovascular, metabolic syndrome and type-2 diabetes. Moreover, immune stimulatory effects of eHsp70 especially in relate to innate/inflammatory immune responses in metabolic syndrome were observed in some studies (García et al. 2013; Ortega 2016). Release of IL-1β and IL-6 in macrophages from obese rats decrease in obese rats and physical exercise with stimulation of eHsp72-induced release of inflammatory cytokines in the obese animals to values closer to those of the healthy lean ones (García et al. 2013).

In addition to immune modulatory effects of exercise that lead to anti-inflammatory effects of exercise training. Immune stimulatory of acute physical exercise introduced as a possible mechanism in positive effects of regular physical exercise in chronic disease (Furrer et al. 2017; Apostolopoulos et al. 2014; Molanouri Shamsi et al. 2017a). It is proposed that exercise with appropriate intensity and duration can induce anti-tumor immune responses especially in pro-inflammatory associated cancers such as colon cancer and postmenopausal breast cancer (Pedersen 2011). As we know a main tumor-directed immune response involving cytolytic CD8+ T cells, Th1 cells and NK cells appears to protect against tumor development and progression (DeNardo and Coussens 2007; Gajewski et al. 2013). It is showed that extracellular Hsp70 can activate NK cells and increase cytotoxic action of NK cells in order to kill tumor cells (Gastpar et al. 2005). Communication between eHsp70 with innate and adaptive immune cells during physical exercise probably can induce anti-tumor immune responses. Moreover, although in some disease such as diabetes immune regulatory effects of exercise related to skeletal muscle released cytokines such as IL-6, this myokine had minor role in cancer protection (Hojman et al. 2011). More studies in this case would response to questions about eHsp70 and cancer and possible role of this chaperokine in immune system and tumor tissue relations in physical exercise.

From another perspective, in some situation such as neurodegenerative disorders, including spinal muscular atrophy, amyotrophic lateral sclerosis, Alzheimer's disease, and Parkinson's disease and also in aging decreased activation of heat shock proteins such as Hsp72 associated with neural degeneration (Adachi et al.

2009; Park et al. 2009). Glial cells have also been shown that to increase Hsp70 for protection to neighboring neurons (Vega et al. 2008). Extracellular Hsp such as Hsp70 have been observed to interact with amyloid β peptides, which are involved in the pathology of Alzheimer's disease (Carnini et al. 2012; Bruinsma et al. 2011; Evans et al. 2006). Luo et al. (2012) suggested that extracellular protect Schwann from hydrogen peroxide induced apoptosis. It is suggested that eHsp70 released during exercise can indeed aid the motoneurons against oxidative and other natures of stress, then the exercise could be used as an important strategy for the treatment of diseases that induce neurodegeneration and dysfunction (Krause and Rodrigues-Krause Jda 2011; Mehrholz et al. 2015; Cho et al. 2010)

From this view, assessment of different exercise intensities on releasing of eHsp70 in chronic disease that positive effects of regular physical exercise was observed about them can be consider for future studies. Also, effects of released eHsp70 following exercise on immune cells can present novel view in possible mechanisms that induce positive effects of exercise in chronic disease.

10.2 Conclusions

Extracellular Hsp7o as a danger signal can activate immune responses. It seems that releasing of eHsp7o following acute physical exercise and acting as a chaperokine would be one of mechanism in immune system activation during and after exercise and can help in exercise training adaptation. Moreover, it is showed that regular physical exercise is effective in the prevention of occurrence and development of numerous chronic diseases. Beneficial effects of regular exercise on immune responses in some disease such as cancer, cardiovascular complications, neuromuscular disorders and diabetes were observed. With considering to chaperokine effects of eHsp7o, this factor can be one of possible mechanisms in protective effects of regular physical exercise in chronic disease.

Acknowledgments This work was supported by the Research Center of Tarbiat Modarres University (TMU), Tehran, Iran. We wish to thank Professor Yaghob Fathoallahy for his kind help and sincere cooperation.

References

Adachi H, Katsuno M, Waza M, Minamiyama M, Tanaka F, Sobue G (2009) Heat shock proteins in neurodegenerative diseases: pathogenic roles and therapeutic implications. Int J Hyperth 25:647–654

Apostolopoulos V, Borkoles E, Polman R, Stojanovska L (2014) Physical and immunological aspects of exercise in chronic diseases. Immunotherapy 6(10):1145–1157

Arisi MF, Chirico EN, Sebeny R, Muthukumaran G, Mu A, De Jonghe BC, Margulies KB, Libonati JR (2017) Myocardial apoptosis and mesenchymal stem cells with acute exercise. Physiol Rep 5(11):e13297

Asea A (2003) Chaperokine-induced signal transduction pathways. Exerc Immunol Rev 9:25–33
Asea A (2005) Stress proteins and initiation of immune response: chaperokine activity of Hsp72. Exerc Immunol Rev 11:34–45
Asea A (2006) Initiation of the immune response by extracellular Hsp72: chaperokine activity of Hsp72. Curr Immunol Rev 2:209–215
Asea A, Kraeft SK, Kurt-Jones EA, Stevenson MA, Chen LB, Finberg RW, Koo GC, Calderwood SK (2000) HSP70 stimulates cytokine production through a CD14-dependant pathway, demonstrating its dual role as a chaperone and cytokine. Nat Med 6:435–442
Asea A, Rehli M, Kabingu E, Boch JA, Bare O, Auron PE, Stevenson MA, Calderwood SK (2002) Novel signal transduction pathway utilized by extracellular HSP70: role of toll-like receptor (TLR)2 and TLR4. J Biol Chem 277:15028–15034
Beckmann RP, Lovett M, Welch WJ (1992) Examining the function and regulation of hsp 70 in cells subjected to metabolic stress. J Cell Biol 117(6):1137–1150
Bishopric NH, Andreka P, Slepak T, Webster KA (2001) Molecular mechanisms of apoptosis in the cardiac myocyte. Curr Opin Pharmacol 1:141–150
Bittencourt A, Porto RR (2017) eHSP70/iHSP70 and divergent functions on the challenge: effect of exercise and tissue specificity in response to stress. Clin Physiol Funct Imaging 37(2):99–105
Booth FW, Roberts CK, Laye MJ (2012) Lack of exercise is a major cause of chronic diseases. Compr Physiol 2(2):1143–1211
Brolinson PG, Elliott D (2007) Exercise and the immune system. Clin Sports Med 26(3):311–319
Bruinsma IB, de Jager M, Carrano A, Versleijen AA, Veerhuis R, Boelens W, Rosemuller AJ, de Waal RM, Verbeek MM (2011) Small heat shock proteins induce a cerebral inflammatory reaction. J Neurosci 31:11992–12000
Bualeong T, Kebir S, Hof D, Goelz L, Graewe M, Ehrentraut SF, Knuefermann P, Baumgarten G, Meyer R, Ehrentraut H (2016) Tlr2 deficiency does not limit the development of left ventricular hypertrophy in a model of transverse aortic constriction induced pressure overload. J Negat Results Biomed 25(15):9
Bukau B, Horwich AL (1998) The Hsp70 and Hsp60 chaperone machines. Cell 92:351–366
Campisi J, Leem TH, Fleshner M (2003) Stress-induced extracellular Hsp72 is a functionally significant danger signal to the immune system. Cell Stress Chaperones 8(3):272–286
Carnini A, Scott LO, Ahrendt E, Proft J, Winkfein RJ, Kim SW, Colicos MA, Braun JE (2012) Cell line specific modulation of extracellular abeta42 by Hsp40. PLoS One 7:e37755
Caux C, Massacrier C, Vanbervliet B, Dubois B, Van Kooten C, Durand I, Banchereau J (1994) Activation of human dendritic cells through CD40 cross-linking. J Exp Med 180:1263–1272
Chiang HL, Terlecky SR, Plant CP, Dice JF (1989) A role for a 70-kilodalton heat shock protein in lysosomal degradation of intracellular proteins. Science 246(4928):382–385
Cho JY, Um HS, Kang EB, Cho IH, Kim CH, Cho JS, Hwang DY (2010) The combination of exercise training and alpha-lipoic acid treatment has therapeutic effects on the pathogenic phenotypes of Alzheimer's disease in NSE/APPsw-transgenic mice. Int J Mol Med 25(3):337–346
Civitarese AE, Hesselink MKC, Russell AP, Ravussin E, Schrauwen P (2005) Glucose ingestion during exercise blunts exercise-induced gene expression of skeletal muscle fat oxidative genes. Am J Phys 289:E1023–E1029
Clayton A, Turkes A, Navabi H, Mason MD, Tabi Z (2005) Induction of heat shock proteins in B-cell exosomes. J Cell Sci 118:3631–3638
Coggan AR (1991) Plasma glucose metabolism during exercise in humans. Sports Med 11(2):102–124
De Maio A (2011) Extracellular heat shock proteins, cellular export vesicles, and the Stress Observation System: a form of communication during injury, infection, and cell damage. Cell Stress Chaperones 16:235–249
De Maio A, Vazquez D (2013) Extracellular heat shock proteins: a new location, a new function. Shock 40(4):239–246
De Nardo DG, Coussens LM (2007) Inflammation and breast cancer. Balancing immune response: crosstalk between adaptive and innate immune cells during breast cancer progression. Breast Cancer Res 9(4):212

Evans CG, Wisen S, Gestwicki JE (2006) Heat shock proteins 70 and 90 inhibit early stages of amyloid beta-(1-42) aggregation in vitro. J Biol Chem 281:33182–33191
Febbraio MA, Koukoulas I (2000) HSP72 gene expression progressively increases in human skeletal muscle during prolonged, exhaustive exercise. J Appl Physiol (1985) 89(3):1055–1060
Febbraio MA, Ott P, Nielsen HB, Steensberg A, Keller C, Krustrup P, Secher NH, Pedersen BK (2002) Exercise induces hepatosplanchnic release of heat shock protein 72 in humans. J Physiol 544:957–962
Febbraio MA, Mesa JL, Chung J, Steensberg A, Keller C, Nielsen HB, Krustrup P, Ott P, Secher NH, Pedersen BK (2004) Glucose ingestion attenuates the exercise-induced increase in circulating heat shock protein 72 and heat shock protein 60 in humans. Cell Stress Chaperones 9(4):390–396
Fleshner M, Johnson JD (2005) Endogenous extra-cellular heat shock protein 72: releasing signal(s) and function. Int J Hyperth 21:457–471
Furrer R, Eisele PS, Schmidt A, Beer M, Handschin C (2017) Paracrine cross-talk between skeletal muscle and macrophages in exercise by PGC-1α-controlled BNP. Sci Rep 7:40789
Gajewski TF, Schreiber H, Fu YX (2013) Innate and adaptive immune cells in the tumor microenvironment. Nat Immunol 14:1014–1022
García JJ, Martín-Cordero L, Hinchado MD, Bote ME, Ortega E (2013) Effects of habitual exercise on the eHsp72-induced release of inflammatory cytokines by macrophages from obese Zucker rats. Int J Sports Med 34:559–564
Gastpar R, Gehrmann M, Bausero MA, Asea A, Gross C, Schroeder JA, Multhoff G (2005) Heat shock protein 70 surface-positive tumor exosomes stimulate migratory and cytolytic activity of natural killer cells. Cancer Res 65:5238–5247
Genth-Zotz S, Bolger AP, Kalra PR, Haehling S, Doehner W, Coats A, Volk HD, Anker SD (2004) Heat shock protein 70 in patients with chronic heart failure: relation to disease severity and survival. Int J Cardiol 96:397–401
Giraldo E, Martin-Cordero L, García JJ, Gehrmann M, Multhoff G, Ortega E (2010) Exercise-induced extracellular 72 kDa heat shock protein (Hsp72) stimulates neutrophil phagocytic and fungicidal capacities via TLR-2. Eur J Appl Physiol 108:217–225
Gleeson M (2006) Immune system adaptation in elite athletes. Curr Opin Clin Nutr Metab Care 9(6):659–665
Gleeson M (2007) Immune function in sport and exercise. J Appl Physiol 103(2):693–699
Gleeson M, Nieman DC, Pedersen BK (2004) Exercise, nutrition and immune function. J Sports Sci 22:115–125
Gleeson M, Bishop NC, Stensel DJ, Lindley MR, Mastana SS, Nimmo MA (2011) The anti-inflammatory effects of exercise: mechanisms and implications for the prevention and treatment of disease. Nat Rev Immunol 11(9):607–615
Gomez-Merino D, Drogou C, Guezennec CY, Burnat P, Bourrilhon C, Tomaszewski A, Milhau S, Chennaoui M (2006) Comparison of systemic cytokine responses after a long distance triathlon and a 100-km run: relationship to metabolic and inflammatory processes. Eur Cytokine Netw 17(2):117–124
Gross C, Hansch D, Gastpar R, Multhoff G (2003) Interaction of heat shock protein 70 peptide with NK cells involves the NK receptor CD94. Biol Chem 384:267–279
Guttman R, Lewis S, Rinzel J (1980) Control of repetitive firing in squid axon membrane as a model for a neuroneoscillator. J Physiol 305:377–395
Hamilton KL, Staib JL, Phillips T, Hess A, Lennon SL, Powers SK (2003) Exercise, antioxidants, and HSP72: protection against myocardial ischemia/reperfusion. Free Radic Biol Med 34(7):800–809
Handschin C, Spiegelman BM (2008) The role of exercise and PGC1α in inflammation and chronic disease. Nature 454(7203):463–469
Hashemi SM, Hassan ZM, Soudi S, Shahabi S (2008) The effect of vaccination with the lysate of heat-shocked tumor cells on nitric oxide production in BALB/c mice with fibrosarcoma tumor. Cell Biol Int 32(7):835–840

Haworth R, Platt N, Keshav S, Hughes D, Darley E, Suzuki H, Kurihara Y, Kodama T, Gordon S (1997) The macrophage scavenger receptor type A is expressed by activated macrophages and protects the host against lethal endotoxic shock. J Exp Med 186:1431–1439

Heck TG, Scholer CM, Bittencourt PIH (2011) Hsp70 expression: does it a novel fatigue signalling factor from immune system to brain? Cell Biochem Funct 29:215–226

Henstridge DC, Febbraio MA, Hargreaves M (2016) Heat shock proteins and exercise adaptations. Our knowledge thus far and the road still ahead. J Appl Physiol 120(6):683–691

Hojman P (2017) Exercise protects from cancer through regulation of immune function and inflammation. Biochem Soc Trans 45(4):905–911

Hojman P, Dethlefsen C, Brandt C, Hansen J, Pedersen L, Pedersen BK (2011) Exercise-induced muscle-derived cytokines inhibit mammary cancer cell growth. Am J Physiol Endocrinol Metab 301(3):E504–E510

Horn P, Kalz A, Lim CL, Pyne D, Saunders P, Mackinnon L, Peake J, Suzuki K (2007) Exercise-recruited NK cells display exercise-associated eHSP-70. Exerc Immunol Rev 13:100–111

Horowitz JF, Klein S (2000) Lipid metabolism during endurance exercise. Am J Clin Nutr 72(2 Suppl):558S–563S

Johnson JD, Fleshner M (2006) Releasing signals, secretory pathways, and immune function of endogenous extracellular heat shock protein 72. J Leukoc Biol 79:425–434

Johnson JD, Campisi J, Sharkey CM, Kennedy SL, Nickerson M, Fleshner M (2005) Adrenergic receptors mediate stress-induced elevations in extracellular Hsp72. J Appl Physiol 99:1789–1795

Knight JA (2012) Physical inactivity: associated diseases and disorders. Ann Clin Lab Sci 42(3):320–337

Kovalchin JT, Wang R, Wagh MS, Azoulay J, Sanders M, Chandawarkar RY (2006) In vivo delivery of heat shock protein 70 accelerates wound healing by up-regulating macrophage-mediated phagocytosis. Wound Repair Regen 14(2):129–137

Krause M, Rodrigues-Krause Jda C (2011) Extracellular heat shock proteins (eHSP70) in exercise: possible targets outside the immune system and their role for neurodegenerative disorders treatment. Med Hypotheses 76(2):286–290

Kujala UM (2004) Evidence for exercise therapy in the treatment of chronic disease based on at least three randomized controlled trials – summary of published systematic reviews. Scand J Med Sci Sports 14(6):339–345

Lancaster GI, Febbraio MA (2005) Mechanisms of stress-induced cellular HSP72 release: implications for exercise-induced increases in extracellular HSP72. Exerc Immunol Rev 11:46–52

Lancaster GI, Halson SL, Khan Q, Drysdale P, Jeukendrup AE, Drayson MT, Gleeson M (2004a) The effects of acute exhaustive exercise and intensified training on type 1/type 2 T cell distribution and cytokine production. Exerc Immunol Rev 10:91–106

Lancaster GI, Moller K, Nielsen B, Secher NH, Febbraio MA, Nybo L (2004b) Exercise induces the release of heat shock protein 72 from the human brain in vivo. Cell Stress Chaperones 9:276–280

Lang CH, Silvis C, Deshpande N, Nystrom G, Frost RA (2003) Endotoxin stimulates in vivo expression of inflammatory cytokines tumor necrosis factor alpha, interleukin-1beta, -6, and high-mobility-group protein-1 in skeletal muscle. Shock 19:538–546

LaVoy EC, Hussain M, Reed J, Kunz H, Pistillo M, Bigley AB, Simpson RJ (2017) T-cell redeployment and intracellular cytokine expression following exercise: effects of exercise intensity and cytomegalovirus infection. Physiol Rep 5(1):e13070

Lee BJ, Clarke ND, Hankey J, Thake CD (2017) Whole body precooling attenuates the extracellular HSP72, IL-6 and IL-10 responses after an acute bout of running in the heat. J Sports Sci 5:1–8

Lehner T, Wang Y, Whittall T, McGowan E, Kelly CG, Singh M (2004) Functional domains of HSP70 stimulate generation of cytokines and chemokines, maturation of dendritic cells and adjuvanticity. Biochem Soc Trans 32:629–632

Lin L, Knowlton AA (2014) Innate immunity and cardiomyocytes in ischemic heart disease. Life Sci 100(1):1–8

Lin Y, Lee H, Berg AH, Lisanti MP, Shapiro L, Scherer PE (2000) The lipopolysaccharide-activated toll-like receptor (TLR)-4 induces synthesis of the closely related receptor TLR-2 in adipocytes. J Biol Chem 275(32):24255–24263

Liu Z, Li X, Qiu L, Zhang X, Chen L, Cao S, Wang F, Meng S (2009) Treg suppress CTL responses upon immunization with HSP gp96. Eur J Immunol 39(11):3110–3120

Locke M, Noble EG (1995) Stress proteins: the exercise response. Can J Appl Physiol 20(2):155–167

Luo X, Tao L, Lin P, Mo X, Chen H (2012) Extracellular heat shock protein 72 protects schwann cells from hydrogen peroxide-induced apoptosis. J Neurosci Res 90:1261–1269

Madden LA, Sandstrom ME, Lovell RJ, McNaughton L (2008) Inducible heat shock protein 70 and its role in preconditioning and exercise. Amino Acids 34:511–516

Marshall HC, Ferguson RA, Nimmo MA (2006) Human resting extracellular heat shock protein 72 concentration decreases during the initial adaptation to exercise in a hot, humid environment. Cell Stress Chaperones 11:129–134

Mathur S, Walley KR, Wang Y, Indrambarya T, Boyd JH (2011) Extracellular heat shock protein 70 induces cardiomyocyte inflammation and contractile dysfunction via TLR2. Circ J 75(10):2445–2452

Matzinger P, Fuchs EJ (1996) Beyond self and non-self: immunity is a conversation, not a war. J NIH Res 8:35–39

Mehrholz J, Kugler J, Storch A, Pohl M, Elsner B, Hirsch K (2015) Treadmill training for patients with Parkinson's disease. Cochrane Database Syst Rev 22(8):CD007830

Mizzen LA, Welch WJ (1988) Characterization of the thermotolerant cell. I. Effects on protein synthesis activity and the regulation of heat-shock protein 70 expression. J Cell Biol 106:1105–1116

Moehler MH, Zeidler M, Wilsberg V, Cornelis JJ, Woelfel T, Rommelaere J, Galle PR, Heike M (2005) Parvovirus H-1-induced tumor cell death enhances human immune response in vitro via increased phagocytosis, maturation, and cross-presentation by dendritic cells. Hum Gene Ther 16:996–1005

Molanouri Shamsi M, Hassan ZH, Gharakhanlou R, Quinn LS, Azadmanesh K, Baghersad L, Isanejad A, Mahdavi M (2014) Expression of interleukin-15 and inflammatory cytokines in skeletal muscles of STZ-induced diabetic rats: effect of resistance exercise training. Endocrine 46(1):60–69

Molanouri Shamsi M, Hassan ZM, Quinn LS, Gharakhanlou R, Baghersad L, Mahdavi M (2015) Time course of IL-15 expression after acute resistance exercise in trained rats: effect of diabetes and skeletal muscle phenotype. Endocrine 49(2):396–403

Molanouri Shamsi M, Mahdavi M, Quinn LS, Gharakhanlou R, Isanegad A (2016) Effect of resistance exercise training on expression of Hsp70 and inflammatory cytokines in skeletal muscle and adipose tissue of STZ-induced diabetic rats. Cell Stress Chaperones 21(5):783–791

Molanouri Shamsi M, Chekachak S, Soudi S, Quinn LS, Ranjbar K, Chenari J, Yazdi MH, Mahdavi M (2017a) Combined effect of aerobic interval training and selenium nanoparticles on expression of IL-15 and IL-10/TNF-α ratio in skeletal muscle of 4T1 breast cancer mice with cachexia. Cytokine 90:100–108

Molanouri Shamsi M, Najedi S, Hassan ZM, Isanejad A, Mahdavi M (2017b) Short term exercise training enhances cell-mediated responses to HSV-1 vaccine in mice. Microb Pathog 110:457–463

Monninkhof EM, Elias SG, Vlems FA, van der Tweel I, Schuit AJ, Voskuil DW, van Leeuwen FE (2007) Physical activity and breast cancer: a systematic review. Epidemiology 18:137–157

Moseley P (2000) Stress proteins and the immune response. Immunopharmacology 48(3):299–302

Motta A, Schmitz C, Rodrigues L, Ribeiro F, Teixeira C, Detanico T, Bonan C, Zwickey H, Bonorino C (2007) Mycobacterium tuberculosis heat-shock protein 70 impairs maturation of dendritic cells from bone marrow precursors, induces interleukin-10 production and inhibits T-cell proliferation in vitro. Immunology 121:462–472

Multhoff G, Botzler C, Wiesnet M, Müller E, Meier T, Wilmanns W, Issels RD (1995) A stress-inducible 72-kDa heat-shock protein (HSP72) is expressed on the surface of human tumor cells, but not on normal cells. Int J Cancer 61(2):272–279

Murase Y, Shimizu K, Tanimura Y, Hanaoka Y, Watanabe K, Kono I, Miyakawa S (2016) Salivary extracellular heat shock protein 70 (eHSP70) levels increase after 59 min of intense exercise and correlate with resting salivary secretory immunoglobulin A (SIgA) levels at rest. Cell Stress Chaperones 21(2):261–269

Nakata N, Kato H, Kogure K (1993) Inhibition of ischaemic tolerance in the gerbil hippocampus by quercetin and anti-heat shock protein-70 antibody. Neuroreport 4:695–698

Nieman DC (1994) Exercise, infection and immunity. Int J Sports Med 15:S131–S141

Noble EG, Milne KJ, Melling CW (2008) Heat shock proteins and exercise: a primer. Appl Physiol Nutr Metab 33:1050–1065

Nocon M, Hiemann T, Muller-Riemenschneider F, Thalau F, Roll S, Willich SN (2008) Association of physical activity with all-cause and cardiovascular mortality: a systematic review and meta-analysis. Eur J Cardiovasc Prev Rehabil 15:239–246

Ogura Y, Naito H, Akin S, Ichinoseki-Sekine N, Kurosaka M, Kakigi R, Sugiura T, Powers SK, Katamoto S, Demirel HA (2008) Elevation of body temperature is an essential factor for exercise-increased extracellular heat shock protein 72 level in rat plasma. Am J Physiol Regul Integr Comp Physiol 294(5):R1600–R1607

Ortega E (2016) The "bioregulatory effect of exercise" on the innate/inflammatory responses. J Physiol Biochem 72(2):361–369

Ortega E, Giraldo E, Hinchado MD, Martínez M, Ibáñez S, Cidoncha A, Collazos ME, García JJ (2006) Role of Hsp72 and norepinephrine in the moderate exercise-induced stimulation of neutrophils' microbicide capacity. Eur J Appl Physiol 98(3):250–255

Panjwani NN, Popova L, Srivastava PK (2002) Heat shock proteins gp96 and hsp70 activate the release of nitric oxide by APCs. J Immunol 168(6):2997–3003

Park SW, Goodpaster BH, Lee JS, Kuller LH, Boudreau R, de Rekeneire N, Harris TB, Kritchevsky S, Tylavsky FA, Nevitt M, Cho YW, Newman AB, Health, Aging, and Body Composition Study (2009) Excessive loss of skeletal muscle mass in older adults with type 2 diabetes. Diabetes Care 32(11):1993–1997

Peake JM, Neubauer O, Walsh NP, Simpson RJ (2017) Recovery of the immune system after exercise. J Appl Physiol 122(5):1077–1087

Peart DJ, McNaughton LR, Midgley AW, Taylor L, Towlson C, Madden LA, Vince RV (2011) Pre-exercise alkalosis attenuates the heat shock protein 72 response to a single-bout of anaerobic exercise. J Sci Med Sport 14(5):435–440

Pedersen BK (2011) Exercise-induced myokines and their role in chronic diseases. Brain Behav Immun 25(5):811–816

Pedersen BK, Bruunsgaard H (1995) How physical exercise influences the establishment of infections. Sports 19:393–400

Pedersen BK, Steensberg A, Fischer C, Keller C, Ostrowski K, Schjerling P (2001) Exercise and cytokines with particular focus on musclederived IL-6. Exerc Immunol Rev 7:18–31

Périard JD, Ruell P, Caillaud C, Thompson MW (2012) Plasma Hsp72 (HSPA1A) and Hsp27 (HSPB1) expression under heat stress: influence of exercise intensity. Cell Stress Chaperones 17(3):375–383

Peters EM, Goetzsche JM, Grobbelaar B, Noakes TD (1993) Vitamin C supplementation reduces the incidence of post-race symptoms of upper respiratory tract infection in ultramarathonrunners. Am J Clin Nutr 57:170–174

Pilegaard H, Keller C, Steensberg A, Helge JW, Pedersen BK, Saltin B, Neufer PD (2002) Influence of pre-exercise muscle glycogen content on exercise-induced transcriptional regulation of metabolic genes. J Physiol 541:261–271

Pillon NJ, Bilan PJ, Fink LN, Klip A (2013) Cross-talk between skeletal muscle and immune cells: muscle-derived mediators and metabolic implications. Am J Physiol Endocrinol Metab 304(5):E453–E465

Radford NB, Fina M, Benjamin IJ, Moreadith RW, Graves KH, Zhao P, Gavva S, Wiethoff A, Sherry AD, Malloy CR, Williams RS (1996) Cardioprotective effects of 70-kDa heat shock protein in transgenic mice. Proc Natl Acad Sci 93:2339–2342

Rodrigues-Krause J, Krause M, O'Hagan C, De Vito G, Boreham C, Murphy C, Newsholme P, Colleran G (2012) Divergence of intracellular and extracellular HSP72 in type 2 diabetes: does fat matter? Cell Stress Chaperones 17(3):293–302

Ronsen O, Pedersen BK, Oritsland TR, Bahr R, Kjeldsen-Kragh J (2001) Leukocyte counts and lymphocyte responsiveness associated with repeated bouts of strenuous endurance exercise. J Appl Physiol 91:425–434

Rubio E, Valenciano AI, Segundo C, Sanchez N, de Pablo F, de la Rosa EJ (2002) Programmed cell death in the neurulating embryo is prevented by the chaperone heat shock cognate 70. Eur J Neurosci 15:1646–1654

Ruell PA, Thompson MW, Hoffman KM, Brotherhood JR, Richards DA (2006) Plasma Hsp72 is higher in runners with more serious symptoms of exertional heat illness. Eur J Appl Physiol 97:732–736

Salo DC, Donovan CM, Davies KJ (1991) HSP70 and other possible heat shock or oxidative stress proteins are induced in skeletal muscle, heart, and liver during exercise. Free Radic Biol Med 11(3):239–246

Scharhag J, George K, Shave R, Urhausen A, Kindermann W (2008) Exercise-associated increases in cardiac biomarkers. Med Sci Sports Exerc 40(8):1408–1415

Shave R, Baggish A, George K, Wood M, Scharhag J, Whyte G, Gaze D, Thompson PD (2010) Exercise-induced cardiac troponin elevation: evidence, mechanisms, and implications. J Am Coll Cardiol 56(3):169–176

Smolka MB, Zoppi CC, Alves AA, Silveira LR, Marangoni S, Pereira-Da-Silva L, Novello JC, Macedo DV (2000) HSP72 as a complementary protection against oxidative stress induced by exercise in the soleus muscle of rats. Am J Physiol Regul Integr Comp Physiol 279(5):R1539–R1545

Sondermann H, Becker T, Mayhew M, Wieland F, Hartl FU (2000) Characterization of a receptor for heat shock protein 70 on macrophages and monocytes. Biol Chem 381(12):1165–1174

Starkie R, Ostrowski SR, Jauffred S, Febbraio M, Pedersen BK (2003) Exercise and IL-6 infusion inhibit endotoxin-induced TNF-alpha production in humans. FASEB J 17(8):884–886

Suzuki K, Peake J, Nosaka K, Okutsu M, Abbiss CR, Surriano R, Bishop D, Quod MJ, Lee H, Martin DT, Laursen PB (2006) Changes in markers of muscle damage, inflammation and HSP70 after an Ironman triathlon race. Eur J Appl Physiol 98:525–534

Theriault JR, Adachi H, Calderwood SK (2006) Role of scavenger receptors in the binding and internalization of heat shock protein 70. J Immunol 177:8604–8611

Triantafilou M, Triantafilou K (2004) Heat-shock protein 70 and heat-shock protein 90 associate with Toll-like receptor 4 in response to bacterial lipopolysaccharide. Biochem Soc Trans 32(Pt 4):636–639

Tuomilehto J, Lindstrom J, Eriksson JG, Valle TT, Hamalainen H, Ilanne-Parikka P, Keinanen-Kiukaanniemi S, Laakso M, Louheranta A, Rastas M, Salminen V, Uusitupa M (2001) Prevention of type 2 diabetes mellitus by changes in lifestyle among subjects with impaired glucose tolerance. N Engl J Med 344:1343–1350

Udono H, Srivastava PK (1993) Heat shock protein 70-associated peptides elicit specific cancer immunity. J Exp Med 178:1391–1396

van Eden W, van der Zee R, Prakken B (2005) Heat-shock proteins induce T-cell regulation of chronic inflammation. Nat Rev Immunol 5(4):318–330

van Hall G (2015) The physiological regulation of skeletal muscle fatty acid supply and oxidation during moderate-intensity exercise. Sports Med 1:S23–S32

Vega VL, Rodriguez-Silva M, Frey T, Gehrmann M, Diaz JC, Steinem C, Multhoff G, Arispe N, De Maio A (2008) Hsp70 translocates into the plasma membrane after stress and is released into the extracellular environment in a membrane-associated form that activates macrophages. J Immunol 180:4299–4307

Walsh RC, Koukoulas I, Garnham A, Moseley PL, Hargreaves M, Febbraio MA (2001) Exercise increases serum Hsp72 in humans. Cell Stress Chaperones 6:386–393

Walters TJ, Ryan KL, Tehrany MR, Jones MB, Paulus LA, Mason PA (1998) HSP70 expression in the CNS in response to exercise and heat stress in rats. J Appl Physiol 84(4):1269–1277

Wang Y, Whittall T, McGowan E, Younson J, Kelly C, Bergmeier LA, Singh M, Lehner T (2005) Identification of stimulating and inhibitory epitopes within the heat shock protein 70 molecule that modulate cytokine production and maturation of dendritic cells. J Immunol 174:3306–3316

Welc SS, Clanton TL (2013) The regulation of interleukin-6 implicates skeletal muscle as an integrative stress sensor and endocrine organ. Exp Physiol 98(2):359–371

Weng D, Song B, Koido S, Calderwood SK, Gong J (2013) Immunotherapy of radioresistant mammary tumors with early metastasis using molecular chaperone vaccines combined with ionizing radiation. J Immunol 191:755–763

Whitham M, Fortes MB (2008) Heat shock protein 72: release and biological significance during exercise. Front Biosci 13:1328–1339

Yu X, Guo C, Yi H, Qian J, Fisher PB, Subjeck JR, Wang XY (2013) A multifunctional chimeric chaperone serves as a novel immune modulator inducing therapeutic antitumor immunity. Cancer Res 73(7):2093–2103

Zhen L, He M, Long M, Blomgran R, Stendahl O (2004) Pathogen-induced apoptotic neutrophils express heat shock proteins and elicit activation of human macrophages. J Immunol 173:6319–6326

Chapter 11
Therapeutic Implications of Heat Shock Proteins in Cancer

Maxim Shevtsov and Gabriele Multhoff

Abstract Evolutionary conserved members of the main heat shock protein families (HSP) are highly overexpressed in many tumor cells where they contribute to protein homeostasis, cytoprotection, cell survival, tumor progression and therapy resistance. Upon cell proliferation and differentiation but also in response to a large variety of different stressors, including anticancer radio- and/or chemotherapy, the synthesis of stress-inducible HSP is further upregulated. Therefore, the inhibition of members of the HSP90, HSP70 and small heat shock proteins (i.e., HSP20, clusterin) in cancer cells has emerged as a promising approach in translational and clinical oncology to break therapy resistance. Moreover, HSP have been found to play key roles in eliciting anti-tumor immune responses by acting as carriers for tumor-specific immunogenic peptides, chaperokines, adjuvants for antigen presentation, or targets for the innate immune system. HSP-based immunotherapies have shown promising results in the therapy of certain subtypes of tumors.

Keywords Adaptive immunity · Cancer vaccine · Clusterin · Geldanamycin · Heat shock proteins · HSP20 · HSP70 · HSP90-families · Immunotherapy · Radicicol

M. Shevtsov (✉)
Center for Translational Cancer Research (TranslaTUM), Radiation Immuno Oncology Group, Technische Universität München (TUM), Munich, Germany

Institute of Cytology of the Russian Academy of Sciences (RAS), St. Petersburg, Russia

Pavlov First Saint Petersburg State Medical University, St. Petersburg, Russia

Polenov Russian Scientific Research Institute of Neurosurgery, St. Petersburg, Russia
e-mail: maxim.shevtsov@tum.de

G. Multhoff
Center for Translational Cancer Research (TranslaTUM), Radiation Immuno Oncology Group, Technische Universität München (TUM), Munich, Germany

A. A. A. Asea, P. Kaur (eds.), *Chaperokine Activity of Heat Shock Proteins*, Heat Shock Proteins 16, https://doi.org/10.1007/978-3-030-02254-9_11

Abbreviations

15-DSG	15-deoxyspergualin
17-AAG	17-(allylamino)-17-demethoxygeldanamycin
17-DMAG	17-desmethoxy-17-*N*,*N*-dimethylaminoethylaminogeldanamycin
ADP	Adenosine diphosphate
Aha1	Activator of Hsp90 ATPase
APCs	Antigen presenting cells
ATP	Adenosine triphosphate
Az	Apoptozole
Cdc37	Cell cycle division protein 37
Clever-1	Common lymphatic endothelial and vascular endothelial receptor-1
CLL	Chronic lymphocytic leukemia
CML	Chronic myelogenous leukemia
EGCG	Epigallocatechin gallate
EGF	Epidermal growth factor
EGFR	Epidermal growth factor receptor
FEEL-1	Fascilin EGF-like scavenger receptor-1
GDA	Geldanamycin
GIST	Gastrointestinal stromal tumors
HCC	Hepatocellular carcinoma
HDAC	Histone deacetylase family
HER2	Human epidermal growth factor receptor 2
HoP	HSP70/HSP90 organizing protein
HSF-1	Heat shock factor 1
HSP	Heat shock protein (family)
IFN-γ	Interferon gamma
IL	Interleukin
IPI-504	17-allylamino-17-demethoxygeldanamycin hydroquinone hydrochloride
LOX-1	Low-density lipoprotein receptor-1
MHC	Major histocompatibility complex
MKT-077	1-ethyl-2-[[3-ethyl-5-(3-methylbenzothiazolin2-yliden)]-4-oxothiazolidin-2-ylidenemethyl] pyridinium chloride
mTOR	Mechanistic target of rapamycin
NEF	Nucleotide exchange factor
NF-κB	Nuclear factor kappa-light-chain-enhancer of activated B cells
NK cells	Natural killer cells
NO	Nitric oxide
PD	Platycodin D
PES	Phenylethynesulfonamide (pifithrin-μ)
RDC	Radicicol
ROS	Reactive oxygen species
sCLU	Small clusterin
SREC-1	Scavenger receptor expressed by endothelial cells-1
Stab-1	Stabilin-1
TLR	Toll-like receptor

11.1 Introduction

Heat shock proteins (HSP) are highly conserved proteins that are present in all nucleated cells (Lindquist 1986). According to their molecular weights HSP can be subdivided into different families: small HSP (i.e., HSP20), HSP40, HSP60, HSP70, HSP90, and HSP110. HSP play key roles in protein homeostasis such as folding/unfolding of proteins, assembly of nascent polypeptides, preventing protein aggregation, formation of multiprotein complexes, transport of proteins across membranes, and protection of cells against various stress-induced apoptosis. Furthermore, some HSP (especially members of the HSP90 family) are important for cell signaling processes and protein trafficking of regulatory kinases (Pearl 2005; Horejsi et al. 2010; Takai et al. 2010), transcription factors (Pratt 1998) and steroid hormone receptors (Pratt et al. 2006). Due to their high metabolic requirements and stressful microenvironment (including hypoxia, low pH, nutritional deficiency) cancer cells often show an upregulated expression of HSP which is associated with poor prognosis and a reduced overall survival of cancer patients. Therefore, the inhibition of HSP provides a promising strategy (Jego et al. 2010) to sensitize therapy-resistant tumor cells to standard therapies such as radio- and/or chemotherapy. Up-to-date Hsp90 inhibitors which are often based on geldanamycin derivatives have demonstrated therapeutic potency in the treatment of lung and breast cancer, and haematological diseases (Jego et al. 2010). Inhibitors of members of the HSP70 and small heat shock protein families have been applied more recently in ongoing clinical trials.

The use of HSP as modulators of the immune response provides another therapeutic approach to fight cancer. HSP with molecular weights of approximately 70 and 90 kDa have the capacity to stimulate anti-tumor immune responses either as carriers for antigenic peptides, which can be cross-presented by major histocompatibility complex (MHC) class I molecules, or as natural immunogens (Srivastava et al. 1986; Ullrich et al. 1986; Srivastava 2002a, b; Li et al. 2002). Depending on the availability of ADP and ATP, members of the HSP70 family have the capacity to bind and release tumor-specific peptides. Following cross-presentation on MHC class I molecules, CD8+ cytotoxic T cell mediated immune responses can be initiated against the tumor. Preclinical models revealed that vaccination with HSP70 or HSP90-chaperoned peptides which were purified from tumor, but not normal cells, are able to mediate specific and protective immunity against autologous tumors. In recent years, a large number of receptors, including the alpha-2 macroglobulin receptor CD91 (Binder et al. 2000a, b), lectin-like oxidized low-density lipoprotein receptor-1 (LOX-1) (Gong et al. 2009; Tosti et al. 2014), scavenger receptor expressed by endothelial cells-1 (SREC-1) (Berwin et al. 2003), toll-like receptors-2/4 (TLRs-2/4) (Yang et al. 2007; Chalmin et al. 2010; Vicencio et al. 2015) with CD14 as a co-receptor (Asea et al. 2000; Vabulas et al. 2002), fasciclin EGF-like, laminin-type EGF-like and link domain-containing scavenger receptor-1 (FEEL-1), common lymphatic endothelial and vascular endothelial receptor-1 (Clever-1), stabilin-1 (Stab-1), and CD40 (Becker et al. 2002) have been identified as potential binders of HSP. Following binding and uptake via these receptors and co-receptors HSP-chaperoned exogenous peptides can be represented on antigen presenting cells (APCs).

Despite their high degree of sequence homology within a family of HSP, certain parts of their amino acid sequence are not conserved and thus can act as classical foreign antigens, which can be recognized by the host's innate and adaptive immune system, especially when presented in a tumor-specific manner (Botzler et al. 1998). In combination with pro-inflammatory cytokines, including interleukin 2 (IL-2), IL-12, or IL-15, the major stress-inducible Hsp70, also termed HSPA1A, or a peptide derived from the oligomerization domain of Hsp70 (aa450-463) has been found to activate the cytolytic, proliferative, and migratory capacity of natural killer (NK) cells (Multhoff et al. 1999). This activation was accompanied by an upregulated expression of activatory C-type lectin receptors such as CD94/NKG2C and NKG2D on a subpopulation of NK cells (Gross et al. 2003a, b, c). Furthermore, an intra-tumoral infusion of recombinant Hsp70 has been shown to increase the infiltration of NK cells and CD8+ cytotoxic T cells into tumors and as a chaperokine the secretion of the pro-inflammatory cytokine interferon gamma (IFN-γ) (Shevtsov et al. 2014a, b, c).

11.2 Inhibitors of Heat Shock Proteins as Anti-tumor Therapeutics

11.2.1 Inhibition of HSP90

The chaperone machinery consisting of the 90-kDa heat shock protein Hsp90 in combination with the co-chaperones Hsp70, Hsp40 and HoP plays an important role in fostering metabolic pathways and the protection of oncogenic factors that are key for tumor growth and progression (Chiosis et al. 2004; Sreedhar et al. 2004; Zhang and Burrows 2004). Like other members of the HSP family, Hsp90 is also dependent upon ATP. In the absence of ATP, the Hsp90 complex cannot mediate protein homeostasis and thus oncogenic factors are released and undergo degradation via the ubiquitin-proteasome pathway (Neckers 2002) (Fig. 11.1). Consequently, interfering with the Hsp90 chaperone cycle has emerged as a promising strategy for targeting a multiplicity of different oncogenic signaling pathways which are critical for tumor growth (Goetz et al. 2003; Picard 2002; Richter and Buchner 2001; Tatokoro et al. 2015). Since high Hsp90 expression levels have been found in the majority of human tumors the application of Hsp90 inhibitors provides a universal therapeutic strategy for many different cancer types (Whitesell and Lindquist 2005; Whitesell et al. 2012).

Presently applied Hsp90 inhibitors can be divided into two major groups: ***(1) inhibitors of the N-terminal ATPase domain*** and ***(2) inhibitors of the C-terminal nucleotide binding domain***. The oldest and best studied inhibitors of the ATPase domain are geldanamycin (GDA) that belong to the group of benzoquinone ansamycins and radicicol (RDC) (Fig. 11.2). Geldanamycin can block the phosphate region of the Hsp90 binding pocket (Roe et al. 1999), whereas, radicicol was found to bind to a region of the ATPase domain that is responsible for binding the adenine ring of ATP to produce hydrogen-bonding interactions within the binding pocket (Schulte

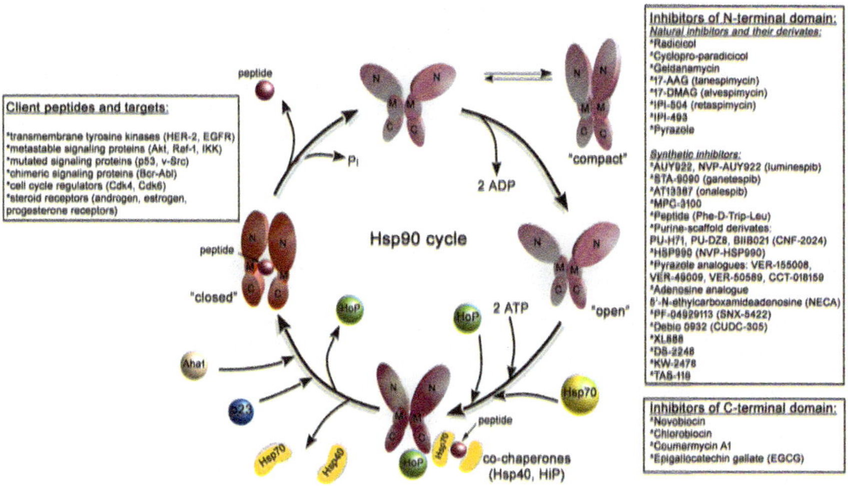

Fig. 11.1 HSP90 chaperone cycle. The chaperone cycle starts with an association of polypeptides with the HSP70/HSP40 complex that is attached to the «open» state of HSP90 via the bridging co-chaperone HoP. HoP promotes peptide transfer from HSP70 to HSP90 preventing HSP90 N-terminal domain association. Following binding of ATP to HSP90, Hop is replaced by p23 and immunophilins producing mature protein complex. Following association of Aha1 to the middle domain of the chaperone ATPase activity of HSP90 is facilitated. As a consequence, ATP hydrolysis stimulates the release of the client protein form the HSP90 complex. *N* N-terminal domain, *C* C-terminal domain, *M* middle domain, *ATP* adenosine triphosphate, *HoP* HSP70/HSP90 organizing protein, *Aha1* activator of HSP90 ATPase

et al. 1998). Although GDA has been found to efficiently inhibit tumor cell growth, a high hepatotoxicity which is unrelated to the Hsp90 inhibition (due to production of superoxide radicals) and a high water insolubility limited its clinical application. Therefore, novel derivates of GDA have been developed with a better water solubility and a lower hepatotoxicity (Supko et al. 1995; Dikalov et al. 1992, 2002). To stabilize the quinone group and to reduce its redox-active potential, stronger electron-donating groups were incorporated into position 17 of GDA. These derivatives were termed 17-(allylamino)-17-demethoxygeldanamycin (17-AAG) and 17-desmethoxy-17-*N,N*-dimethylaminoethylamino-geldanamycin (17-DMAG).

17-AAG is one of the best studied Hsp90 inhibitors that has been shown to suppress tumor growth in a number of cancer cell lines in vitro and in clinical trials (Taldone et al. 2008; Modi et al. 2007). Since the binding affinity of 17-AAG to tumor cell-derived Hsp90 is significantly higher than to that of normal cells this inhibitor shows a tumor-selective targeting (Kamal et al. 2003). Despite this tumor-selectivity, several clinical phase I and II trials applying 17-AAG as a single agent in metastatic pancreatic cancer (Pedersen et al. 2015), other solid tumors (Saif et al. 2013; Iyer et al. 2012; Hubbard et al. 2011; Tse et al. 2008; Ramalingam et al. 2008; Weigel et al. 2007; Bagatell et al. 2007; Banerji et al. 2005), refractory acute myeloid leukemia (Walker et al. 2013), breast cancer (Gartner et al. 2012), ovarian cancer (Hendrickson et al. 2012), metastatic melanoma (Pacey et al. 2012; Solit et al.

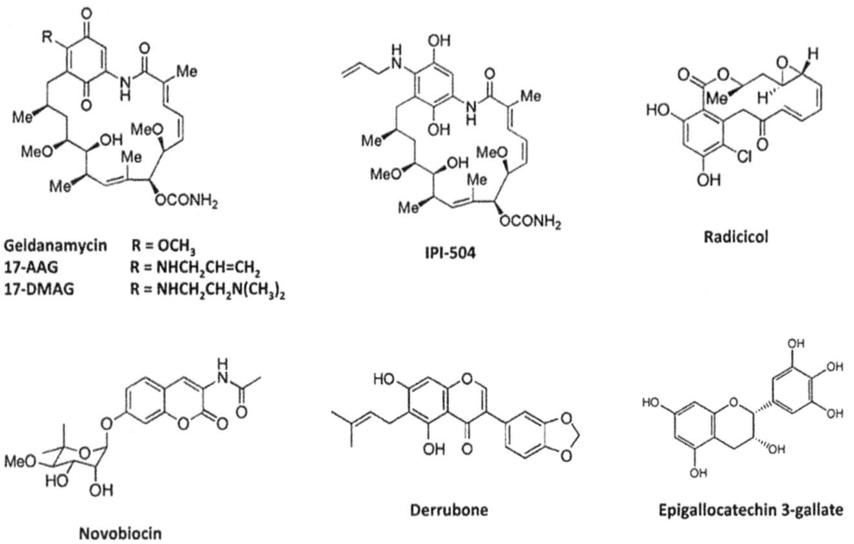

Fig. 11.2 Chemical structures of the naturally occurring Hsp90 inhibitors and its derivatives (Source: Li et al. 2009a, b)

2008a, b), multiple myeloma (Richardson et al. 2010), metastatic prostate cancer (Heath et al. 2008a, b), and renal cell carcinoma (Ronnen et al. 2006) failed to show significant improvement in clinical outcome. In contrast, Modi et al. demonstrated in a phase II clinical trial that a combination of 17-AAG with trastuzumab provided an improved clinical outcome in patients with HER2-positive, metastatic breast cancer who were refractory to a therapy with trastuzumab alone (Modi et al. 2011). These data indicate that an improved patient stratification and the combination of 17-AAG with other therapies might increase the potency of 17-AAG in cancer therapy in the future.

17-DMAG (alvespimycin) has been found to be more potent in its in vivo anti-tumor activity than 17-AAG (Egorin et al. 2002; Yu et al. 2017). In phase I and II clinical trials 17-DMAG has shown a higher oral bioavailability, lower cytotoxicity and an increased stability than 17-AAG (Hollingshead et al. 2005; Ronnen et al. 2006). The improved anti-tumor activity of 17-DMAG could be explained in part by its capability to promote ROS generation in tumor cells as an off-target effect (Kim et al. 2017). The activity of 17-DMAG has been demonstrated in patients with HER2+ metastatic breast cancer and ovarian cancer (Jhaveri et al. 2012), solid tumors (Aregbe et al. 2012; Pacey et al. 2011; Kummar et al. 2010), and acute myeloid leukemia (Lancet et al. 2010).

Therapeutic efficacy of IPI-504 (17-allylamino-17-demethoxygeldanamycin hydroquinone hydrochloride) (retaspimycin), an analog of 17-AAG, has been shown in leukemic mouse models (Peng et al. 2007), in phase I and II clinical trials with philadelphia chromosome-positive chronic myelogenous leukemia (CML) patients and a patient with metastatic GIST tumor after failure of an imatinib and sunitinib therapy (Wagner et al. 2013) (Fig. 11.3). Subsequent studies with patients

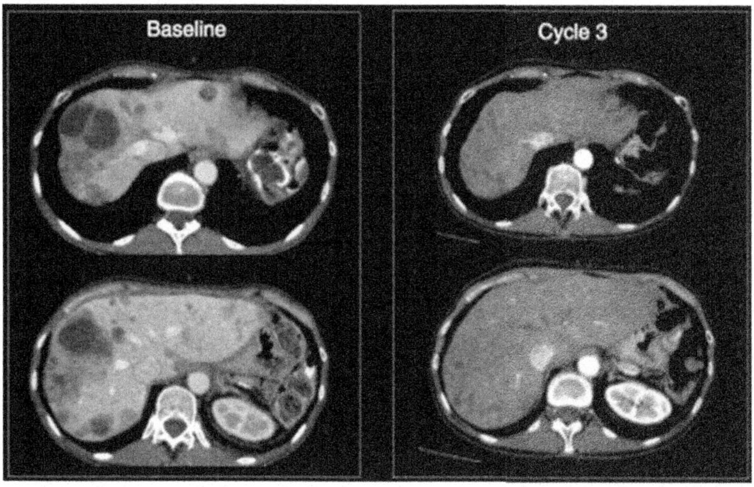

Fig. 11.3 Outcome of patient with metastatic GIST treated with IPI-504. CT scans from a patient treated with 400 mg/m^2 IPI-504 twice weekly for 2 weeks after failure of an imatinib and sunitinib therapy (Source: Wagner et al. 2013)

bearing metastatic HER2+ breast cancer (Modi et al. 2013), multiple myeloma (Siegel et al. 2011), and castration-resistant prostate cancer (Oh et al. 2011) failed to show significant anti-tumor activity. Due to its chemical and metabolic instability radicicol displayed anti-tumor activity in vitro but did not show any therapeutic potency in in vivo studies (Pearl et al. 2008; Soga et al. 2003). Subsequent chemical modifications of radicicol (e.g., cycloproparadicicol) improved its therapeutic activity in preclinical studies (Yamamoto et al. 2003).

Additional Hsp90 inhibitors, such as diarylisoxazole (Zhang et al. 2017), purine-scaffold derivates (Chiosis 2006), pyrazole analogues (Cheung et al. 2005), and three core amino acid peptides (Phe-D-Trp-Leu) (Orosz et al. 2006) are presently under investigation in vitro and in vivo. One of the pyrazole analogues that have entered clinical trials (Brough et al. 2008) is NVP-AUY922. Preliminary data demonstrated a limited efficacy with low toxicity of NVP-AUY922 in patients with pancreatic cancer (Renouf et al. 2016), gastrointestinal stromal tumors (GIST) (Bendell et al. 2016), lung cancer (Johnson et al. 2015), solid tumors (Doi et al. 2014), multiple myeloma (Seggewiss-Bernhardt et al. 2015). Encouraging clinical responses were detected with NVP-AUY922 in therapy-refractory non-Hodgkin lymphoma (Oki et al. 2015) (Fig. 11.4) and HER2+ positive metastatic breast cancer in combination with trastuzumab (Kong et al. 2016) (Fig. 11.5).

Synthetic molecules that were developed to avoid side effects attributed to the benzoquinone group in GDA-based agents demonstrated promising preclinical anti-tumor effects (Sharp et al. 2007a, b). Several small molecule inhibitors of Hsp90, including AT13387, MPC3100, STA9090, XL888 and NVP-AUY922 have been investigated in clinical trials (Hong et al. 2013a, b). 16% of patients with advanced solid tumors who received NVP-AUY922 showed a disease stabilization and 9% of these patients showed a partial tumor response (Sessa et al. 2013). Promising results

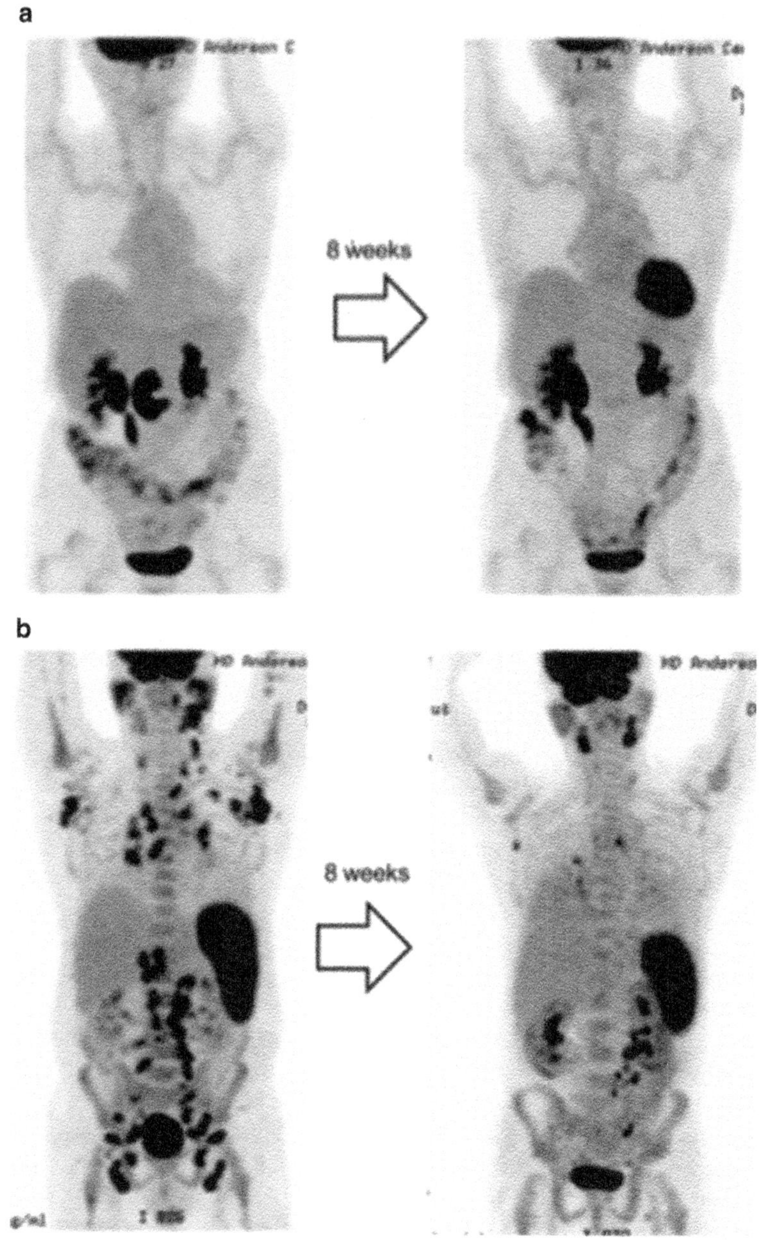

Fig. 11.4 PET scans after 8 weeks of treatment with NVP-AUY922. (**a**) A 74-year-old female with diffuse large B-cell lymphoma with seven prior treatment regimens. The patient achieved complete response after two cycles of NVP-AUY922. Therapy was discontinued after four cycles due to toxicity. The response is ongoing and the patient has remained in complete remission for more than 24 months (**b**) A 60-year-old male with angioimmunoblastic lymphoma with ten prior treatment regimens of NVP-AUY922. The patient achieved partial response but the response was only short-lived <1 month (Source: Oki et al. 2015)

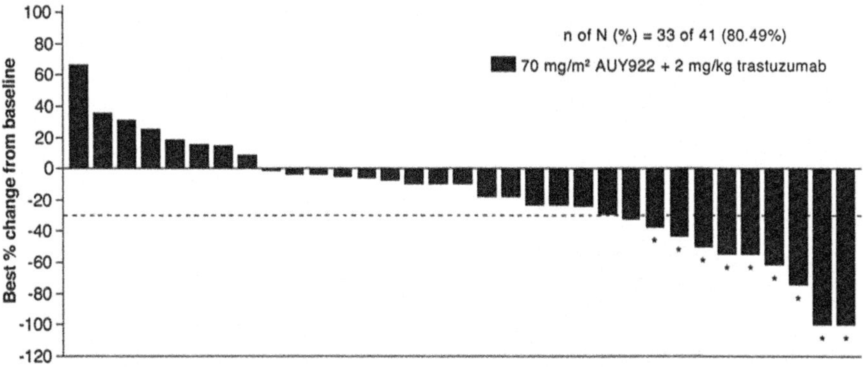

Fig. 11.5 Best change from baseline (%; computed tomography response) in target lesions among evaluable patients who received 70 mg/m^2 NVP-AUY922 + 2 mg/kg trastazumab. *Confirmed Response Evaluation Criteria in Solid Tumors response (RECIST). 'n' represents number of patients with a baseline and at least one post-baseline tumor assessment in target lesions by investigator and 'N' represents total number of patients treated with NVP-AUY922 70 mg/m^2 plus trastuzumab (Source: Kong et al. 2016)

were also observed in a NVP-AUY922 single-agent or combined phase II trial in patients with EGFR overexpressing NSCLC (Samuel et al. 2010) and HER2+ breast cancer (Schroder et al. 2011). A combination of NVP-AUY922 and the EGFR inhibitor erlotinib showed responses in a phase II trial in patients with EGFR-mutant lung cancer who had acquired resistance to EGFR tyrosine kinase inhibitors (Johnson et al. 2013).

Another group of Hsp90 inhibitors comprises purine-based compounds including PU-H71, CNF-2024 (BIIB021) and PU-DZ8. These inhibitors compete with ATP/ADP for the ATP-binding site in the N-terminal domain of Hsp90 which in turn reduces its chaperoning activity. BIIB021 was shown to bind to the ATP-binding pocket of Hsp90 and thus down-regulated HER2 in several in vitro experiments and human xenograft tumor mouse models (Lundgren et al. 2009). BIIB021 decreases the viability of Hodgkin lymphoma cells through an inhibition of the NF-κB pathway and an activation of the activatory NK cell receptor NKG2D on lymphoma cells (Boll et al. 2009). A first clinical trial demonstrated activity of BIIB021 in chronic lymphocytic leukemia (CLL) (Castro et al. 2005; Hong et al. 2013a, b).

SNX-5422 a derivative of 2-aminobenzamide, was successfully tested preclinically in multiple myeloma (Okawa et al. 2009) and breast cancer (Fadden et al. 2010) mouse models. A partial response was observed in a patient with transformed lymphoma and a prolonged stabilization of disease was seen in a patient with multiple myeloma in a phase I trial using SNX-5422 for the treatment of hematological malignancies (Reddy et al. 2013) (Fig. 11.6). Alternative inhibitors of Hsp90 that are presently not yet in clinical trials include 5′-N-ethylcarboxamideadenosine (NECA) (adenosine analogue) (Soldano et al. 2003), 3,4-diarylpyrazole CCT-018159 (Cheung et al. 2005), pyrazole amide analogue VER-49009 and VER-50589 (Sharp et al. 2007a, b).

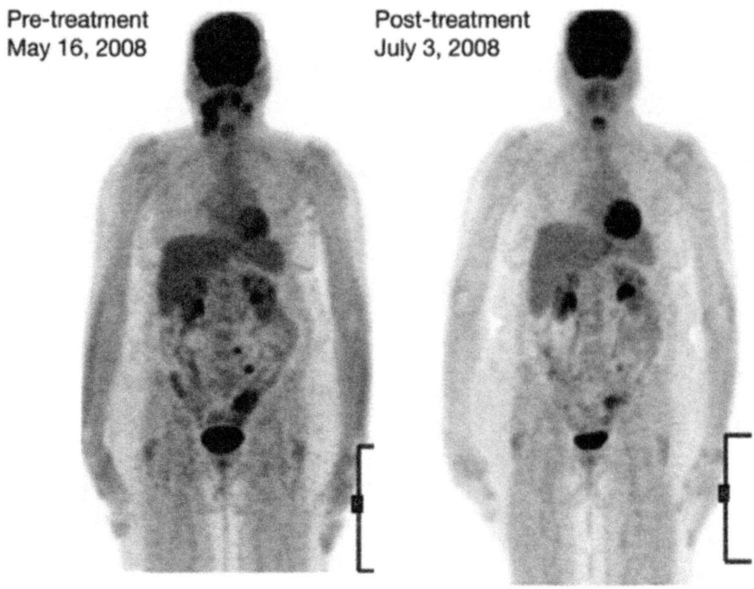

Fig. 11.6 Computed Tomography (CT)/Positron-Emission Tomography (PET) scan in a patient with transformed lymphoma, showing a partial tumor response after two cycles of treatment (Source: Reddy et al. 2013)

The most prominent inhibitor of the C-terminal domain of Hsp90 is novobiocin, an antibiotic isolated from *Streptomyces* (Marcu et al. 2000a, b). A number of preclinical studies indicate that novobiocin destabilizes the interaction of oncogenic client proteins with Hsp90 and thus induces its degradation via the ubiquitin-proteasome pathway (McConkey and Zhu 2008; Marcu et al. 2000a, b). Due to its low therapeutic efficacy further chemical modifications were necessary to improve the potency of novobiocin (e.g., chlorobiocin, coumermycin A1). However, none of these novel compounds have yet been tested in clinical trials (Marcu et al. 2000a, b; Burlison and Blagg 2006). Epigallocatechin gallate (EGCG), the primary flavonoid component of green tea, also binds to the C-terminal domain of Hsp90 and thereby prevents its dimerization (Yin et al. 2009). However, due to the poor chemical and metabolic stability and low bioavailability of EGCG this compound has not been tested in clinical trials.

Unfortunately, the therapeutic efficacy of many Hsp90 inhibitors in preclinical studies could not be confirmed in human trials (Ramalingam et al. 2015; Solit et al. 2008a, b; Johnson et al. 2015; Cercek et al. 2014; Heath et al. 2008a, b). One reason for this negative clinical outcome could be based on an acquired drug resistance of tumor cells, as shown previously for glioblastoma, lung and melanoma cells lines after long-term treatment with VER50589 and NVP-AUY922 (McCollum et al. 2008; Gaspar et al. 2009). In another study, the group of Chai et al. (2017) demonstrated that 17-AAG-resistant tumor cell lines (i.e., advanced mesenchymal human cancer cell line MDA-MB-435, human breast cancer cell line MDA-MB-231) showed cross-resistance to second generation benzoquinone ansamycin-based

Hsp90 inhibitors, such as CCT018159, VER50589, NVP-AUY922 and structurally different Hsp90 inhibitors such as radicicol. Molecular analysis of resistant tumor cells revealed an altered expression pattern of members of the histone deacetylase family (HDAC 1, 5, 6). A subsequent treatment of the cells with pan-HDAC inhibitors (TSA, LBH589) and the class II HDAC-inhibitor SNDX275 that affect the acetylation profile of Hsp90 could re-sensitize these tumor cells towards 17-AAG and 17-DMAG (Chai et al. 2017). Previously, it was reported that hyperacetylation of lysine residues of Hsp90 can impact its association with co-chaperones and ATP binding, and thus results in a reduced chaperone activity that in turn causes an increased degradation of Hsp90 client proteins (Rao et al. 2008). Moreover, hyperacetylation enhances the binding of 17-AAG to Hsp90 and thus might further inhibit its chaperoning function (Rao et al. 2008). The inhibition of HDAC6 which was identified as the main deacetylase for Hsp90 (Scroggins et al. 2007) results in an increased acetylation of Hsp90 and subsequently decreases its chaperoning activity (Kramer et al. 2014).

Although numerous Hsp90 inhibitors have already been tested in preclinical and clinical trials, the specificity of theses inhibitors towards the four HSP90 family paralogs, cytosolic Hsp90α (inducible/major form) and Hsp90β (constitutive/minor form), 94-kDa glucose-regulated protein (Grp94) in the endoplasmic reticulum, and tumor necrosis factor receptor associated protein 1 (Trap1) in mitochondrial matrix) (Maloney and Workman 2002) has not yet been studied in detail. The specificity towards paralogs could influence the efficacy and side effects of Hsp90 inhibitors. Thus in the study of Liu et al. it was demonstrated that NECA (5′-N-ethylcarboxamidoadenosine) preferably binds cytosolic Hsp90 but to a lesser extent to Grp94 (Liu and Street 2016).

Exposure of cancer cells to the Hsp90 inhibitor 17-AAG was shown to increase resistance to chemotherapy due to an increased expression of other pro-survival chaperones such as Hsp70. The binding of ansamycin agents to Hsp90 induces the nuclear localization and trimerization of heat shock factor 1 (HSF-1), a transcription factor that in turn induces the synthesis of Hsp70 (Kaur and Ralhan 2000; Whitesell et al. 2003). Therefore, a simultaneous application of inhibitors directed against both chaperones, Hsp90 and Hsp70 might be able to improve therapeutic efficacy. Several studies have demonstrated an overexpression of Hsp70 upon Hsp90 inhibition as an off-target effect (Mayor-López et al. 2014; Gaspar et al. 2009). As a consequence, the reduction of the intracellular Hsp70/Hsc70 levels by siRNA interference in combination with Hsp90 inhibitors has been shown to increase its anti-tumor effects (Guo et al. 2005). In subsequent experiments combinations of small-molecule kinase inhibitor D11 with 17-AAG have been shown to enhance cytotoxicity and autophagy of human glioblastoma cell lines U87-MG and M059K (Schaefer et al. 2017). In conclusion, the combination of different HSP inhibitors might have the capacity to overcome resistance of tumor cells that is induced by the heat shock response (Powers et al. 2008; Mayer et al. 2009).

Future strategies could exploit the inhibition of co-chaperones that are necessary for the interaction of Hsp90 with its client proteins. The cell cycle division protein 37 (Cdc37) acting as an adaptor for loading of tyrosine kinases to the Hsp90 complex could be considered as a potential target (Vaughan et al. 2006; Smith et al.

2008). Recently platycodin D (PD) was identified as a possible inhibitor of the interaction of Hsp90 with Cdc37. Intriguingly, this inhibition did not induce the production of Hsp70 (Li et al. 2017). Moreover, the combination of PD with the mTOR inhibitor everolimus proved to be therapeutically beneficial in the treatment of cancer (Li et al. 2017).

11.2.2 Inhibition of HSP70

HSP70 represents a family of stress proteins with a molecular weight of approximately 70 kDa with at least 13 different members (Kampinga et al. 2009). According to their subcellular localization the members of this family can be subdivided into endoplasmatic reticulum-residing Grp70, mitochondrial mtHsp70, and cytosolic Hsc70/Hsp70 that are constitutively expressed or stress-inducible. HSP70 support protein synthesis and refolding, assembly of nascent polypeptides, prevent protein aggregation and assist protein trafficking and degradation (Koldewey et al. 2017; Radons 2016) (Fig. 11.7). HSP70 also interfere with both, extrinsic and intrinsic apoptotic pathways (Lee et al. 2005; Park et al. 2001, 2002). Moreover, HSP70 play essential roles in protein degradation through the ubiquitin-proteasome and autophagy pathway (Fernández-Fernández et al. 2017).

HSP70 consist of two functional regions: (i) N-terminal domain that is responsible for the ATP hydrolysis and (ii) C-terminal domain including the EEVD motif that is responsible for peptide binding (Vogel et al. 2006). The major role of members of the HSP70 family is the maintenance of protein homeostasis including oncogenic client proteins. Therefore, strategies that aimed to inhibit HSP70 resulted in increased tumor cell death and sensitivity to stress (Johnston and Kucey 1988; Riabowol et al. 1988). The major stress-inducible Hsp70 is frequently overexpressed in many tumor cell types due to high metabolic stress induced by a fast tumor proliferation. First approaches to inhibit Hsp70 aimed at the transcription factor HSF1 by using the biolflavonoid quercetin (Hosokawa et al. 1992), KNK437 (benzylidene lactam compound) (Yokota et al. 2000; Davenport et al. 2010), triazole nucleoside analog (Xia et al. 2012). Despite a delayed tumor progression after inhibition of HSF1, the major limitation of this approach was the non-specific targeting of HSF1 in tumor and normal cells. Therefore, novel approaches aim to identify agents that can affect specific functions of Hsp70 in tumor cells. Current approaches that aim to inhibit HSP70 address ***the N-terminal domain, C-terminal domain*** and the ***interaction of HSP70 with its co-chaperones***.

11.2.2.1 Inhibitors of the N-Terminal Domain of HSP70

Disruption of the interaction of ATP with the N-terminal domain of Hsp70 affects the binding of the chaperone with its client proteins and thus affects the protein homeostasis. 15-deoxyspergualin (15-DSG) was the first compound that was shown

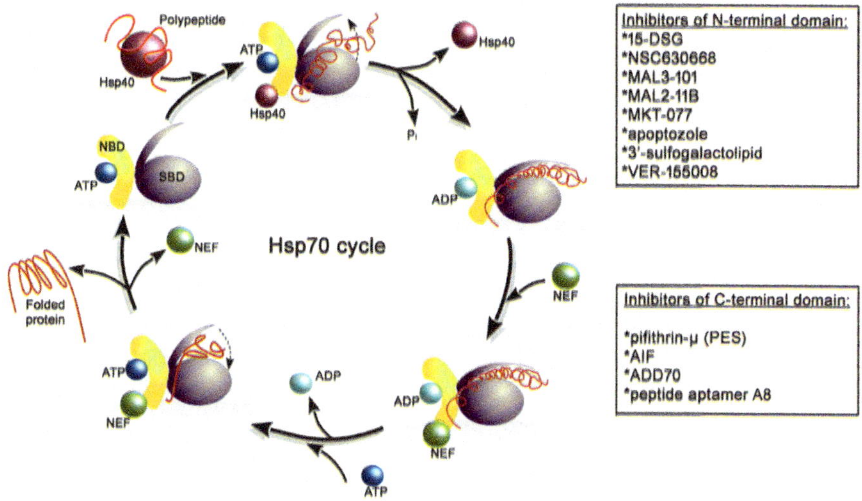

Fig. 11.7 HSP70 chaperone cycle. At the first step of the chaperone cycle the HSP40 with a binded client peptide co-operates with HSP70. Peptide is transferred to the substrate-binding domain (SBD) of the ATP-bound form of HSP70. Following the hydrolysis of the ATP to APD and P_i stimulated by HSP40 more stable interaction between SBD and polypeptide is formed. Interaction of nucleotide exchange factor (NEF) with the HSP70-peptide complex induces the exchange of the ADP to ATP that consequently facilitates the release of the peptide from SBD

to impact the ATPase activity of Hsp70 (Nadeau et al. 1994). Subsequently structurally related to 15-DSG other compounds (i.e., dihydropyrimidine NSC630668, MAL3-101, MAL2-11B) were shown to block protein dislocation and thus reduce cell proliferation (Fewell et al. 2001; Wisen et al. 2010). The main limitation of these substances is based on a limited understanding of the exact molecular mechanism of their mode of action. Other newly identified compounds (i.e., azure C, methylene blue, myricetin) have not yet been assessed for their Hsp70 targeting specificity (Jinwal et al. 2009).

MKT-077 (1-ethyl-2-[[3-ethyl-5-(3-methylbenzothiazolin 2-yliden)]-4-oxothiazolidin-2-ylidenemethyl] pyridinium chloride) another inhibitor of the N-terminal domain of Hsp70 has demonstrated high anti-tumor activity by inhibiting mtHsp70. However, in clinical trials (Koya et al. 1996) MKT-077 resulted in severe renal dysfunctions in patients with solid tumors (Propper et al. 1999; Britten et al. 2000). Presumably chemical modifications of the drug could reduce its renal toxicity for future clinical trials.

For the small molecular inhibitor apoptozole (Az) (Williams et al. 2008) and adamantyl sulfogalactosyl ceramide (3′sulfogalactolipid) (Whetstone and Lingwood 2003) also anti-tumor effects have been described, although the precise molecular mechanism of action have not yet been elucidated (Powers et al. 2010). A recent study of Park et al. demonstrated that Az binds to the ATPase domain of Hsc70 and thus promotes leukemia cells death through a caspase-dependent apoptotic pathway (Park et al. 2017).

Adenosine-derived VER-155008 is a promising compound which selectively binds certain isoforms of Hsp70. The compound VER-155008 alone demonstrated a high anti-tumor activity in breast and HCT116 colon cancer cells, but also increases the activity of the Hsp90 inhibitor 17-AAG (Massey et al. 2010).

11.2.2.2 Inhibitors of the C-Terminal Domain

2-phenylethynesulfonamide (PES) (pifithrin-μ) has been shown to co-operate with the substrate-binding domain of Hsp70 and therefore affects the association of the chaperone with ist client protein. This subsequently results in an aggregation of misfolded proteins and a destabilization of lysosomal membranes (Leu et al. 2009). Moreover, pifithrin-μ also can interfere with the Hsp90/Hsp70 chaperone complex which in turn results in further protein aggregation (Leu et al. 2011).

Another inhibitor derived from AIF derived decoy for Hsp70 (ADD70) was shown to suppress the chaperoning function of Hsp70 (Gurbuxani et al. 2003). Subsequent experiments in B16F10 melanoma and rat colon cancer models demonstrated that ADD70 induces a delay in tumor growth and sensitizes cancer cells to chemotherapeutic agents (Schmitt et al. 2006). The aptamer peptide A8 specifically targets Hsp70 (but not Hsc70 or Hsp90) and shows a chemosensitizing effect in combination with cisplatin (Rerole et al. 2011).

11.2.2.3 Inhibitors of the Interaction of HSP70 with Co-chaperones

Affecting co-chaperones of Hsp70 (especially Hsp40 homologs) results in a disruption of the chaperone function. D-peptide that specifically targets bacterial DnaJ downregulates Hsp70-dependent protein refolding (Bischofberger et al. 2003). Other compounds including MAL2-11B (Wright et al. 2009), pyrimidotriazinediones (Yi and Regan 2008), HSP110DE9 (Dorard et al. 2011) were also shown to influence the interaction with Hsp70. Moreover inhibition of the Hsp70 nucleotide exchange factor Hsp110 with a dominant negative mutant (HSP110DE9) sensitizes tumor cells to chemotherapy (Dorard et al. 2011).

11.2.2.4 Inhibitors of Small Heat Shock Proteins

Hsp27 and small clusterin (sCLU) represent stress-activated molecular chaperones that are involved in cytoprotective transcriptional pathways and thus are important for tumor cell survival. They represent ATP-dependent molecular chaperones that are characterized by α-crystallin domain (conserved sequence of 80–100 amino acids) placed between the C- and N-terminal domain (Jakob et al. 1993).

A recent study with apartorsen (OGX-427) a compound targeting Hsp27 demonstrated tolerability and reduction in tumor biomarkers in patients with prostate and ovarian cancer (Zoubeidi and Gleave 2012). A phase I dose-escalation study with OGX-427 in patients with castration-resistant prostate cancer and other advanced cancers also showed tolerability and clinical responses that justified further phase II clinical trials (Chi et al. 2016).

More clinical data are available for inhibitors of sCLU inhibitors. Phase I and II clinical trials with OGX-011 an inhibitor of sCLU in localized prostate cancer (Chi et al. 2005, 2009, 2010), metastatic breast cancer (Chia et al. 2009), and NSCLC (Laskin et al. 2012) clearly demonstrated anti-tumor efficacy. However, a recently published phase III clinical trial (SYNERGY trial; ClinicalTrials.gov number NCT01188187) using clusterin inhibitors in combination with docetaxel and prednisone did not show significant benefit in patients with metastatic castration-resistant prostate cancer compared to a treatment with docetaxel and prednisone alone (Chi et al. 2017).

11.3 HSP-Based Vaccines

Earlier Old et al. isolated from mouse fibrosarcomas and identified a glycoprotein with a molecular weight of 96 kDa (gp96), which was shown to act as a tumor rejection antigen (Srivastava et al. 1986). In subsequent studies it was demonstrated that gp96 represents an ER-residing member of the HSP90 family with ATPase activity (Li and Srivastava 1993). Other molecular chaperones including the members of the HSP70 family also were shown to possess immunomodulatory activity, when tumor-specific antigens were bound (Udono and Srivastava 1993, Udono et al. 1994). To purify HSP-peptide complexes from tumor cells a novel purification method based on ADP-affinity chromatography was proposed since application of ATP sepharose columns could decrease the amount of the immunogenic peptides due to the internal ATPase activity in the column (Srivastava 1997; Peng et al. 1997).

Chaperoned by gp96 and Hsp70 immunogenic peptides were shown not only to enhance specific anti-tumor immune response but also can be applied for infectious diseases treatment (Srivastava 1994; Srivastava and Liu 2003; SenGupta et al. 2004; Blachere et al. 1997). Belli et al. demonstrated that the immunogenicity of HSP-based vaccines was significantly increased following the oxidative stress (Belli et al. 2002), indicating the fact that the repertoire and the amount of immunogenic peptides might differ in non-stressed vs stressed cells On the other hand, due to the high expression of the stress-inducible members of HSP70 and HSP90 families more immunogenic peptides could be chaperoned and subsequently processed in the antigen-presenting cells. In the studies by Srivastava et al. it was demonstrated that HSP70/HSP90-chaperoned peptides complexes are tumor-specific and could elicit CD8+ cytotoxic T cell-mediated immunity (Srivastava 1994). To elicit an effective immunotherapy in preclinical tumor models the presence of M1 macrophages, CD8+ cytotoxic and CD4+ T helper cells in the effector phase is of high importance (Udono et al. 1994). Chaperoned by HSP exogenous antigens through the endogenous pathway can be presented on MHC I class molecules bypassing presentation by MHC class II antigens (Suto and Srivastava 1995). The described process of HSP-mediated switch of peptides is termed antigen cross-presentation (Suto and Srivastava 1995; Noessner et al. 2002; Murshid et al. 2012).

For a long time the exact receptors what are involved in the uptake of the HSP-peptide complexes by antigen-presenting cells have not been identified until recently. Thus Binder et al. at first described the interaction of the chaperone-peptide complexes as specific and saturable indicating the receptor-ligand interactions (Binder et al. 2000a, b). Later the same group showed that CD91, low-density lipoprotein-related protein, is a receptor for peptides complexed with HSP70 and HSP90 families as well as for calreticulin and alpha-2 macroglobulin (Basu et al. 2001; Binder et al. 2000a, b). Up-to-date several receptors that are involved in binding, uptake and signaling of peptide complexes by APCs (Arnold-Schild et al. 1999) were identified including SREC-1, FEEL-1, Clever-1, Stab-1 (Gong et al. 2009; Berwin et al. 2003; Murshid et al. 2011, 2015), TLRs-2/4, and their co-factors CD14 (Asea et al. 2000, 2002; Vabulas et al. 2002), LOX-1 (Tosti et al. 2014), and CD40 (Becker et al. 2002).

Intriguingly, as was shown by Chandawarkar et al., HSP-peptide complexes can elicit anti-tumor immune responses when being applied at low doses (Chandawarkar et al. 1999). Indeed many-fold higher dose of HSP70/HSp90-peptide complexes demonstrated immunosuppressive activities. Therefore, induction of the immunotolerance by gp96-peptide complexes demonstrated a therapeutic potential in management of encephalomyelitis and autoimmune diabetes in animal models (Chandawarkar et al. 2004; Binder et al. 2012). TLR4 and TLR2 were shown to be associated with Hsp70 mediated immunotolerance. Thus TLR2/MyD88 signaling pathway was described as one of the protection mechanisms following ischemic reperfusion injury when activated by the exosomal fraction of Hsp70 (Vicencio et al. 2015). The TLR4/ERK1,2/p38/MAPK pathway in myeloid-derived suppressor cells was shown to induce pStat3-mediated immunosuppression (Chalmin et al. 2010).

Therapeutic potency of the HSP and/or HSP70/HSP90-peptide based vaccines initiated several clinical trials in clinical oncology (Oki and Younes 2004) including studies in glioblastoma (Bloch et al. 2014), pancreatic carcinoma (Maki et al. 2007), renal cell carcinoma (Wood 2008), gastric carcinoma (Hertkorn et al. 2002; Heike et al. 2000), chronic myeloid leukemia (Hammond 2004), late stage melanoma (Janetzki et al. 2000; Srivastava 2006; Castelli et al. 2004; Pilla et al. 2006; Testori et al. 2008; Schadendorf et al. 2006; Nestle et al. 1998), and metastatic colon carcinoma (Mazzaferro et al. 2003) (Table 11.1). Although these studies demonstrated certain immunological responses only in a minority group of patients clinical responses (CR) were reported.

11.4 HSP70 as an Immunomodulator

Molecular chaperone Hsp70 has been shown to be overexpressed in various tumor cells (including colorectal, breast, glioblastomas, sarcomas, prostate and pancreatic carcinomas, squamous cell carcinomas of the head and neck, hematological malignancies) but not on corresponding normal cells (Gehrmann et al. 2003; Multhoff et al. 1997a, b). Membrane-bound Hsp70 (mHsp70) has been identified on freshly

Table 11.1 Phase I–III clinical trials using HSP-based vaccines

HSP vaccine	Tumor entity	Study	References
Gp96	Late stage melanoma	Pilot	Janetzki et al. (2000)
Gp96 + sPD-1	Malignant melanoma	Phase I–III	Testori et al. (2008), Schadendorf et al. (2006), and Nestle et al. (1998)
Gp96 + GM-CSF	Malignant melanoma	Phase I–II	Castelli et al. (2004)
Gp96 + GM-CSF +IFNy	Malignant melanoma	Phase I–II	Pilla et al. (2006)
Hsp70	Malignant melanoma	Phase I	Castelli et al. (2004)
Gp96	Metastatic colon carcinoma	Phase I	Mazzaferro et al. (2003)
Gp96	Gastric carcinoma	Phase I	Hertkorn et al. (2002) and Heike et al. (2000)
Gp96	Pancreatic carcinoma	Phase I	Maki et al. (2007)
Gp96	Hodgkin lymphoma	Phase I	Hammond (2004)
Hsp70	Chronica lymphatic leukemia	Phase I	Li et al. (2009a, b)
Hsp70	Advanced solid tumors	Pilot	Guzhova et al. (2013)
Gp96	Glioblastoma	Phase I–II	Bloch et al. (2014)
Hsp70	Glioblastoma	Phase I	Guzhova et al. (2013)
Hsp70-activated NK cells	Colon carcinoma, NSCLC	Phase I–II	Krause et al. (2004) and Specht et al. (2015)
Hsp70	HIV	Phase I	SenGupta et al. (2004)
Hsp70 mRNA	HCC	Phase I	Maeda et al. (2015)

isolated tumor biopsies by tumor cells surface iodination/biotinylation (Multhoff et al. 1995; Shin et al. 2003) and by cmHsp70.1 monoclonal antibody employing flow cytometry analysis (Stangl et al. 2011). Recently this method was described for the indirect assessment of the mHsp70 and exosomal Hsp70 (Breuninger et al. 2014; Ciocca and Calderwood 2005). Membrane-associated Hsp70 tumor phenotype was shown to correlate with the aggressiveness of tumors including the tumor invasion and metastatic activity (Multhoff et al. 1997a, b; Gastpar et al. 2005; Murakami et al. 2015). However, it was shown that NK cells can efficiently kill the mHsp70 positive tumor cells (Multhoff et al. 2001). Employing protein/peptide and antibody blocking assays CD94, C-type lectin receptor, was shown to specifically interact with the 14-mer TKD peptide in the structure of the mHsp70 on tumor cells thus resulting in the subsequent activation of NK cells (Biassoni 2009; Pegram et al. 2011; Sullivan et al. 2007; Borrego et al. 2006). In the studies by Gross et al. it was demonstrated that co-incubation of NK cells with purified Hsp70 or Hsp70-derived TKD peptide (in the presence of low dose IL-2) resulted in the enhanced cytotoxicity and migratory activity of NK cells towards mHsp70-positive cancer cells (Gross et al. 2003a, b, c). The schematic representation of the Hsp70 immunomodulatory activity is shown in Fig. 11.8.

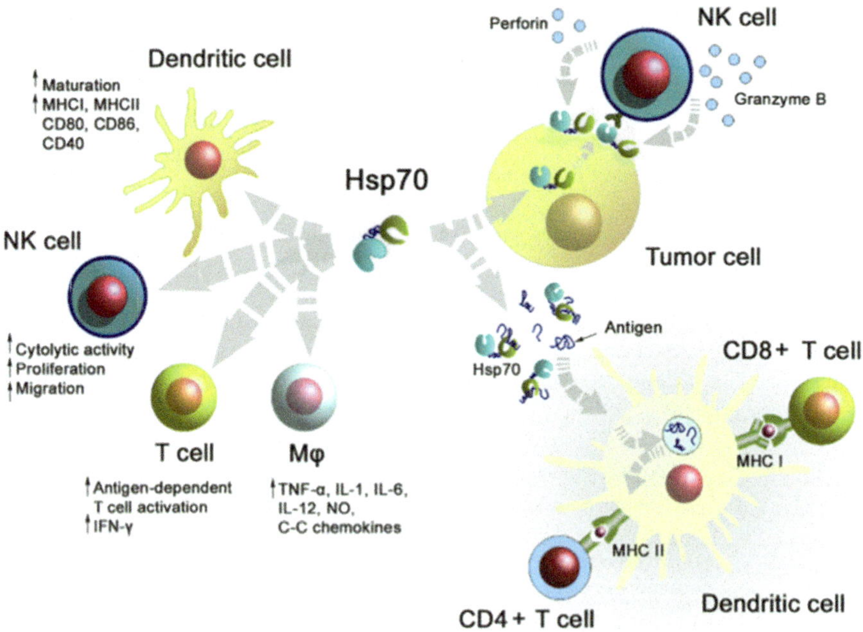

Fig. 11.8 Major immune modulatory functions of Hsp70 either alone or bound to exosomes or in combination with tumor-derived peptides. *IFN-γ* interferon gamma, *IL* interleukin, *Mϕ* macrophages, *MHC* major histocompatibility complex *NK cells* natural killer cells, *NO* nitric oxide

Hsp70 could function as a chaperokine inducing the cytolytic, proliferative and migratory activity of NK cells, maturation of DCs, activation of T cells that results in production of pro-inflammatory cytokines. At the same time the mHsp70 plays role of the tumor-specific antigen which is targeted by the preactivated NK cells. Furthermore, the members of HSP70 and HSP90 families were shown to enhance uptake, processing and presentation of the tumor-specific antigens on the MHC I and II class molecules to the CD8+ cytotoxic T lymphocytes and CD4+ T helper cells, respectively.

By Gross et al. it was shown that following activation of the NK cells the latter produce a serine protease granzyme B, which is directly interacting with mHsp70 inducing the granzyme-B-mediated apoptosis (Gross et al. 2003a, b, c). Stimulation of the NK cells with Hsp70-derived TKD peptide and low dose IL-2 significantly increased the production of the granzyme B (Gross et al. 2003a, b). Subsequent studies demonstrated that the tumor cells which do express lower amounts of mHsp70 are not lysed by granzyme B, indicating the role of the granzyme B-mediated apoptosis in tumor lysis (Gross et al. 2003b).

Based on the promising results of the in vitro and in vivo studies first-in-man phase I trial of the ex vivo TKD/IL-2 stimulated autologous NK cells in metastasized colorectal and NSCLC patients demonstrated safety and tolerability (Krause et al. 2004). Subsequent phase II trial has been launched in NSCLC patients to assess the therapeutic efficacy of the Hsp70 peptide TKD + IL-2-activated autologous NK cells (Specht et al. 2015).

11.5 Chaperokine Function of HSP

HSP, particularly the major stress-inducible Hsp70, were shown to function as a cytokines (i.e., chaperokines), initiating both, innate and adaptive immune responses (Elsner et al. 2007; Takemoto et al. 2010; Guzhova et al. 2013; Asea 2006). At the same time due to the involvement in the tumor antigens uptake, processing and presentation by APCs these HSP act as classical chaperones. In the studies by Shevtsov et al. it was shown that exogenous Hsp70 uptaken by cancer cells can induce the translocation of its own intracellular analogue to the cell membrane, thus sensitizing the tumor cells towards cytolytic activity of the NK cells (Shevtsov et al. 2014a, b, c). Furthermore, extracellular Hsp70 could also induce the secretion of the intracellular chaperone in complex with tumor peptides into the extracellular milieu, thus promoting the adaptive immune responses (Shevtsov et al. 2014a, b, c). Subsequent in vivo experiments in the model of intracranial C6 glioma in rats demonstrated that single intratumoral injection of the Hsp70 or prolonged delivery (via the osmotic subcutaneous pump) significantly delayed the tumor progression (as shown by the magnetic resonance imaging) and increased the overall survival of animals. The therapeutic potency of the injected Hsp70 was based on the generation of the effective innate and adaptive anti-tumor immune responses (Shevtsov et al. 2014a, b, c). Other treatment modalities including the upregulation of *hsp70.1* gene were shown to have a significant therapeutic activity (Shevtsov et al. 2014a, b, c; Geng et al. 2006; Rafiee et al. 2001; Nishikawa et al. 2010; Ito et al. 2004). Combination with other therapeutic modalities might further improve the anti-tumor activity of Hsp70. Thus combined therapy of local hyperthermia employing magnetite cationic liposomes (MCLs) with intratumoral injection of Hsp70 in the model of subcutaneous melanoma in mice resulted in significant delay of tumor growth (Ito et al. 2004). In the recent studies by Shevtsov et al. it was shown that superparamagnetic iron oxide nanoparticles (SPIONs) functionalized with Hsp70 could efficiently deliver tumor peptides into the DCs in comparison to non-coated SPIONs or soluble Hsp70 thus eliciting an efficient CD8+ cytotoxic anti-tumor immune response in animals (Shevtsov et al. 2015). Furthermore, several clinical trials confirmed that administration of Hsp70 as a single agent or in combination with other therapies could induce the anti-tumor immune responses (Shevtsov et al. 2014a, b, c; Li et al. 2009a, b; Maeda et al. 2015). In the first-in-man phase I trial intratumoral injection recombinant human Hsp70 in neuro-oncological patients (n = 12) resulted in partial response (PR) in one patient and complete response (CR) in another patient (Shevtsov et al. 2014a, b, c). Promising results were reported in another clinical phase I trial in non-resectable or recurrent HCV-related hepatocellular carcinoma (HCC) patients when Hsp70 mRNA (Hsp70-DCs) was applied (Maeda et al. 2015). Thus 7 out of 12 patients were reported to have either a CR or stable disease (SD), indicating the efficacy of the proposed therapy. Presumably, multimodal anti-cancer therapies based on conventional therapies (including radio- and/or chemotherapy, hyperthermia) in combination with HSP-based therapies can provide a treatment option for future clinical studies (Schildkopf et al. 2011).

11.6 Conclusions

Molecular chaperones, particularly members of the HSP90, HSP70 and small heat shock protein families (i.e., Hsp27, clusterin) are abundantly expressed in tumor cells and contribute to tumorogenesis and cell survival during radio- and/or chemotherapy. Much effort in recent decades was dedicated to the search for potent inhibitors of these chaperones that could be applied in cancer therapy. Recent clinical trials using inhibitors against members of the HSP70 and HSP90 families demonstrate clinical responses especially in patients with lymphoma and breast cancer. Due to compensatory effects of Hsp90 inhibitors towards an increased expression of Hsp70 in tumor cells it appears beneficial to combine several HSP inhibitors in cancer therapy. Especially members of the HSP70 and HSP90 families have been found to elicit protective anti-tumor immunity in preclinical models and in tumor patients either alone or in complex with tumor-derived peptides. HSP and HSP70/HSP90-peptide complexes can act as classical tumor-specific foreign antigens for the innate immune system, chaperokines, and adjuvants that facilitate uptake, processing, and presentation of tumor-specific antigens which are cross-presented by APCs to CD8+ cytotoxic T lymphocytes. Patient-derived NK cells can regain the capacity to recognize highly aggressive tumor cells presenting Hsp70 on their plasma membrane after stimulation with Hsp70 peptide TKD + IL-2.

Acknowledgements The study was supported by the Alexander von Humboldt Stiftung, by a grant from Russian Foundation for Basic Research 15-08-08148A, in part by a grant of the Russian Science Foundation 14-50-00068 and by the Federal Agency of Scientific Organizations, Russia and by a grant of the Deutsche Forschungsgemeinschaft (DFG SFB824/3, STA 1520/1-1, MAP), BMBF (01GU0823, 02NUK038A) and BMWi (AiF project), Germany. This work was supported by the german Research Foundation (DFG) and the Technische Universität München within the funding programme Open Access Publishing. The authors are grateful to Nan-Jong Lee for preparation of the Figures.

References

Aregbe AO, Sherer EA, Egorin MJ, Scher HI, Solit DB, Ramanathan RK, Ramalingam S, Belani CP, Ivy PS, Bies RR (2012) Population pharmacokinetic analysis of 17-dimethylaminoethyl amino-17-demethoxygeldanamycin (17-DMAG) in adult patients with solid tumors. Cancer Chemother Pharmacol 70(1):201–205

Arnold-Schild D, Hanau D, Spehner D, Schmid C, Rammensee HG, de la SH et al (1999) Cutting edge: receptor-mediated endocytosis of heat shock proteins by professional antigen-presenting cells. J Immunol 162:3757–3760

Asea A (2006) Initiation of the immune response by extracellular Hsp72: chaperokine activity of Hsp72. Curr Immunol Rev 2:209–215

Asea A, Kraeft SK, Kurt-Jones EA, Stevenson MA, Chen LB, Finberg RW et al (2000) HSP70 stimulates cytokine production through a CD14-dependant pathway, demonstrating its dual role as a chaperone and cytokine. Nat Med 6:435–442

Asea A, Rehli M, Kabingu E, Boch JA, Bare O, Auron PE et al (2002) Novel signal transduction pathway utilized by extracellular HSP70: role of TLR2 and TLR4. J Biol Chem 277:15028–15034

Bagatell R, Gore L, Egorin MJ, Ho R, Heller G, Boucher N, Zuhowski EG, Whitlock JA, Hunger SP, Narendran A, Katzenstein HM, Arceci RJ, Boklan J, Herzog CE, Whitesell L, Ivy SP, Trippett TM (2007) Phase I pharmacokinetic and pharmacodynamics study of 17-N-allylamino-17-demethoxygeldanamycin in pediatric patients with recurrent or refractory solid tumors: a pediatric oncology experimental therapeutics investigators consortium study. Clin Cancer Res 13(6):1783–1788

Banerji U, O'Donnell A, Scurr M, Pacey S, Stapleton S, Asad Y, Simmons L, Maloney A, Raynaud F, Campbell M, Walton M, Lakhani S, Kaye S, Workman P, Judson I (2005) Phase I pharmacokinetic and pharmacodynamic study of 17-allylamino, 17-demethoxygeldanamycin in patients with advanced malignancies. J Clin Oncol 23(18):4152–4161

Basu S, Binder RJ, Ramalingam T, Srivastava PK (2001) CD91 is a common receptor for heat shock proteins gp96, hsp90, hsp70, and calreticulin. Immunity 14:303–313

Becker T, Hartl FU, Wieland F (2002) CD40, an extracellular receptor for binding and uptake of Hsp70-peptide complexes. J Cell Biol 158:1277–1285

Belli F, Testori A, Rivoltini L, Maio M, Andreola G, Sertoli MR et al (2002) Vaccination of metastatic melanoma patients with autologous tumor-derived heat shock protein gp96-peptide complexes: clinical and immunologic findings. J Clin Oncol 20:4169–4180

Bendell JC, Bauer TM, Lamar R, Joseph M, Penley W, Thompson DS, Spigel DR, Owera R, Lane CM, Earwood C, Burris HA 3rd. (2016) A phase 2 study of the Hsp90 inhibitor AUY922 as treatment for patients with refractory gastrointestinal stromal tumors. Cancer Investig 34(6):265–270

Berwin B, Hart JP, Rice S, Gass C, Pizzo SV, Post SR et al (2003) Scavenger receptor-A mediates gp96/GRP94 and calreticulin internalization by antigen-presenting cells. EMBO J 22:6127–6136

Biassoni R (2009) Human natural killer receptors, co-receptors, and their ligands. Curr Protoc Immunol., Chapter 14:Unit14.10. https://doi.org/10.1002/0471142735.im1410s84

Binder RJ, Han DK, Srivastava PK (2000a) CD91: a receptor for heat shock protein gp96. Nat Immunol 1:151–155

Binder RJ, Harris ML, Menoret A, Srivastava PK (2000b) Saturation, competition, and specificity in interaction of heat shock proteins (hsp) gp96, hsp90, and hsp70 with CD11b+ cells. J Immunol 165:2582–2587

Binder RJ, Zhou YJ, Messmer MN, Pawaria S (2012) CD91-dependent modulation of immune responses by heat shock proteins: a role in autoimmunity. Autoimmune Dis 2012:863041

Bischofberger P, Han W, Feifel B, Schonfeld HJ, Christen P (2003) D-peptides as inhibitors of the DnaK/DnaJ/GrpE chaperone system. J Biol Chem 278:19044–19047

Blachere NE, Li Z, Chandawarkar RY, Suto R, Jaikaria NS, Basu S et al (1997) Heat shock protein-peptide complexes, reconstituted in vitro, elicit peptide-specific cytotoxic T lymphocyte response and tumor immunity. J Exp Med 186:1315–1322

Bloch O, Crane CA, Fuks Y, Kaur R, Aghi MK, Berger MS et al (2014) Heat-shock protein peptide complex-96 vaccination for recurrent glioblastoma: a phase II, single-arm trial. Neuro-Oncology 16:274–279

Boll B et al (2009) Heat shock protein 90 inhibitor BIIB021 (CNF2024) depletes NF-kappaB and sensitizes Hodgkin's lymphoma cells for natural killer cell-mediated cytotoxicity. Clin Cancer Res 15:5108–5116

Borrego F, Masilamani M, Marusina AI, Tang X, Coligan JE (2006) The CD94/NKG2 family of receptors: from molecules and cells to clinical relevance. Immunol Res 35:263–278

Botzler C, Li G, Issels RD, Multhoff G (1998) Definition of extracellular localized epitopes of Hsp70 involved in an NK immune response. Cell Stress Chaperones 3:6–11

Breuninger S, Erl J, Knape C, Gunther S, Regel I, Rodel F et al (2014) Quantitative analysis of liposomal heat shock protein 70 (Hsp70) in the blood of tumor patients using an novel lipHsp70 ELISA. J Clin Cell Immunol 5:2–10

Britten CD, Rowinsky EK, Baker SD, Weiss GR, Smith L, Stephenson J et al (2000) A phase I and pharmacokinetic study of the mitochondrial-specific rhodacyanine dye analog MKT 077. Clin Cancer Res 6:42–49

Brough PA, Aherne W, Barril X et al (2008) 4,5-diarylisoxazole Hsp90 chaperone inhibitors: potential therapeutic agents for the treatment of cancer. J Med Chem 51:196–218
Burlison JA, Blagg BS (2006) Synthesis and evaluation of coumermycin A1 analogues that inhibit the Hsp90 protein folding machinery. Org Lett 8:4855–4858
Castelli C, Rivoltini L, Rini F, Belli F, Testori A, Maio M et al (2004) Heat shock proteins: biological functions and clinical application as personalized vaccines for human cancer. Cancer Immunol Immunother 53:227–233
Castro JE et al (2005) ZAP-70 is a novel conditional heat shock protein 90 (Hsp90) client: inhibition of Hsp90 leads to ZAP-70 degradation, apoptosis, and impaired signaling in chronic lymphocytic leukemia. Blood 106:2506–2512
Cercek A, Shia J, Gollub M, Chou JF, Capanu M, Raasch P, Reidy-Lagunes D, Proia DA, Vakiani E, Solit DB et al (2014) Ganetespib, a novel Hsp90 inhibitor in patients with KRAS mutated and wild type, refractory metastatic colorectal cancer. Clin Colorectal Cancer 13:207–212
Chai RC, Vieusseux JL, Lang BJ, Nguyen CH, Kouspou MM, Britt KL, Price JT (2017) Histone deacetylase activity mediates acquired resistance towards structurally diverse HSP90 inhibitors. Mol Oncol 11(5):567–583
Chalmin F, Ladoire S, Mignot G, Vincent J, Bruchard M, Remy-Martin JP et al (2010) Membrane-associated Hsp72 from tumor-derived exosomes mediates STAT3-dependent immunosuppressive function of mouse and human myeloid-derived suppressor cells. J Clin Invest 120:457–471
Chandawarkar RY, Wagh MS, Srivastava PK (1999) The dual nature of specific immunological activity of tumor-derived gp96 preparations. J Exp Med 189:1437–1442
Chandawarkar RY, Wagh MS, Kovalchin JT, Srivastava P (2004) Immune modulation with high-dose heat-shock protein gp96: therapy of murine autoimmune diabetes and encephalomyelitis. Int Immunol 16:615–624
Cheung KM et al (2005) The identification, synthesis, protein crystal structure and in vitro biochemical evaluation of a new 3,4-diarylpyrazole class of Hsp90 inhibitors. Bioorg Med Chem Lett 15:3338–3343
Chi KN, Eisenhauer E, Fazli L, Jones EC, Goldenberg SL, Powers J et al (2005) A phase I pharmacokinetic and pharmacodynamic study of OGX-011, a 2-methoxyethyl antisense oligonucleotide to clusterin, in patients with localized prostate cancer. J Natl Cancer Inst 97:1287–1296
Chi KN, Hotte SJ, Yu E, Tu D, Eigl BJ, Tannock I et al (2009) Mature results of a randomized phase II study of OGX-011 in combination with docetaxel/prednisone versus docetaxel/prednisone in patients with metastatic castration-resistant prostate cancer. J Clin Oncol 27 Abstract 5012
Chi KN, Hotte SJ, Yu EY, Tu D, Eigl BJ, Tannock I et al (2010) Randomized phase II study of docetaxel and prednisone with or without OGX-011 in patients with metastatic castration-resistant prostate cancer. J Clin Oncol 28:4247–4254
Chi KN, Yu EY, Jacobs C, Bazov J, Kollmannsberger C, Higano CS et al (2016) A phase I dose-escalation study of apatorsen (OGX-427), an antisense inhibitor targeting heat shock protein 27 (Hsp27), in patients with castration-resistant prostate cancer and other advanced cancers. Ann Oncol 27(6):1116–1122
Chi KN, Higano CS, Blumenstein B, Ferrero JM, Reeves J, Feyerabend S et al (2017) Custirsen in combination with docetaxel and prednisone for patients with metastatic castration-resistant prostate cancer (SYNERGY trial): a phase 3, multicentre, open-label, randomised trial. Lancet Oncol 18(4):473–485
Chia S, Dent S, Ellard S, Ellis PM, Vandenberg T, Gelmon K et al (2009) Phase II trial of OGX-011 in combination with docetaxel in metastatic breast cancer. Clin Cancer Res 15(2):708–713
Chiosis G (2006) Discovery and development of purine-scaffold Hsp90 inhibitors. Curr Top Med Chem 6:1183–1191
Chiosis G, Vilenchik M, Kim J, Solit D (2004) Hsp90: the vulnerable chaperone. Drug Discov Today 9:881–888
Ciocca DR, Calderwood SK (2005) Heat shock proteins in cancer: diagnostic, prognostic, predictive, and treatment implications. Cell Stress Chaperones 10:86–103

Davenport EL, Zeisig A, Aronson LI, Moore HE, Hockley S, Gonzalez D et al (2010) Targeting heat shock protein 72 enhances Hsp90 inhibitor-induced apoptosis in myeloma. Leukemia 24:1804–1807

Dikalov S, Rumyantseva GV, Piskunov AV, Weiner LM (1992) Role of quinone-iron (III) interaction in NADPH-dependent enzymatic generation of hydroxyl radicals. Biochemistry 31:8947–8953

Dikalov S, Landmesser U, Harrison DG (2002) Geldanamycin leads to superoxide formation by enzymatic and non-enzymatic cycling. J Biol Chem 277:25480–25485

Doi T, Onozawa Y, Fuse N, Yoshino T, Yamazaki K, Watanabe J, Akimov M, Robson M, Boku N, Ohtsu A (2014) Phase I dose-escalation study of the HSP90 inhibitor AUY922 in Japanese patients with advanced solid tumors. Cancer Chemother Pharmacol 74(3):629–636

Dorard C, de Thonel A, Collura A, Marisa L, Svrcek M, Lagrange A et al (2011) Duval, expression of a mutant HSP110 sensitizes colorectal cancer cells to chemotherapy and improves disease prognosis. Nat Med 17:1283–1289

Egorin MJ, Lagattuta TF, Hamburger DR, Covey JM, White KD, Musser SM, Eiseman JL (2002) Pharmacokinetics, tissue distribution, and metabolism of 17-(dimethylaminoethylamino)-17-demethox-ygeldanamycin (NSC 707545) in CD2F1 mice and fischer 344 rats. Cancer Chemother Pharmacol 49:7–19

Elsner L, Muppala V, Gehrmann M, Lozano J, Malzahn D, Bickeboller H et al (2007) The heat shock protein Hsp70 promotes mouse NK cell activity against tumors that express inducible NKG2D ligands. J Immunol 179:5523–5533

Fadden P et al (2010) Application of chemoproteomics to drug discovery: identification of a clinical candidate targeting hsp90. Chem Biol 17:686–694

Fernández-Fernández, M.R., Gragera, M., Ochoa-Ibarrola, L.., (2017) Quintana-Gallardo, L., Valpuesta, J.M. (2017) Hsp70 – a master regulator in protein degradation. FEBS Lett. https://doi.org/10.1002/1873-3468.12751. [Epub ahead of print] Review

Fewell SW, Day BW, Brodsky JL (2001) Identification of an inhibitor of hsc70- mediated protein translocation and ATP hydrolysis. J Biol Chem 276:910–914

Gartner EM, Silverman P, Simon M, Flaherty L, Abrams J, Ivy P, Lorusso PM (2012) A phase II study of 17-allylamino-17-demethoxygeldanamycin in metastatic or locally advanced, unresectable breast cancer. Breast Cancer Res Treat 131(3):933–937

Gaspar N, Sharp SY, Pacey S, Jones C, Walton M, Vassal G, Eccles S, Pearson A, Workman P (2009) Acquired resistance to 17-allylamino-17-demethoxygeldanamycin (17-AAG, tanespimycin) in glioblastoma cells. Cancer Res 69:1966–1975

Gastpar R, Gehrmann M, Bausero MA, Asea A, Gross C, Schroeder JA et al (2005) Heat shock protein 70 surface-positive tumor exosomes stimulate migratory and cytolytic activity of natural killer cells. Cancer Res 65:5238–5247

Gehrmann M, Schmetzer H, Eissner G, Haferlach T, Hiddemann W, Multhoff G (2003) Membrane-bound heat shock protein 70 (Hsp70) in acute myeloid leukemia: a tumor specific recognition structure for the cytolytic activity of autologous NK cells. Haematologica 88:474–476

Geng H, Zhang GM, Xiao H, Yuan Y, Li D, Zhang H et al (2006) HSP70 vaccine in combination with gene therapy with plasmid DNA encoding sPD-1 overcomes immune resistance and suppresses the progression of pulmonary metastatic melanoma. Int J Cancer 118:2657–2664

Goetz MP, Toft DO, Ames MM, Erlichman C (2003) The hsp90 chaperone complex as a novel target for cancer therapy. Ann Oncol 14:1169–1176

Gong J, Zhu B, Murshid A, Adachi H, Song B, Lee A et al (2009) T cell activation by heat shock protein 70 vaccine requires TLR signaling and scavenger receptor expressed by endothelial cells-1. J Immunol 183:3092–3098

Gross C, Hansch D, Gastpar R, Multhoff G (2003a) Interaction of heat shock protein 70 peptide with NK cells involves the NK receptor CD94. Biol Chem 384:267–279

Gross C, Koelch W, DeMaio A, Arispe N, Multhoff G (2003b) Cell surface-bound heat shock protein 70 (Hsp70) mediates perforin-independent apoptosis by specific binding and uptake of granzyme B. J Biol Chem 278:41173–41181

Gross C, Schmidt-Wolf IG, Nagaraj S, Gastpar R, Ellwart J, Kunz-Schughart LA et al (2003c) Heat shock protein 70-reactivity is associated with increased cell surface density of CD94/CD56 on primary natural killer cells. Cell Stress Chaperones 8:348–360

Guo F, Rocha K, Bali P, Pranpat M, Fiskus W, Boyapalle S et al (2005) Abrogation of heat shock protein 70 induction as a strategy to increase antileukemia activity of heat shock protein 90 inhibitor 17-allylamino-demethoxy geldanamycin. Cancer Res 65:10536–10544

Gurbuxani S, Schmitt E, Cande C, Parcellier A, Hammann A, Daugas E et al (2003) Heat shock protein 70 binding inhibits the nuclear import of apoptosis-inducing factor. Oncogene 22:6669–6678

Guzhova IV, Shevtsov MA, Abkin SV, Pankratova KM, Margulis BA (2013) Intracellular and extracellular Hsp70 chaperone as a target for cancer therapy. Int J Hyperth 29:399–408

Hammond DB (2004) What ASCO does for its members. J Clin Oncol 22:2251–2252

Heath EI, Hillman DW, Vaishampayan U, Sheng S, Sarkar F, Harper F, Gaskins M, Pitot HC, Tan W, Ivy SP (2008a) A phase II trial of 17-allylamino-17-demethoxygeldanamycin in patients with hormone-refractory metastatic prostate cancer. Clin Cancer Res 14:7940–7946

Heath EI, Hillman DW, Vaishampayan U, Sheng S, Sarkar F, Harper F, Gaskins M, Pitot HC, Tan W, Ivy SP, Pili R, Carducci MA, Erlichman C, Liu G (2008b) A phase II trial of 17-allylamino-17-demethoxygeldanamycin in patients with hormone-refractory metastatic prostate cancer. Clin Cancer Res 14(23):7940–7946

Heike M, Hertkorn C, Reitsma DJ, Kommoss F, Simiantonaki N, Woelfel T et al (2000) Pilot trial of vaccination with autologous tumor-derived GP96 heat shock protein-peptide complex (HSPPC-96) in patients after surgery for gastric carcinoma. Proc Am Soc Clin Oncol 19:1825

Hendrickson AE, Oberg AL, Glaser G, Camoriano JK, Peethambaram PP, Colon-Otero G, Erlichman C, Ivy SP, Kaufmann SH, Karnitz LM, Haluska P (2012) A phase II study of gemcitabine in combination with tanespimycin in advanced epithelial ovarian and primary peritoneal carcinoma. Gynecol Oncol 124(2):210–215

Hertkorn C, Lehr A, Wölfel T, Junginger T, Dippold WG, Galle PR et al (2002) Phase I trial of vaccination with autologous tumor-derived gp96 in patients after surgery for gastric cancer. Proc Am Soc Clin Oncol 21:30a

Hollingshead M, Alley M, Burger AM et al (2005) In vivo anti-tumor efficacy of 17-DMAG (17-dimethylaminoethylamino-17-demethoxygeldanamycin hydrochloride), a water-soluble geldanamycin derivative. Cancer Chemother Pharmacol 56:115–125

Hong D, Said R, Falchook G, Naing A, Moulder S, Tsimberidou AM, Galluppi G, Dakappagari N, Storgard C, Kurzrock R, Rosen LS (2013a) Phase I study of BIIB028, a selective heat shock protein 90 inhibitor, in patients with refractory metastatic or locally advanced solid tumors. Clin Cancer Res 19(17):4824–4831

Hong DS et al (2013b) Targeting the molecular chaperone heat shock protein 90 (HSP90): lessons learned and future directions. Cancer Treat Rev 39:375–387

Horejsi Z, Takai H, Adelman CA, Collis SJ, Flynn H, Maslen S, Skehel JM, de Lange T, Boulton SJ (2010) CK2 phospho-dependent binding of R2TP complex to TEL2 is essential for mTOR and SMG1 stability. Mol Cell 39:839–850

Hosokawa N, Hirayoshi K, Kudo H, Takechi H, Aoike A, Kawai K, Nagata K (1992) Inhibition of the activation of heat shock factor in vivo and in vitro by flavonoids. Mol Cell Biol 12:3490–3498

Hubbard J, Erlichman C, Toft DO, Qin R, Stensgard BA, Felten S, Ten Eyck C, Batzel G, Ivy SP, Haluska P (2011) Phase I study of 17-allylamino-17 demethoxygeldanamycin, gemcitabine and/or cisplatin in patients with refractory solid tumors. Investig New Drugs 29(3):473–480

Ito A, Matsuoka F, Honda H, Kobayashi T (2004) Anti-tumor effects of combined therapy of recombinant heat shock protein 70 and hyperthermia using magnetic nanoparticles in an experimental subcutaneous murine melanoma. Cancer Immunol Immunother 53:26–32

Iyer G, Morris MJ, Rathkopf D, Slovin SF, Steers M, Larson SM, Schwartz LH, Curley T, DeLaCruz A, Ye Q, Heller G, Egorin MJ, Ivy SP, Rosen N, Scher HI, Solit DB (2012) A phase I trial of docetaxel and pulse-dose 17-allylamino-17-demethoxygeldanamycin in adult patients with solid tumors. Cancer Chemother Pharmacol 69(4):1089–1097

Jakob U, Gaestel M, Engel K, Buchner J (1993) Small heat shock proteins are molecular chaperones. J Biol Chem 268:1517–1520

Janetzki S, Palla D, Rosenhauer V, Lochs H, Lewis JJ, Srivastava PK (2000) Immunization of cancer patients with autologous cancer-derived heat shock protein gp96 preparations: a pilot study. Int J Cancer 88:232–238

Jego G, Hazoume A, Seigneuric R, Garrido C (2010) Targeting heat shock proteins in cancer. Cancer Lett. https://doi.org/10.1016/j.canlet.2010.10.014

Jhaveri K, Miller K, Rosen L, Schneider B, Chap L, Hannah A, Zhong Z, Ma W, Hudis C, Modi S (2012) A phase I dose-escalation trial of trastuzumab and alvespimycin hydrochloride (KOS-1022; 17 DMAG) in the treatment of advanced solid tumors. Clin Cancer Res 18(18):5090–5098

Jinwal UK, Miyata Y, Koren J 3rd, Jones JR, Trotter JH, Chang L et al (2009) Chemical manipulation of hsp70 ATPase activity regulates tau stability. J Neurosci 29:12079–12088

Johnson ML et al (2013) A phase II study of HSP90 inhibitor AUY922 and erlotinib (E) for patients (pts) with EGFR-mutant lung cancer and acquired resistance (AR) to EGFR tyrosine kinase inhibitors (EGFR TKIs). J Clin Oncol 31(Suppl):8036

Johnson ML, Yu HA, Hart EM, Weitner BB, Rademaker AW, Patel JD, Kris MG, Riely GJ (2015) Phase I/II study of HSP90 inhibitor AUY922 and erlotinib for EGFR-mutant lung cancer with acquired resistance to epidermal growth factor receptor tyrosine kinase inhibitors. J Clin Oncol 33:1666–1673

Johnston RN, Kucey BL (1988) Competitive inhibition of hsp70 gene expression causes thermosensitivity. Science 242:1551–1554

Kamal A, Thao L, Sensintaffar J, Zhang L, Boehm MF, Fritz LC et al (2003) A high-affinity conformation of Hsp90 confers tumour selectivity on Hsp90 inhibitors. Nature 425:407–410

Kampinga HH, Hageman J, Vos MJ, Kubota H, Tanguay RM, Bruford EA, Cheetham ME, Chen B, Hightower LE (2009) Guidelines for the nomenclature of the human heat shock proteins. Cell Stress Chaperones 14:105–111

Kaur J, Ralhan R (2000) Induction of apoptosis by abrogation of HSP70 expression in human oral cancer cells. Int J Cancer 85:1–5

Kim JG, Lee SC, Kim OH, Kim KH, Song KY, Lee SK, Choi BJ, Jeong W, Kim SJ (2017) HSP90 inhibitor 17-DMAG exerts anticancer effects against gastric cancer cells principally by altering oxidant-antioxidant balance. Oncotarget. 2017 Apr 10. https://doi.org/10.18632/oncotarget.17007 [Epub ahead of print]

Koldewey P, Horowitz S, Bardwell JCA (2017) Chaperone-client interactions: non-specificity engenders multifunctionality. J Biol Chem 292(29):12010–12017

Kong A, Rea D, Ahmed S, Beck JT, López López R, Biganzoli L, Armstrong AC, Aglietta M, Alba E, Campone M, Hsu Schmitz SF, Lefebvre C, Akimov M, Lee SC (2016) Phase 1B/2 study of the HSP90 inhibitor AUY922 plus trastuzumab in metastatic HER2-positive breast cancer patients who have progressed on trastuzumab-based regimen. Oncotarget 7(25):37680–37692

Koya K, Li Y, Wang H, Ukai T, Tatsuta N, Kawakami M et al (1996) MKT-077, a novel rhodacyanine dye in clinical trials, exhibits anticarcinoma activity in preclinical studies based on selective mitochondrial accumulation. Cancer Res 56:538–543

Kramer OH, Mahboobi S, Sellmer A (2014) Drugging the HDAC6-HSP90 interplay in malignant cells. Trends Pharmacol Sci 35:501–509

Krause SW, Gastpar R, Andreesen R, Gross C, Ullrich H, Thonigs G et al (2004) Treatment of colon and lung cancer patients with ex vivo heat shock protein 70-peptide-activated, autologous natural killer cells: a clinical phase I trial. Clin Cancer Res 10:3699–3707

Kummar S, Gutierrez ME, Gardner ER, Chen X, Figg WD, Zajac-Kaye M, Chen M, Steinberg SM, Muir CA, Yancey MA, Horneffer YR, Juwara L, Melillo G, Ivy SP, Merino M, Neckers L, Steeg PS, Conley BA, Giaccone G, Doroshow JH, Murgo AJ (2010) Phase I trial of 17-dimethylaminoethylamino-17-demethoxygeldanamycin (17-DMAG), a heat shock protein inhibitor, administered twice weekly in patients with advanced malignancies. Eur J Cancer 46(2):340–347

Lancet JE, Gojo I, Burton M, Quinn M, Tighe SM, Kersey K, Zhong Z, Albitar MX, Bhalla K, Hannah AL, Baer MR (2010) Phase I study of the heat shock protein 90 inhibitor alvespimy-

cin (KOS-1022, 17-DMAG) administered intravenously twice weekly to patients with acute myeloid leukemia. Leukemia 24(4):699–705
Laskin JJ, Nicholas G, Lee C, Gitlitz B, Vincent M, Cormier Y et al (2012) Phase I/II trial of custirsen (OGX-011), an inhibitor of clusterin, in combination with a gemcitabine and platinum regimen in patients with previously untreated advanced non-small cell lung cancer. J Thorac Oncol 7(3):579–586
Lee JS, Lee JJ, Seo JS (2005) HSP70 deficiency results in activation of c-Jun N- terminal kinase, extracellular signal-regulated kinase, and caspase-3 in hyperosmolarity-induced apoptosis. J Biol Chem 280:6634–6641
Leu JI, Pimkina J, Frank A, Murphy ME, George DL (2009) A small molecule inhibitor of inducible heat shock protein 70. Mol Cell 36:15–27
Leu JI, Pimkina J, Pandey P, Murphy ME, George DL (2011) HSP70 inhibition by the small-molecule 2-phenylethynesulfonamide impairs protein clearance pathways in tumor cells. Mol Cancer Res 9:936–947
Li Z, Srivastava PK (1993) Tumor rejection antigen gp96/grp94 is an ATPase: implications for protein folding and antigen presentation. EMBO J 12:3143–3151
Li Z, Menoret A, Srivastava P (2002) Roles of heat-shock proteins in antigen presentation and cross-presentation. Curr Opin Immunol 14:45–51
Li JL, Liu HL, Zhang XR, Xu JP, Hu WK, Liang M et al (2009a) A Phase I trial of intra-tumoral administration of recombinant oncolytic adenovirus overexpressing HSP70 in advanced solid tumor patients. Gene Ther 16:376–382
Li Y, Zhang T, Sun D (2009b) New developments in HSp90 inhibitors as anti-cancer therapeutics: mechanisms, clinical perspective and more potential. Drug Resist Updat 12(1–2):17–27
Li T, Chen X, Dai XY, Wei B, Weng QJ, Chen X, Ouyang DF, Yan R, Huang ZJ, Jiang HL, Zhu H, Lu JJ (2017) Novel Hsp90 inhibitor platycodin D disrupts Hsp90/Cdc37 complex and enhances the anticancer effect of mTOR inhibitor. Toxicol Appl Pharmacol 330:65–73
Lindquist S (1986) The heat-shock response. Annu Rev Biochem 55:1151–1191
Liu S, Street TO (2016) 5′-N-ethylcarboxamidoadenosine is not a paralog-specific Hsp90 inhibitor. Protein Sci 25(12):2209–2215
Lundgren K et al (2009) BIIB021, an orally available, fully synthetic small-molecule inhibitor of the heat shock protein Hsp90. Mol Cancer Ther 8:921–929
Maeda Y, Yoshimura K, Matsui H, Shindo Y, Tamesa T, Tokumitsu Y et al (2015) Dendritic cells transfected with heat-shock protein 70 messenger RNA for patients with hepatitis C virus-related hepatocellular carcinoma: a Phase I dose escalation clinical trial. Cancer Immunol Immunother 64:1047–1056
Maki RG, Livingston PO, Lewis JJ, Janetzki S, Klimstra D, Desantis D et al (2007) A phase I pilot study of autologous heat shock protein vaccine HSPPC-96 in patients with resected pancreatic adenocarcinoma. Dig Dis Sci 52:1964–1972
Maloney A, Workman P (2002) Hsp90 as a new therapeutic target for cancer therapy: the story unfolds. Expert Opin Biol Ther 2:3–24
Marcu MG, Chadli A, Bouhouche I, Catelli M, Neckers LM (2000a) The heat shock protein 90 antagonist novobiocin interacts with a previously unrecognized ATP-binding domain in the carboxyl terminus of the chaperone. J Biol Chem 275:37181–37186
Marcu MG, Schulte TW, Neckers L (2000b) Novobiocin and related coumarins and depletion of heat shock protein 90-dependent signaling proteins. J Natl Cancer Inst 92:242–248
Massey AJ, Williamson DS, Browne H, Murray JB, Dokurno P, Shaw T et al (2010) A novel, small molecule inhibitor of Hsc70/Hsp70 potentiates Hsp90 inhibitor induced apoptosis in HCT116 colon carcinoma cells. Cancer Chemother Pharmacol 66:535–545
Mayer MP, Prodromou C, Frydman J (2009) The Hsp90 mosaic: a picture emerges. Nat Struct Mol Biol 16:2–6
Mayor-López L, Tristante E, Carballo-Santana M, Carrasco-García E, Grasso S, García-Morales P et al (2014) Comparative study of 17-AAG and NVP-AUY922 in pancreatic and colorectal cancer cells: are there common determinants of sensitivity? Transl Oncol Elsevier 7:590–604

Mazzaferro V, Coppa J, Carrabba MG, Rivoltini L, Schiavo M, Regalia E et al (2003) Vaccination with autologous tumor-derived heat-shock protein gp96 after liver resection for metastatic colorectal cancer. Clin Cancer Res 9:3235–3245

McCollum AK, TenEyck CJ, Stensgard B, Morlan BW, Ballman KV, Jenkins RB, Toft DO, Erlichman C (2008) P-glycoprotein-mediated resistance to Hsp90-directed therapy is eclipsed by the heat shock response. Cancer Res 68:7419–7427

McConkey DJ, Zhu K (2008) Mechanisms of proteasome inhibitor action and resistance in cancer. Drug Resist Updat 11:164–179

Modi S, Stopeck AT, Gordon MS, Mendelson D, Solit DB, Bagatell R et al (2007) Combination of trastuzumab and tanespimycin (17-AAG, KOS-953) is safe and active in trastuzumab-refractory HER-2 overexpressing breast cancer: a phase I dose-escalation study. J Clin Oncol Am Soc Clin Oncol 25:5410–5417

Modi S, Stopeck A, Linden H, Solit D, Chandarlapaty S, Rosen N, D'Andrea G, Dickler M, Moynahan ME, Sugarman S, Ma W, Patil S, Norton L, Hannah AL, Hudis C (2011) HSP90 inhibition is effective in breast cancer: a phase II trial of tanespimycin (17-AAG) plus trastuzumab in patients with HER2-positive metastatic breast cancer progressing on trastuzumab. Clin Cancer Res 17(15):5132–5139

Modi S, Saura C, Henderson C, Lin NU, Mahtani R, Goddard J, Rodenas E, Hudis C, O'Shaughnessy J, Baselga J (2013) A multicenter trial evaluating retaspimycin HCL(IPI-504) plus trastuzumab in patients with advanced or metastatic HER2-positive breast cancer. Breast Cancer Res Treat 139(1):107–113

Multhoff G, Botzler C, Wiesnet M, Muller E, Meier T, Wilmanns W et al (1995) A stress-inducible 72-kDa heat-shock protein (HSP72) is expressed on the surface of human tumor cells, but not on normal cells. Int J Cancer 61:272–279

Multhoff G, Botzler C, Jennen L, Schmidt J, Ellwart J, Issels R (1997a) Heat shock protein 72 on tumor cells: a recognition structure for natural killer cells. J Immunol 158:4341–4350

Multhoff G, Mizzen L, Winchester CC, Milner CM, Wenk S, Eissner G et al (1997b) Heat shock protein 70 (Hsp70) stimulates proliferation and cytolytic activity of natural killer cells. Exp Hematol 27:1627–1636

Multhoff G, Mizzen L, Winchester CC, Milner CM, Wenk S, Eissner G, Kampinga HH, Laumbacher B, Johnson J (1999) Heat shock protein 70 (Hsp70) stimulates proliferation and cytolytic activity of natural killer cells. Exp Hematol. 27(11):1627–1636 PubMed PMID: 10560910

Multhoff G, Pfister K, Gehrmann M, Hantschel M, Gross C, Hafner M et al (2001) A 14-mer Hsp70 peptide stimulates natural killer (NK) cell activity. Cell Stress Chaperones 6:337–344

Murakami N, Kuhnel A, Schmid TE, Ilicic K, Stangl S, Braun IS et al (2015) Role of membrane Hsp70 in radiation sensitivity of tumor cells. Radiat Oncol 10:149

Murshid A, Theriault J, Gong J, Calderwood SK (2011) Investigating receptors for extracellular heat shock proteins. Methods Mol Biol 787:289–302

Murshid A, Gong J, Calderwood SK (2012) The role of heat shock proteins in antigen cross presentation. Front Immunol 3:63

Murshid A, Borges TJ, Calderwood SK (2015) Emerging roles for scavenger receptor SREC-I in immunity. Cytokine 75:256–260

Nadeau K, Nadler SG, Saulnier M, Tepper MA, Walsh CT (1994) Quantitation of the interaction of the immunosuppressant deoxyspergualin and analogs with Hsc70 and Hsp90. Biochemistry 33:2561–2567

Neckers L (2002) Hsp90 inhibitors as novel cancer chemotherapeutic agents. Trends Mol Med 8:S55–S61

Nestle FO, Alijagic S, Gilliet M, Sun Y, Grabbe S, Dummer R et al (1998) Vaccination of melanoma patients with peptide- or tumor lysate-pulsed dendritic cells. Nat Med 4:328–332

Nishikawa M, Otsuki T, Ota A, Guan X, Takemoto S, Takahashi Y et al (2010) Induction of tumor-specific immune response by gene transfer of Hsp70-cell-penetrating peptide fusion protein to tumors in mice. Mol Ther 18:421–428

Noessner E, Gastpar R, Milani V, Brandl A, Hutzler PJ, Kuppner MC et al (2002) Tumor-derived heat shock protein 70 peptide complexes are cross-presented by human dendritic cells. J Immunol 169:5424–5432

Oh WK, Galsky MD, Stadler WM, Srinivas S, Chu F, Bubley G, Goddard J, Dunbar J, Ross RW (2011) Multicenter phase II trial of the heat shock protein 90 inhibitor, retaspimycin hydrochloride (IPI-504), in patients with castration-resistant prostate cancer. Urology 78(3):626–630

Okawa Y et al (2009) SNX-2112, a selective Hsp90 inhibitor, potently inhibits tumor cell growth, angiogenesis, and osteoclastogenesis in multiple myeloma and other hematologic tumors by abrogating signaling via Akt and ERK. Blood 113:846–855

Oki Y, Younes A (2004) Heat shock protein-based cancer vaccines. Expert Rev Vaccines 3:403–411

Oki Y, Younes A, Knickerbocker J, Samaniego F, Nastoupil L, Hagemeister F, Romaguera J, Fowler N, Kwak L, Westin J (2015) Experience with HSP90 inhibitor AUY922 in patients with relapsed or refractory non-Hodgkin lymphoma. Haematologica 100(7):e272–e274

Orosz A, Szabo A, Szeman G et al (2006) Novel nontoxic heat shock protein 90 inhibitors having selective antiproliferative effect. Int J Biochem Cell Biol 38:1352–1362

Pacey S, Wilson RH, Walton M, Eatock MM, Hardcastle A, Zetterlund A, Arkenau HT, Moreno-Farre J, Banerji U, Roels B, Peachey H, Aherne W, de Bono JS, Raynaud F, Workman P, Judson I (2011) A phase I study of the heat shock protein 90 inhibitor alvespimycin (17-DMAG) given intravenously to patients with advanced solid tumors. Clin Cancer Res 17(6):1561–1570

Pacey S, Gore M, Chao D, Banerji U, Larkin J, Sarker S, Owen K, Asad Y, Raynaud F, Walton M, Judson I, Workman P, Eisen T (2012) A phase II trial of 17-allylamino, 17-demethoxygeldanamycin (17-AAG, tanespimycin) in patients with metastatic melanoma. Investig New Drugs 30(1):341–349

Park HS, Lee JS, Huh SH, Seo JS, Choi EJ (2001) Hsp72 functions as a natural inhibitory protein of c-Jun N-terminal kinase. EMBO J 20:446–456

Park HS, Cho SG, Kim CK, Hwang HS, Noh KT, Kim MS et al (2002) Heat shock protein hsp72 is a negative regulator of apoptosis signal-regulating kinase 1. Mol Cell Biol 22:7721–7730

Park SH, Kim WJ, Li H, Seo W, Park SH, Kim H et al (2017) Anti-leukemia activity of a Hsp70 inhibitor and its hybrid molecules. Sci Rep 7(1):3537

Pearl LH (2005) Hsp90 and Cdc37–a chaperone cancer conspiracy. Curr Opin Genet Dev 15:55–61

Pearl LH, Prodromou C, Workman P (2008) The Hsp90 molecular chaperone: an open and shut case for treatment. Biochem J 410:439–453

Pedersen KS, Kim GP, Foster NR, Wang-Gillam A, Erlichman C, McWilliams RR (2015) Phase II trial of gemcitabine and tanespimycin (17AAG) in metastatic pancreatic cancer: a Mayo Clinic Phase II Consortium study. Investig New Drugs 33(4):963–968

Pegram HJ, Andrews DM, Smyth MJ, Darcy PK, Kershaw MH (2011) Activating and inhibitory receptors of natural killer cells. Immunol Cell Biol 89:216–224

Peng P, Menoret A, Srivastava PK (1997) Purification of immunogenic heat shock protein 70-peptide complexes by ADP-affinity chromatography. J Immunol Methods 204:13–21

Peng C, Brain J, Hu Y et al (2007) Inhibition of heat shock protein 90 prolongs survival of mice with BCR-ABL-T315I-induced leukemia and suppresses leukemic stem cells. Blood 110:678–685

Picard D (2002) Heat-shock protein 90, a chaperone for folding and regulation. Cell Mol Life Sci 59:1640–1648

Pilla L, Patuzzo R, Rivoltini L, Maio M, Pennacchioli E, Lamaj E et al (2006) A phase II trial of vaccination with autologous, tumor-derived heat-shock protein peptide complexes Gp96, in combination with GM-CSF and interferon-alpha in metastatic melanoma patients. Cancer Immunol Immunother 55:958–968

Powers MV, Clarke PA, Workman P (2008) Dual targeting of HSC70 and HSP72 inhibits HSP90 function and induces tumor-specific apoptosis. Cancer Cell 14:250–262

Powers MV, Jones K, Barillari C, Westwood I, van Montfort RL, Workman P (2010) Targeting HSP70: the second potentially druggable heat shock protein and molecular chaperone? Cell Cycle 9:1542–1550

Pratt WB (1998) The hsp90-based chaperone system: involvement in signal transduction from a variety of hormone and growth factor receptors. Proc Soc Exp Biol Med 217:420–434

Pratt WB, Morishima Y, Murphy M, Harrell M (2006) Chaperoning of glucocorticoid receptors. Handb Exp Pharmacol:111–138

Propper DJ, Braybrooke JP, Taylor DJ, Lodi R, Styles P, Cramer JA et al (1999) Phase I trial of the selective mitochondrial toxin MKT077 in chemo-resistant solid tumours. Ann Oncol 10:923–927
Radons J (2016) The human HSP70 family of chaperones: where do we stand? Cell Stress Chaperones 21(3):379–404
Rafiee M, Kanwar JR, Berg RW, Lehnert K, Lisowska K, Krissansen GW (2001) Induction of systemic anti-tumor immunity by gene transfer of mammalian heat shock protein 70.1 into tumors in situ. Cancer Gene Ther 8:974–981
Ramalingam SS, Egorin MJ, Ramanathan RK, Remick SC, Sikorski RP, Lagattuta TF, Chatta GS, Friedland DM, Stoller RG, Potter DM, Ivy SP, Belani CP (2008) A phase I study of 17-allyl amino-17-demethoxygeldanamycin combined with paclitaxel in patients with advanced solid malignancies. Clin Cancer Res 14(11):3456–3461
Ramalingam S, Goss G, Rosell R, Schmid-Bindert G, Zaric B, Andric Z, Bondarenko I, Komov D, Ceric T, Khuri F et al (2015) A randomized phase II study of ganetespib, a heat shock protein 90 inhibitor, in combination with docetaxel in second-line therapy of advanced non-small cell lung cancer (GALAXY-1). Ann Oncol 26:1741–1748
Rao R, Fiskus W, Yang Y, Lee P, Joshi R, Fernandez P, Mandawat A, Atadja P, Bradner JE, Bhalla K (2008) HDAC6 inhibition enhances 17-AAG–mediated abrogation of hsp90 chaperone function in human leukemia cells. Blood 112:1886–1893
Reddy N, Voorhees PM, Houk BE, Brega N, Hinson JM Jr, Jillela A (2013) Phase I trial of the HSP90 inhibitor PF-04929113 (SNX5422) in adult patients with recurrent, refractory hematologic malignancies. Clin Lymphoma Myeloma Leuk 13(4):385–391
Renouf DJ, Hedley D, Krzyzanowska MK, Schmuck M, Wang L, Moore MJ (2016) A phase II study of the HSP90 inhibitor AUY922 in chemotherapy refractory advanced pancreatic cancer. Cancer Chemother Pharmacol 78(3):541–545
Rerole AL, Gobbo J, De Thonel A, Schmitt E, Pais de Barros JP, Hammann A et al (2011) Peptides and aptamers targeting HSP70: a novel approach for anticancer chemotherapy. Cancer Res 71:484–495
Riabowol KT, Mizzen LA, Welch WJ (1988) Heat shock is lethal to fibroblasts microinjected with antibodies against hsp70. Science 242:433–436
Richardson PG, Badros AZ, Jagannath S, Tarantolo S, Wolf JL, Albitar M, Berman D, Messina M, Anderson KC (2010) Tanespimycin with bortezomib: activity in relapsed/refractory patients with multiple myeloma. Br J Haematol 150(4):428–437
Richter K, Buchner J (2001) Hsp90: chaperoning signal transduction. J Cell Physiol 188:281–290
Roe SM, Prodromou C, O'Brien R, Ladbury JE, Piper PW, Pearl LH (1999) Structural basis for inhibition of the Hsp90 molecular chaperone by the anti-tumor antibiotics radicicol and geldanamycin. J Med Chem 42:260–266
Ronnen EA, Kondagunta GV, Ishill N, Sweeney SM, Deluca JK, Schwartz L, Bacik J, Motzer RJ (2006) A phase II trial of 17-(allylamino)-17-demethoxygeldanamycin in patients with papillary and clear cell renal cell carcinoma. Investig New Drugs 24:543–546
Saif MW, Erlichman C, Dragovich T, Mendelson D, Toft D, Burrows F, Storgard C, Von Hoff D (2013) Open-label, dose-escalation, safety, pharmacokinetic, and pharmacodynamic study of intravenously administered CNF1010(17-(allylamino)-17-demethoxygeldanamycin [17-AAG]) in patients with solid tumors. Cancer Chemother Pharmacol 71(5):1345–1355
Samuel TA et al (2010) AUY922, a novel HSP90 inhibitor: final results of a first-in-human study in patients with advanced solid malignancies. J Clin Oncol 28(Suppl):2528
Schadendorf D, Ugurel S, Schuler-Thurner B, Nestle FO, Enk A, Brocker EB et al (2006) Dacarbazine (DTIC) versus vaccination with autologous peptide-pulsed dendritic cells (DC) in first-line treatment of patients with metastatic melanoma: a randomized phase III trial of the DC study group of the DeCOG. Ann Oncol 17:563–570
Schaefer S, Svenstrup TH, Guerra B (2017) The small-molecule kinase inhibitor D11 counteracts 17-AAG-mediated up-regulation of HSP70 in brain cancer cells. PLoS One 12(5):e0177706
Schildkopf P, Frey B, Ott OJ, Rubner Y, Multhoff G, Sauer R et al (2011) Radiation combined with hyperthermia induces Hsp70-dependent maturation of dendritic cells and release of pro-inflammatory cytokines by dendritic cells and macrophages. Radiother Oncol 101:109–115

Schmitt E, Maingret L, Puig PE, Rerole AL, Ghiringhelli F, Hammann A et al (2006) Heat shock protein 70 neutralization exerts potent anti-tumor effects in animal models of colon cancer and melanoma. Cancer Res 66:4191–4197

Schroder CP et al (2011) Use of biomarkers and imaging to evaluate the treatment effect of AUY922, an HSP90 inhibitor, in patients with HER2+ or ER+ metastatic breast cancer. J Clin Oncol 29(Suppl):e11024

Schulte TW, Akinaga S, Soga S, Sullivan W, Stensgard B, Toft D, Neckers LM (1998) Antibiotic radicicol binds to the N-terminal domain of Hsp90 and shares important biological activities with geldanamycin. Cell Stress Chaperones 3:100–108

Scroggins BT, Robzyk K, Wang D, Marcu MG, Tsutsumi S, Beebe K, Cotter RJ, Felts S, Toft D, Karnitz L et al (2007) An acetylation site in the middle domain of Hsp90 regulates chaperone function. Mol Cell 25:151–159

Seggewiss-Bernhardt R, Bargou RC, Goh YT, Stewart AK, Spencer A, Alegre A, Bladé J, Ottmann OG, Fernandez-Ibarra C, Lu H, Pain S, Akimov M, Iyer SP (2015) Phase 1/1B trial of the heat shock protein 90 inhibitor NVP-AUY922 as monotherapy or in combination with bortezomib in patients with relapsed or refractory multiple myeloma. Cancer 121(13):2185–2192

SenGupta D, Norris PJ, Suscovich TJ, Hassan-Zahraee M, Moffett HF, Trocha A et al (2004) Heat shock protein-mediated cross-presentation of exogenous HIV antigen on HLA class I and class II. J Immunol 173:1987–1993

Sessa C et al (2013) First-in-human phase I dose-escalation study of the HSP90 inhibitor AUY922 in patients with advanced solid tumors. Clin Cancer Res 19:3671–3680

Sharp SY et al (2007a) In vitro biological characterization of a novel, synthetic diaryl pyrazole resorcinol class of heat shock protein 90 inhibitors. Cancer Res 67:2206–2216

Sharp SY et al (2007b) Inhibition of the heat shock protein 90 molecular chaperone in vitro and in vivo by novel, synthetic, potent resorcinylic pyrazole/isoxazole amide analogues. Mol Cancer Ther 6:1198–1211

Shevtsov MA, Kim AV, Samochernych KA, Romanova IV, Margulis BA, Guzhova IV et al (2014a) Pilot study of intra-tumoral injection of recombinant heat shock protein 70 in the treatment of malignant brain tumors in children. Onco Targets Ther 7:1071–1081

Shevtsov MA, Komarova EY, Meshalkina DA, Bychkova NV, Aksenov ND, Abkin SV et al (2014b) Exogenously delivered heat shock protein 70 displaces its endogenous analogue and sensitizes cancer cells to lymphocytes-mediated cytotoxicity. Oncotarget 5:3101–3114

Shevtsov MA, Pozdnyakov AV, Mikhrina AL, Yakovleva LY, Nikolaev BP, Dobrodumov AV et al (2014c) Effective immunotherapy of rat glioblastoma with prolonged intra-tumoral delivery of exogenous heat shock protein Hsp70. Int J Cancer 135:2118–2128

Shevtsov MA, Nikolaev BP, Yakovleva LY, Parr MA, Marchenko YY, Eliseev I et al (2015) 70-kDa heat shock protein coated magnetic nanocarriers as a nanovaccine for induction of anti-tumor immune response in experimental glioma. J Control Release 220:329–340

Shin BK, Wang H, Yim AM, Le Naour F, Brichory F, Jang JH et al (2003) Global profiling of the cell surface proteome of cancer cells uncovers an abundance of proteins with chaperone function. J Biol Chem 278:7607–7616

Siegel D, Jagannath S, Vesole DH, Borello I, Mazumder A, Mitsiades C, Goddard J, Dunbar J, Normant E, Adams J, Grayzel D, Anderson KC, Richardson P (2011) A phase 1 study of IPI-504 (retaspimycin hydrochloride) in patients with relapsed or relapsed and refractory multiple myeloma. Leuk Lymphoma 52(12):2308–2315

Smith JR, Clarke PA, de Billy E, Workman P (2008) Silencing the cochaperone CDC37 destabilizes kinase clients and sensitizes cancer cells to HSP90 inhibitors. Oncogene. https://doi.org/10.1038/onc.2008.380

Soga S, Shiotsu Y, Akinaga S, Sharma SV (2003) Development of radicicol analogues. Curr Cancer Drug Targets 3:359–369

Soldano KL, Jivan A, Nicchitta CV, Gewirth DT (2003) Structure of the N-terminal domain of GRP94. Basis for ligand specificity and regulation. J Biol Chem 278:48330–48338

Solit DB, Osman I, Polsky D, Panageas KS, Daud A, Goydos JS, Teitcher J, Wolchok JD, Germino FJ, Krown SE, Coit D, Rosen N, Chapman PB (2008a) Phase II trial of 17-allylamino-

17-demethoxygeldanamycin in patients with metastatic melanoma. Clin Cancer Res 14(24):8302–8307

Solit DB, Osman I, Polsky D, Panageas KS, Daud A, Goydos JS, Teitcher J, Wolchok JD, Germino FJ, Krown SE (2008b) Phase II trial of 17-allylamino-17-demethoxygeldanamycin in patients with metastatic melanoma. Clin Cancer Res 14:8302–8307

Specht HM, Ahrens N, Blankenstein C, Duell T, Fietkau R, Gaipl US et al (2015) Heat shock protein 70 (Hsp70) peptide activated natural killer (NK) cells for the treatment of patients with non-small cell lung cancer (NSCLC) after radiochemotherapy (RCTx) – from preclinical studies to a clinical phase II trial. Front Immunol 6:162

Sreedhar AS, Soti C, Csermely P (2004) Inhibition of Hsp90: a new strategy for inhibiting protein kinases. Biochim Biophys Acta 1697:233–242

Srivastava PK (1994) Heat shock proteins in immune response to cancer: the fourth paradigm. Experientia 50:1054–1060

Srivastava PK (1997) Purification of heat shock protein-peptide complexes for use in vaccination against cancers and intracellular pathogens. Methods 12:165–171

Srivastava P (2002a) Interaction of heat shock proteins with peptides and antigen presenting cells: chaperoning of the innate and adaptive immune responses. Annu Rev Immunol 20:395–425

Srivastava PK (2002b) Roles of heat-shock proteins in innate and adaptive immunity. Nat Rev Immunol 2:185–194

Srivastava PK (2006) Therapeutic cancer vaccines. Curr Opin Immunol 18:201–205

Srivastava IK, Liu MA (2003) Gene vaccines. Ann Intern Med 138:550–559

Srivastava PK, DeLeo AB, Old LJ (1986) Tumor rejection antigens of chemically induced sarcomas of inbred mice. Proc Natl Acad Sci U S A 83:3407–3411

Stangl S, Gehrmann M, Riegger J, Kuhs K, Riederer I, Sievert W et al (2011) Targeting membrane heat-shock protein 70 (Hsp70) on tumors by cmHsp70.1 antibody. Proc Natl Acad Sci U S A 108:733–738

Sullivan LC, Clements CS, Beddoe T, Johnson D, Hoare HL, Lin J et al (2007) The heterodimeric assembly of the CD94-NKG2 receptor family and implications for human leukocyte antigen-E recognition. Immunity 27:900–911

Supko JG, Hickman RL, Grever MR, Malspeis L (1995) Preclinical pharmacologic evaluation of geldanamycin as an anti-tumor agent. Cancer Chemother Pharmacol 36:305–315

Suto R, Srivastava PK (1995) A mechanism for the specific immunogenicity of heat shock protein-chaperoned peptides. Science 269:1585–1588

Takai H, Xie Y, de Lange T, Pavletich NP (2010) Tel2 structure and function in the Hsp90-dependent maturation of mTOR and ATR complexes. Genes Dev 24:2019–2030

Takemoto S, Nishikawa M, Guan X, Ohno Y, Yata T, Takakura Y (2010) Enhanced generation of cytotoxic T lymphocytes by heat shock protein 70 fusion proteins harboring both CD8(+) T cell and CD4(+) T cell epitopes. Mol Pharm 7:1715–1723

Taldone T, Gozman A, Maharaj R, Chiosis G (2008) Targeting Hsp90: small-molecule inhibitors and their clinical development. Curr Opin Pharmacol 8:370–374

Tatokoro M, Koga F, Yoshida S, Kihara K (2015) Heat shock protein 90 targeting therapy: state of the art and future perspective. EXCLI J 14:48–58

Testori A, Richards J, Whitman E, Mann GB, Lutzky J, Camacho L et al (2008) Phase III comparison of vitespen, an autologous tumor-derived heat shock protein gp96 peptide complex vaccine, with physician's choice of treatment for stage IV melanoma: the C-100-21 Study Group. J Clin Oncol 26:955–962

Tosti G, Cocorocchio E, Pennacchioli E, Ferrucci PF, Testori A, Martinoli C (2014) Heat-shock proteins-based immunotherapy for advanced melanoma in the era of target therapies and immunomodulating agents. Expert Opin Biol Ther 14:955–967

Tse AN, Klimstra DS, Gonen M, Shah M, Sheikh T, Sikorski R, Carvajal R, Mui J, Tipian C, O'Reilly E, Chung K, Maki R, Lefkowitz R, Brown K, Manova-Todorova K, Wu N, Egorin MJ, Kelsen D, Schwartz GK (2008) A phase 1 dose-escalation study of irinotecan in combination with 17-allylamino-17-demethoxygeldanamycin in patients with solid tumors. Clin Cancer Res 14(20):6704–6711

Udono H, Srivastava PK (1993) Heat shock protein 70-associated peptides elicit specific cancer immunity. J Exp Med 178:1391–1396

Udono H, Levey DL, Srivastava PK (1994) Cellular requirements for tumor-specific immunity elicited by heat shock proteins: tumor rejection antigen gp96 primes CD8+ T cells in vivo. Proc Natl Acad Sci U S A 91:3077–3081

Ullrich SJ, Robinson EA, Law LW, Willingham M, Appella E (1986) A mouse tumor-specific transplantation antigen is a heat shock-related protein. Proc Natl Acad Sci U S A 83:3121–3125

Vabulas RM, Ahmad-Nejad P, Ghose S, Kirschning CJ, Issels RD, Wagner H (2002) HSP70 as endogenous stimulus of the toll/interleukin-1 receptor signal pathway. J Biol Chem 277:15107–15112

Vaughan CK, Gohlke U, Sobott F et al (2006) Structure of an Hsp90-Cdc37-Cdk4 complex. Mol Cell 23:697–707

Vicencio JM, Yellon DM, Sivaraman V, Das D, Boi-Doku C, Arjun S et al (2015) Plasma exosomes protect the myocardium from ischemia-reperfusion injury. J Am Coll Cardiol 65:1525–1536

Vogel M, Bukau B, Mayer MP (2006) Allosteric regulation of Hsp70 chaperones by a proline switch. Mol Cell 21:359–367

Wagner AJ, Chugh R, Rosen LS, Morgan JA, George S, Gordon M, Dunbar J, Normant E, Grayzel D, Demetri GD (2013) A phase I study of the HSP90 inhibitor retaspimycin hydrochloride (IPI-504) in patients with gastrointestinal stromal tumors or soft-tissue sarcomas. Clin Cancer Res 19(21):6020–6029

Walker AR, Klisovic R, Johnston JS, Jiang Y, Geyer S, Kefauver C, Binkley P, Byrd JC, Grever MR, Garzon R, Phelps MA, Marcucci G, Blum KA, Blum W (2013) Pharmacokinetics and dose escalation of the heat shock protein inhibitor 17-allyamino-17-demethoxygeldanamycin in combination with bortezomib in relapsed or refractory acute myeloid leukemia. Leuk Lymphoma 54(9):1996–2002

Weigel BJ, Blaney SM, Reid JM, Safgren SL, Bagatell R, Kersey J, Neglia JP, Ivy SP, Ingle AM, Whitesell L, Gilbertson RJ, Krailo M, Ames M, Adamson PC (2007) A phase I study of 17-allylaminogeldanamycin in relapsed/refractory pediatric patients with solid tumors: a Children's Oncology Group study. Clin Cancer Res 13(6):1789–1793

Whetstone H, Lingwood C (2003) 3′Sulfogalactolipid binding specifically inhibits Hsp70 ATPase activity in vitro. Biochemistry 42(6):1611–1617

Whitesell L, Lindquist SL (2005) HSP90 and the chaperoning of cancer. Nat Rev Cancer 5:761–772

Whitesell L, Bagatell R, Falsey R (2003) The stress response: implications for the clinical development of hsp90 inhibitors. Curr Cancer Drug Targets 3:349–358

Whitesell L, Santagata S, Lin NU (2012) Inhibiting HSP90 to treat cancer: a strategy in evolution. Curr Mol Med 12:1108–1124

Williams DR, Ko SK, Park S, Lee MR, Shin I (2008) An apoptosis-inducing small molecule that binds to heat shock protein 70. Angew Chem Int Ed Eng 47:7466–7469

Wisen S, Bertelsen EB, Thompson AD, Patury S, Ung P, Chang L et al (2010) Binding of a small molecule at a protein–protein interface regulates the chaperone activity of hsp70–hsp40. ACS Chem Biol 5:611–622

Wood C (2008) A vaccine for renal cancer. Lancet 372:1460–1461

Wright CM, Seguin SP, Fewell SW, Zhang H, Ishwad C, Vats A et al (2009) Inhibition of simian virus 40 replication by targeting the molecular chaperone function and ATPase activity of T antigen. Virus Res 141(1):71–80

Xia Y, Liu Y, Rocchi P, Wang M, Fan Y, Qu F, Iovanna JL, Peng L (2012) Targeting heat shock factor 1 with a triazole nucleoside analog to elicit potent anticancer activity on drug-resistant pancreatic cancer. Cancer Lett 318:145–153

Yamamoto K, Garbaccio RM, Stachel SJ et al (2003) Total synthesis as a resource in the discovery of potentially valuable anti-tumor agents: cycloproparadicicol. Angew Chem Int Ed Eng 42:1280–1284

Yang Y, Liu B, Dai J, Srivastava PK, Zammit DJ, Lefrancois L et al (2007) Heat shock protein gp96 is a master chaperone for toll-like receptors and is important in the innate function of macrophages. Immunity 26:215–226

Yi F, Regan L (2008) A novel class of small molecule inhibitors of Hsp90. ACS Chem Biol 3:645–654

Yin Z, Henry EC, Gasiewicz TA (2009) (−)-Epigallocatechin-3-gallate is a novel Hsp90 inhibitor. Biochemistry 48:336–345

Yokota S, Kitahara M, Nagata K (2000) Benzylidene lactam compound, KNK437, a novel inhibitor of acquisition of thermotolerance and heat shock protein induction in human colon carcinoma cells. Cancer Res 60:2942–2948

Yu S, Cai X, Wu C, Liu Y, Zhang J, Gong X, Wang X, Wu X, Zhu T, Mo L, Gu J, Yu Z, Chen J, Thiery JP, Chai R, Chen L (2017) Targeting HSP90-HDAC6 regulating network implicates precision treatment of breast cancer. Int J Biol Sci 13(4):505–517

Zhang H, Burrows F (2004) Targeting multiple signal transduction pathways through inhibition of Hsp90. J Mol Med 82:488–499

Zhang C, Wang X, Liu H, Zhang M, Geng M, Sun L, Shen A, Zhang A (2017) Design, synthesis and pharmacological evaluation of 4,5-diarylisoxazols bearing amino acid residues within the 3-amido motif as potent heat shock protein 90 (Hsp90) inhibitors. Eur J Med Chem 125:315–326

Zoubeidi A, Gleave M (2012) Small heat shock proteins in cancer therapy and prognosis. Int J Biochem Cell Biol 44(10):1646–1656

Part II
Physiological Responses

Chapter 12
Physical Exercise and Heat Shock Proteins

Amin Isanejad and Hamid Amini

Abstract Molecular chaperones consist of several highly conserved families of proteins, many of which consist of heat shock proteins. Heat shock proteins (HSP) are a group of cytoprotective proteins critical in the maintenance of protein and cellular homeostasis and protect the cell against further insults. HSP also can actively release to circulation and function as chaperokine. It has become suggested that they induced or activated with acute exercise or after chronic exercise training. It seems that a combination of thermal, oxidative and metabolic stress and muscle damage involved in exercise-induced HSP. These chaperones can play important role in biological response and adaptation by exercise. It is well known that the levels of certain molecular chaperones are elevated during stress to provide protection to the cell. For example stress response get disturbed in some conditions such as, cancer, diabetes and Alzheimer's, furthermore in conditions that may stimulated disease such as aging. In the present chapter we discussed the role of exercise on these classes of chaperones in biology of exercise and human diseases. Exercise training, by improving the level of HSP, can help to amplify the stress response. These observations provided an opportunity to exploit protective role of HSP health and physical performance. Therefore, identifying the biological effects of these chaperons in exercise can lead to a greater understanding of the mechanisms of exercise in medicine and exercise adaptation biology.

Keywords Chaperones · HSP · Muscle adaptation · Exercise training · ROS · Aging

A. Isanejad (✉)
Immunoregulation Research Center, Shahed University, Tehran, Iran

Sport Sciences Department, Shahed University, Tehran, Iran
e-mail: A.isanezhad@shahed.ac.ir

H. Amini
Department of Physical Education and Sports Sciences, Yadegar-e-Imam Khomeini (RAH) Shahre Rey Branch, Islamic Azad University, Tehran, Iran

A. A. A. Asea, P. Kaur (eds.), *Chaperokine Activity of Heat Shock Proteins*, Heat Shock Proteins 16, https://doi.org/10.1007/978-3-030-02254-9_12

Abbreviations

ADT	Androgen deprivation therapy
BiP	Immunoglobulin binding protein
CCT	Chaperonin containing TCP-1 beta subunit
CK	Creatine kinase
CRP	C reactive protein
ER	Endoplasmic reticulum
ERK	Extracellular signal-regulated kinase
FOXOs	Forkhead box protein O1
GRP75	Glucose-regulated protein 75
GS	Glycogen synthase
GSK3αm	Glycogen synthase kinases α
HSF	Heat shock factor
HSF1	Heat shock factor1
HSP	Heat shock proteins
IKK	Inhibitor of NF-kb kinase
IL-6	Interleukin-6
IRE1a	Inositol-requiring transmembrane kinase/endonuclease-1
JNK	c-Jun NH2-terminal kinase
NFkB	Nuclear factor kappa-light-chain-enhancer of activated B cells
NGEMP	Nuclear genes encoding mitochondrial proteins
NP22	Neuronal protein 22
P13K/AKT	Phosphoinositide 3-kinase-Akt
p38 MAPK	P38 mitogen-activated protein kinases
PGC-1	Peroxisome proliferator-activated receptor gamma coactivator 1-alpha
RNS	Reactive nitrogen species
ROS	Reactive oxygen species
SIRT1	Silent mating type information regulation 2 homolog
SOD	Superoxide dismutase
TNF-α	Tumor necrosis factor alpha
TRL	Toll like receptor
UPR	Unfolded protein response

12.1 Introduction

Physiologically, stress can be viewed as any disruption to homeostasis and can exist at the level of the cell, tissue, organ, organ system or whole organism. At the cellular level, the rapid change in expression of a family of highly conserved chaperons such as heat shock proteins (HSP) are one of the key responses to stress (Kregel 2002). Heat shock response (HSR) is one of the important cellular endogenous components that allow the body to cope with stressful conditions including metabolic stress (Richter et al. 2010; Dimauro et al. 2016). This response involves a set of highly conserved proteins called heat shock proteins (HSP) known also as molecular chaperones and other proteins essential for protection and recovery from stress (Richter et al. 2010; Dimauro et al. 2016). The primary functions of molecular chaperones are to bind selectively to nascent polypeptides and partially folded intermediates, preventing their irreversible aggregation and misfolding and to facilitate protein refolding. As a chaperone, HSPs are therefore considered to perform important 'housekeeping' functions within the cell and are critical for efficient protein turnover (Fehrenbach and Northoff 2001). HSP are a group of proteins with critical role in reverse or inhibit denaturation or unfolding of cellular proteins from chronically and acutely stressful conditions such as high temperature. These also involved in the control of cell signaling, modulation of immune response and chronic disease conditions (Richter et al. 2010; Dimauro et al. 2016; Shamsi et al. 2016). At first, it was believed that these proteins reacted only to heat stress but it now known to be induced by different kinds of stress such as exercise. The members of the HSP family are primarily classified according to their molecular weight including HSP90, HSP70 family, mitochondrial HSP60 and the small HSP such as HSP27 and αB-crystalin. HSP27 is localized in the cytosol in the unstressed cell and is translocate to within or around the nucleus following stress (Arrigo and Welch 1987). In addition to the role of HSP27 in cell growth and transformation (Welch 1992), it has been suggested to protection against thermal (Locke 1997) and oxidative stress (Escobedo et al. 2004). HSP72 as a chaperon also play a direct role in protecting skeletal muscle from contraction- induced damage (Koh 2002). αB-crystallin functions as a molecular chaperone to either prevent aggregation of denatured proteins or to facilitate refolding upon the removal of stress (Jakob et al. 1993). It has also recently been demonstrated that αB-crystallin maintains skeletal muscle myosin ATPase activity and prevents its aggregation under heat shock stress (Melkani et al. 2006). Therefore, these small HSP can play an important physiological role which functions to protect the skeletal muscle during stressful insults, such as exercise, in an attempt to maintain muscle performance. HSP60 is located primarily within the muscle mitochondrial matrix that plays an important role in normal cell function by facilitating the correct folding and assembly of proteins as they enter the mitochondria and facilitating protein transport across intracellular membrane (Hood 2001). The most highly conserved of HSPs and the most widely studied to date are members of the HSP70 family. Four major isoforms of the HSP70 family have been identified in mammalian cells, the most prominent of which include a cognate isoform (referred to as HSC70 or HSP73) and an inducible isoform (referred to as HSP70 or HSP72) (Morton et al. 2007). HSP70 is considered to be primarily stress inducible (Millarski and Morimoto 1989). HSP70 is the key stress protein involved in cytoprotection (Locke and Noble 2002). Also HSP70 can protect the cells and tissues from thermal stress (Katschinski 2004) and damaging contractions (Isanejad et al. 2015; McArdle et al. 2004). In this regard, elevated levels of HSP70 in cells or tissues by exercise may therefore permit increased tolerance to the biochemical and physiological stresses induced by exercise (Locke 1997).HSP70 level is higher in slow twitch muscle fiber (type I) than fast twitch fibers (II) (O'Neill et al. 2006) because of increased rate of protein turnover that is characteristic of these fibers (Obinata et al. 1981). Another member of HSP70 family is HSC70 that constitutively expressed in the cytoplasm of unstressed cells at higher levels than HSP70 (Morton et al. 2007). In this way, HSC70 is only slightly increased after stressful condition such as exercise (McArdle et al. 2001). The remaining forms of HSP70 family are GRP75 and GRP78 that located in the sarcoplasmic/endoplasmic reticulum and mitochondria respectively (Pelham 1986). Mattson et al. (2000) demonstrated an increased expression of GRP75 in rodent plantaris muscle following 8 weeks of endurance exercise training; they suggested a possible role in facilitating increased mitochondrial protein import and folding during exercise-induced mitochondrial biogenesis. HSP90 is also a highly conserved family of proteins which represents 1–2% of total cellular protein under resting conditions (Katschinski 2004). The HSP90 family consists of three proteins: the glucose regulated protein GRP94 (located in the endoplasmic reticulum) and the two closely related cytoplasmic isoforms of HSP90α and HSP90β (Morton et al. 2007). Although HSP90 is one of the most abundant proteins in unstressed mammalian cells, it is also increasingly expressed in response to stress (Welch 1992). An increased production of HSP during or following stressful insults which functions to restore cellular homeostasis and to provide cytoprotection against subsequent periods of normally damaging stresses (McArdle et al. 2004; Maglara et al. 2003; Suzuki et al. 2000). HSPs were originally found to be induced in cells exposed to sub-lethal heat shock (Tissieres et al. 1974). However, now we know that HSP can expressed in response to a large variety of other stressors including ischemia (Marber et al. 1995), protein degradation (Chiang et al. 1989), hypoxia (Guttman et al. 1980), acidosis (Weitzel et al. 1985), oxidative stress (Adrie et al. 2000), increased intracellular calcium (Welch et al. 1983) and energy depletion (Sciandra and Subjeck 1983) in varied tissues. Given that several of the above stressors are characteristic of the homeostatic perturbations during acute or chronic exercise. There are numerus evidences showing that exercise induce the expression of chprokines such as HSP in animal tissues such as heart (Salo et al. 1991; Skidmore et al. 1995), brain (Walters et al. 1998), liver (Salo et al. 1991) and skeletal muscle (Isanejad et al. 2015; Locke et al. 1990; Salo et al. 1991; Skidmore et al. 1995; Hernando and Manso 1997). It has also been demonstrated that endurance and resistance exercise can up-regulate HSP expression in human skeletal muscle (Puntschart et al. 1996; Walsh et al. 2001; Milne and Noble 2002). This chapter focuses on chaperokine activity of HSP in response to physical exercise, next we explained the possible cellular and molecular mechanisms responsible to HSP response to exercise. Finally, we discuss about the role of exercise on the chaperokine activity of HSP in some conditions such as aging, Alzheimer and diabetes diseases.

12.2 Heat Shock Proteins Response to Exercise

12.2.1 Acute Exercise and Chaperones (HSP) Response

Molecular chaperones also referred as heat shock proteins (HSP) are a category of proteins that are involved in protein import, folding, unfolding and remodeling (Beckmann et al. 1990). They help newly synthesized proteins to reach their native fold; mediate either assembly of complexes or unfolding and disassembly while protecting them from various types of stress such as exercise and protein aggregation (Bukau et al. 2006). In this respect, there are several studies that have been shown the expression of HSP as chaperones in response to acute physical exercise. For example it has been seen the positive effects of acute exercise on HSP72 in serum (Walsh et al. 2001), plasma (Steensberg et al. 2006) and skeletal muscle (Walsh et al. 2001), also increased expression of hsp70 mRNA in skeletal muscle (Isanejad et al. 2015; Silver et al. 2012; Morton et al. 2007; Febbraio 2000) and myocardial (Melling et al. 2007) and increase the hsp70 protein in peripheral blood leukocytes (Fehrenbach et al. 2000). Folkesson et al. (2008) demonstrated that HSP27 was accumulated after an acute resistance exercise (8-10 repeats in 70% of 1RM) in myofibrillar structures that can act as chaperones to remodeling the injured muscle fibers. In contrast, Benini et al. (2015) found that acute resistance exercise did not induce a significant effect on HSP 70 and HSP 60, despite increased levels of creatine kinase (CK) and IL-6. Also, Feasson et al. (2002) observed the higher levels of HSP27 and αB-crystallin after downhill running and for 14 days. This suggest that HSP can play an integral role in the maintenance and remodelling of myofibrillar structures following exercise induced muscle damage (Feasson et al. 2002). It has been proposed that during or after damaging exercise sessions the inflammatory response that occurs in the days following exercise may also contribute to increases in muscle HSP levels as invading phagocytic cells contain relatively high levels of HSP (Khassaf et al. 2003). Also non-damaging exercises can regulate the muscle HSP expression such as HSP70 and HSP60 (Khassaf et al. 2003; Jackson et al. 2004; Vasilaki et al. 2006). Repeated bouts of exercise can affect the HSP response to exercise. In this respect, Thompson et al. (2003) have investigated the HSP response to acute bout of resistance exercise that are damaging in nature. They observed lower level of HSP27 and HSP72 after second bout of resistance exercise. Some investigator have been examined the stress response following downhill running (Isanejad et al. 2015; Feasson et al. 2002; Thompson et al. 2003). Recently, we also demonstrated that downhill running increased the expression of HSP72 in sedentary rats; however it didn't increase in trained animals (Isanejad et al. 2015). On the other side, there are several studies suggested that trained muscle may respond to a bout of exercise with a faster and larger expression of HSP thereby helping the cell to recover quicker (Campisi et al. 2003; Gonzalez et al. 2000). These data appear to represent an adaptive mechanism of the stress response whereby basal levels of stress proteins may be down regulated following repeated bouts of exercise. It has been well documented that trained subjects may exhibit a lower stress response to exercise, given that the adaptations by long exercise training function to reduce the degree of homeostatic unbalance that occurs during a given exercise stress (Nething et al. 2004; Liu et al. 2006). Therefore, the baseline level of fitness can modulate the responses of HSP to acute endurance exercise. In this regard, it had been demonstrated that lower levels of fitness, induced faster and larger responses of HSP60 and HSP70 to exercise, in the way that, these responses were slower in fit subjects (Morton et al. 2009a, b). Also, long intensive exercise causes greater heat shock responses in compared with short moderate exercise (Morton et al. 2009a, b).

Morton et al. (2009a, b) suggested that the ATP depletions, muscle damage, increased oxidative stress and metabolic disturbance are involved in the intense and long-term exercise induced production of HSP (Morton et al. 2009a, b). Interestingly, they suggested that increased both systemic and local muscle temperature per se have no effect on exercise-induced production of HSPs. It suggests that non-heat-stress factors associated with muscle contractile activity are importance in mediating this response. In acute exercise condition, Heat shock proteins bind to the unfolded and denatured proteins and dissociate from HSF-1. This allows HSF1 to form trimers and translocate to the cell nucleus and activate transcription (Henstridge et al. 2016). The acute bout of exercise can stimulate HSP expression, especially hsp70 and hsp72, in various organs of the body depends on the duration and intensity of exercise and levels of fitness (Liu et al. 2000; Locke et al. 1990; Locke and Noble 2002; Skidmore et al. 1995). Therefore, the different responses of HSP to exercise is dependent on variable including training statues/recent activity levels (Feasson et al. 2002) thermal adaptation (Lyashko et al. 1994), energy availability/muscle glycogen content (Febbraio and Koukoulas 2000), gender (Paroo et al. 1999) and age (Vasilaki et al. 2002).

12.2.2 *Chronic Exercise*

Generally, higher levels of heat shock proteins during/after acute exercise mostly play a role to regain cellular hemostasis and better recovery, however increased content of HSP following chronic exercise likely represent adaptations that serve to maintain homeostatic balance during a given stress (Noble 2002). Also, the higher basal content of HSP in trained subjects may be a pivotal component of cellular and molecular mechanism by which exercise confers protection to cells and tissues against a variety of stressors (Locke 1997). Regular exercise training can improve overall health and protect against common chronic diseases including coronary artery disease, diabetes mellitus, and cancers and so on (Hopkins et al. 2012). When the body engages in exercise training several times a week or more frequently, some physiological systems, including musculoskeletal; cardiovascular or respiratory systems, undergoes specific adaptations that increase the body's efficiency and functional capacity (Hopkins et al. 2012). One of the exercise-related adaptations is increased the body's capacity for response to stress stimuli. A wealth of data suggests that various types of chronic exercise training including both aerobic or resistance training are accompanied by an increase in HSP, such as HSP72, HSP27, GRP75 and HSP70 variety of tissues, especially skeletal muscles (Naito et al. 2000; Gjovaag and Dahl 2006; Liu et al. 1999, 2000; Paulsen et al. 2012; Isanejad et al. 2015; Gonzalez et al. 2000). For example Ogura et al. (2006) showed that 9-week of sprint-interval training increase HSP72 levels in plantaris and soleus muscles.

Furthermore, Liu et al. (1999) by investigating the effects of 4 weeks training in elite rowers, indicated that HSP response to training is related to the total exercise volume. In another study, they suggested that hsp70 response is dependent on exercise intensity (Liu et al. 2000). Exercise mode is another important factor involving in exercise induced HSP (Liu et al. 2004).

The exercise-related adaptations can be due to changes in mRNA expression of HSP and in result increased protein levels of it's after each exercise training session (Hawley et al. 2014). Consequently, it would be expected that exercise training over long time can lead to generalized increase HSP levels. In trained people, HSP induction blunted in response to acute exercise training (Smolka et al. 2000) and mRNA and proteins expression of its decrease immediately after exercise training (Liu et al. 2004). It can be concluded that upregulation of the HSP through chronic exercise is a critical component of the cellular and molecular adaptation by regular training. This adaptation may help the body to deal with unaccustomed stress (Henstridge et al. 2016; Noble et al. 2008; Fabbraio et al. 2000). Some studies demonstrated that exercise can lead to mount a greater and/or faster stress response following an acute bout of stress (Isanejad et al. 2015; Gonzalez et al. 2000; Fehrenbach et al. 2000). It has been suggested that trained people have more ability to recover and regain cellular homeostasis following an acute exercise stress, possibility due to increased HSP synthesis (Isanejad et al. 2015; Gonzalez et al. 2000; Fehrenbach et al. 2000; Campisi et al. 2003). Effects exercise training, as a non-pharmacological approach, on cells endogenous protective system is an exciting area in science relating to health. (Locke 2002). Consequently, given their cytoprotective properties, the HSP response to periods of exercise training provides an exciting research area with both obvious performance and health implications.

12.2.3 HSP Responses to Acute Stress Following Chronic Training

It is well known that cellular and molecular adaptation following chronic exercise training can cause to better response to acute bout of stress (Petriz et al. 2017; Gonzalez et al. 2000). Gonzalez et al. (2000) demonstrated that trained rodents had higher level of HSP72 following an acute bout of moderate running on treadmill in compare with sedentary animals. Also, Fehernbach et al. (2002) indicated that in leukocytes of trained people exposed to thermal shock, gene expression of HSP70 is significantly higher. Also, Campisi et al. (2003) found that habitual physical activity facilitates stress-induction of HSP72 in brain, peripheral and immune tissues. Consistent with these findings, we recently observed a higher level of HSP72 in trained animals (Isanejad et al. 2015). Taken together, it has been suggested that trained people may have a more ability to recover and regain cellular homeostasis following an acute bout exercise or other forms of stress, probably due to the increased level of HSP. On the other hand, it is possible that HSP responses in trained subjects diminished when facing to stressor conditions such as an acute exercise, due to appropriate endogenous defense systems to combat any homeostatic disruptions. Smolka et al. (2000) observed that following an acute exercise, production of HSP70 in soleus muscle of trained animals is lower than sedentary animals. They also showed that exercise training lead to increased activity of antioxidant enzymes of catalase and glutathione reductase in trained animals in compare to sedentary animals. Nething et al. (2004) observed that in trained people, an acute bout of high intensity resistance training or low intensity endurance training, didn't lead to any increase in muscle HSP70 content Therefore, it might propose that HSP is the secondary defense system.

12.3 Possible Mechanisms in Exercise-Induced HSP

Stress responses are triggered by various stressors such as temperature; oxidative stress and mechanical stimuli, metabolic changes, and cytokines production (Noble et al. 2008). These stimuli may modulate the effects of exercise on the levels of HSP (Morton et al. 2009a, b); however, it is difficult to find a precise mechanism that is responsible for induce these exercise-induced stress response. In following sections, we study the role possible mechanisms involving in exercise-induced HSP.

12.3.1 Temperature

Approximately two thirds of the energy used by skeletal muscle to produce contractions is converted to thermal energy, which lead to elevation in temperature of the contracting muscle and in core temperature of whole body (Morton et al. 2007). In this respect, elevated core and muscle temperature is a possible factor for higher levels of HSP following exercise (Lancaster and Febbraio 2005; Liu and Steinacker 2001; Liu et al. 2006). Kim et al. (2004) demonstrated that exercise induced upregulation of HSP70 is dependent to the absolute body temperature. Both in vitro and in vivo showed that increased cell temperature to 42 °C, HSP70 content of muscle of animal's models is increased. (Morton et al. 2009a, b; Maglara et al. 2003; Oishi et al. 2002, 2003). In general, ability to survive and adapt to thermal stress is a fundamental requirement of cellular surveillance, and key heat shock proteins (HSP) involved in these responses (Samali and Orrenius 1998). Even in some mammalian (for example human), considerable variations in core temperature can occur during severe environmental stress, such as exercise training. The role of thermal stimuli in activation of exercise-induced stress response is provided from a number of experimental studies. Some studies have shown that increased temperature in the skeletal muscles lead to increased expression of several HSP including HSP70 and HSP72 (Oishi et al. 2002; Kim et al. 2004; Morton et al. 2009a, b). Also, it observed that in the absence of increased temperature, the response to thermal stress in myocardium is decreased (Harris and Starnes 2001). Higher in temperature induced by exercise also lead to increased ROS (Zuo et al. 2000), therefore it is potent activators of the exercise-induced HSP.

In contrast, Morton et al. (2007) observed that the thermal stress alone does not lead to the exercise-induced HSP expression (Morton et al. 2007). In addition, it has been proposed that up -regulation of HSP70 content in skeletal and cardiac muscle were independent of exercise-increased core temperature (Skidmore et al. 1995; Staib et al. 2007). These contradictory findings may arise from different response of tissue to stress, for example cardiac tissue response is more dependent on increased temperature than skeletal muscle (Skidmore et al. 1995, Staib et al. 2007). Some studies recently suggested that heat stress can act as an effective post-exercise pathway to enhance mitochondrial adaptations in skeletal muscle (Tamura et al. 2014). Nnuclear-encoded mitochondrial proteins and mitochondrial transcription factors including p38 MAPK and mTOR, as precursor proteins, can be translated and synthesized in the cytosol and then transported into the mitochondria and refolded by HSP60 and HSP72 (Tamura et al. 2014). HSP (e.g. HSP60 and HSP72) by refolding these factors, can involve in the mitochondrial adaptations, which probably differ from adaptation of exercise training. These data suggested that exercise may affects HSP expression may be independent to elevated temperature by exercise. Therefore other possible mechanism should be considered.

12.3.2 Reactive Oxygen Species (ROS)

Free radicals and reactive oxygen species as 'injurious molecules' have capacity to promote cellular damage. These are involved in the pathogenesis of a variety of diseases (Powers and Hamilton 2002). Exercise-induced expression of reactive oxygen species (ROS) may be play pivotal role in up-regulating HSP expression (Fittipaldi et al. 2014). In addition, ROS produced by exercise can activate the HSF-1(Jackson 2005). In a biological context, ROS are formed as a natural byproduct of the normal metabolism of oxygen. It has important roles in cell signaling and homeostasis. However, during the environmental stress (e.g. intensive exercise training), ROS levels can increase dramatically (Jackson et al. 2004). During exercise, the skeletal muscle is the main source of ROS and RNS. These molecules play an important role, not only in the skeletal muscle adaptation to exercise, but also in the muscle plasticity that occurs during a period of prolonged inactivity (Fittipaldi et al. 2014). It seems that ROS in normal levels cause to transient and reversible oxidation proteins, on the other side in high levels it can cause to mechanical disruption of protein structure and secondary inflammatory processes. In this condition, stress response is induced (Fittipaldi et al. 2014). Milne and Noble 2002) also showed that 15 min of non-damage isometric exercises increased the release of superoxide from the muscle into the extracellular space and increased the content of HSP70 in soleus and extensor digitorum longus muscles (Milne and Noble 2002). The ability of the ROS in driving Hsp70 expression in blood cells after endurance exercise has been widely documented in several studies (Fehrenbach et al. 2000; Niess et al. 1999, 2002). It has been suggested that higher level of HSP70 may be due to oxidative damage to the cellular membrane (Dimauro et al. 2016). It seen that misfolding of proteins due to ROS, is one of the main mechanisms involving in the exercise-induced HSP expression. However, cellular signaling pathways like p38 MAPK/JNK/ERK and P13K/AKT can activated the HSP expression but the involving these mechanisms in the exercise-induced adaptations in humans is only speculative (Dimauro et al. 2016). Although ROS can stimulate HSP, but it has been suggested that antioxidants can reduce HSP' stimulation in immune cells exposed to heat (Gorman et al. 1999). In addition, this occurs in cells that are directly exposed to ROS (Khassaf et al. 2003). Moreover, antioxidants supplementation can reduce the release of HSP into the circulation, as well as the blunted the expression of HSP mRNA and protein in skeletal muscle following exercise (Khassaf et al. 2003; Fischer et al. 2006; Jackson et al. 2004). Also, in some studies, the effects of exogenous antioxidants such as vitamin C, E, and beta-carotene on HSP have been investigated following exercise training (Khassaf et al. 2003; Fischer et al. 2006). Khassaf et al. (2003) indicated that vitamin C, vitamin E and beta-carotene supplementation can reduce the HSP70 content of the lateralis muscle following a one-leg bike exercise (Khassaf et al. 2003). Also in this study, it showed that vitamin C can reduce the HSP72 muscle response to one-step ergometer training (45 min at 70% maximal oxygen consumption). In the absence of vitamin C, this response increased significantly after exercise. In another study, Fischer et al. (2006) evaluated the effects of exercise and vitamin E on HSP72 expression. After 28 days of supplementation, subjects completed 3 h of constriction extension of the knee extension at 50% intensity. They showed that the serum concentration of HSP72 after exercise was negatively correlated with plasma vitamin E levels (Fischer et al. 2006). These data suggest that ROS produced by exercise can reduce because of increasing tissue antioxidant capacity; therefore the ROS play an important role in exercise-induced HSP. Regarding these results, it seems that up-regulation of heat shock protein in order to modify the effects of ROS, in addition to improving cellular function can play an important role in preventing diseases.

Metabolic stress is a physiological process that occurs during exercise in response to low energy charge. It leads to metabolite accumulation [lactate, phosphate inorganic (Pi) and ions of hydrogen (H+)] in muscle cells and other tissue (Noble et al. 2008). Some researchers have suggested that metabolic stress has an important impact on hormonal release, hypoxia, cell swelling and production of reactive oxygen species (ROS) (Hernando and Manso 1997); McArdle et al. 2001; Salo et al. 1991). In this regards, ATP depletion (Benjamin et al. 1992) and acidosis can increase the content of HSP (Weitzel et al. 1985). Benjamin et al. (1992) have shown that ATP depletion up to 30%, as well as reducing the pH from 7.3 to 6, lead to stimulate the HSF-1 (Benjamin et al. 1992). These authors also suggest that the depletion of acid-dependent ATP was one of the most important metabolic factors for expressing HSP during exercise. In addition, Chang et al. (2001) showed that decreased levels of ATP after ischemic injury lead to increase the HSF-1 binding to DNA and its transcription activation (a critical step in activating signaling cascades) in myocardium (Chang et al. 2001). However, it is unlikely that during exercise training, ATP levels will drop to these levels. Also, some studies have shown that the reduction in available glycogen is one of the factors involving in the exercise-produce HSP in human skeletal muscle (Febbraio and Koukoulas 2000; Febbraio et al. 2002). Febbraio and Koukoulas (2000) observed that increased expression of the HSP72 gene in the muscles during an endurance cycling was associated with a decrease in glycogen levels in the muscles (Febbraio and Koukoulas 2000). Febbraio et al. (2002) by using the two-legged exhaustive knee extensor protocol observed that the gene expression of the HSP72 after exercise only occurred in muscle that had glycogen discharged 24 h before exercise (Febbraio et al. 2002). Also, Tupling et al. (2007) observed that the peak stimulation of HSP70 in isomeric contraction was higher in the muscle that more glycogen was depletion from that during the training (Tupling et al. 2007). After exercise, skeletal muscle immediately restores its degraded glycogen content through dephosphorylation and activating the glycogen synthase (GS) enzyme (Nielsen and Wojtaszewski 2004). To activate the GS and replenish muscle glycogen reserves, deactivation GSK3α is necessary (a kinase protein responsible for phosphorylation and deactivation of GS) (Markuns et al. 1999). Subsequently, HSF1 phosphorylated by GSK3α. This can lead to the attachment of a series of proteins, the removal of the transcription factor, and limiting its transcriptional ability (Noble et al. 2008).

Interestingly, the HSF1 transcription activity was stopped (Noble et al. 2008; Melling et al. 2006) after an exercise that led to an increase in ERK1/2, with the inhibition of protein kinase A (a circular AMP dependent kinase protein). Melling et al. (2006) also observed that activation of ERK1/2, despite suppressing HSF1 activity, has no effect on nuclear localization or the ability connect of HSF1 (Melling et al. 2006). It seems that HSF1 phosphorylation by ERK1/2 is independent process from protein damage or the accumulation of abnormal proteins, and is preferably activate the network of cell-signaling cascades and exercise-related factors that regulate the production of HSP (Noble et al. 2008). Although these studies suggest the potential role of glycogen availability in exercise-induced HSP expression, but it seems difficult to detect the exact mechanisms of this relation. Moreover, in some studies didn't found any association. For example, in a study, increased levels of HSP70, HSP60, and aB-crystallin in lateralis and gastrocnemius muscles in response to exercise training was independent of internal and external access to glycogen (Morton et al. 2009a, b). These data suggested that metabolic stress play an important role in HSP responses to exercise.

12.3.3 Mitochondria Biogenesis

One of adaptations of long-term endurance training is mitochondrial biogenesis. In mitochondrial biogenesis, with increase mass and number of mitochondrial, production of ATP as a response to greater energy expenditure is increased.

Mitochondrial biogenesis as a major adaptation of skeletal muscle to exercise training is induced by numerous signaling pathways that respond to metabolic, mechanical, and hypoxic stress. In muscle contraction, this stress is generated within the myocyte (Barone et al. 2016). Hsp60 is one of factors involving in this adaptation. Hsp60 mainly act as a chaperon. Hsp60 play a role in mitochondrial protein import and macromolecular assembly. It may facilitate the correct folding of imported proteins, and may also prevent misfolding and promote the refolding and proper assembly of unfolded polypeptides generated under stress conditions in the mitochondrial matrix. HSP60 has been shown to be up-regulated in the skeletal muscle of rodents following exercise training (Mattson et al. 2000) presumably to facilitate increased mitochondrial protein import and assist in exercise-induced mitochondrial biogenesis. In addition mitochondrial, it also was found in cytosolic and extracellular environment (Cappello et al. 2008). Hsp60 could be involved in many of molecular pathways originated by exercise training such as mitochondrial biogenesis. Some studies showed increase Hsp60 coincides with the increased citrate synthase activity (an enzyme localized in the mitochondrial matrix) in rat muscle after endurance training (Mattson et al. 2000). Also Barone et al. (2016) has been shown that concomitantly with the increased levels of Hsp60 released in the blood stream of trained mice, mitochondrial copy number and the expression of three isoforms of PGC1α increased (Barone et al. 2016). PGC1α is a dominant regulator of oxidative metabolism, acting as a transcriptional co-activator of nuclear receptors and other transcription factors regulating mitochondrial biogenesis (Powers et al. 2011). In particular, it have suggested that up-regulation of Hsp60 lead to protect mitochondrial proteins against damage and to facilitate protein import and folding in exercise-induced mitochondrial biogenesis (Barone et al. 2016). A large body of evidence now exists supporting a role for mitochondria dysfunction and oxidative damage in pathogenesis age related disease such as Alzheimer's (Lin and Beal 2006; Castellani et al. 2002; Hashimoto et al. 2003). Therefore exercise training by inducing of HSP might enhance cell resistance to stress and prevent disease.

12.3.4 The Role of Inflammation on the HSPs Response to Exercise

Inflammation has previously been demonstrated to be involved in variety of disease such as metabolic disorders, cancer (Grivennikov et al. 2010), vascular diseases (Sullivan et al. 2000), Alzheimer's (Tuppo and Arias 2005). Inflammation also has been proposed as positive regulator of HSP expression (Stephanou et al. 1998). On the other hand, cytokine produced during exercise might induce stress responses by exercise (Liu and Steinacker 2001). Regarding the increased secretion of inflammatory cytokine such as IL-6, due to exercise (Akira et al. 1993), it appears that the invasion of monocytes into the muscle to help eliminate the myofibrillar micro-injuries and regeneration is one of the main causes of increased levels of IL-6 may induces the up-regulation of HSP following the damaging exercise (Thompson et al. 2001). Generally, it was thought that inflammatory response to exercise was a consequence of an immune response resulting from exercise-induced muscle damage (Malm 2001). However, recently it has been suggested that exercise induced increased level of circulatory IL-6 following non-damaging exercise protocols (Febbraio et al. 2003). Furthermore, Starkie et al. (2000) indicated that circulatory monocytes are not responsible in elevated levels of plasma IL-6. It is well known that a muscular IL-6 level following exercise is the main sources of high circulatory levels of IL-6 (Steensberg et al. 2000). Interestingly, muscle-derived IL-6 may play pivotal role in activation some cellular stress response such as signaling molecule. Also it can lead to a significant increase in HSP72 gene expression (Febbraio et al. 2002). Febbraio and colleagues (2002) suggested that IL-6 may activate HSP expression via activation activator of transcription 3 (STAT3). Therefore, inflammatory cytokine such as IL-6 may act as upstream to induction of HSP expression by exercise.

12.4 Chaperokine Activity of HSP; The Role of Exercise

Chaperokine, is a term recently coined to better describe the unique function of HSP as both chaperone and cytokine (Asea 2006). HSP were thought to be cytoplasmic proteins. However, it has been proposed that they can be released in to extra-cellular space and affects the function of other cells (Tytell 2005). Physical exercise as stress can release HSP such as HSP70 to extracellular milieu (Walsh et al. 2001). The contracting skeletal muscle is not the tissue source of the exercise-induced increase in the circulating HSP72 concentration (Walsh et al. 2001). Furthermore, Febbraio et al. (2002) have been shown the hepatosplanchnic tissues as source of eHSP72. They concluded that exercise induced release of HSP72 from liver via a specific exocytose pathway regardless of any tissue damage. Therefore, according present data exercise induced high level of circulatory HSP72 and that this increase is not the result of tissue damage (Lancaster et al. 2005).

eHSP72 has been suggested to act as a danger signal to the immune system. eHSP72 also have an immunological function (Asea et al. 2000a; Fleshner et al. 2003). HSP70 can bind to immune cells such as NK (Multhoff et al. 1997), dendritic cells (Asea et al. 2002) and macrophages (Sondermann et al. 2000). eHSP72 have a subsequent effects on cytokine cascade and resulting inflammatory response (Asea et al. 2000a, b). There are numerus evidences showing that strenuous exercise can increased the concentration of eHSP72 (Walsh et al. 2001; Lancaster et al. 2004;

Fehrenbach et al. 2005) and return to baseline levels within 24 h (Walsh et al. 2001). Marshall et al. (2006) observed the lower level of eHSP70 after repeated bout of exercise. They concluded that, this change was possibly due to an enhanced role in the immune response. Lancaster et al. (2004) also observed that the brain is able to release Hsp72 in response to physical exercise. Recently we also demonstrated that trained rats (after 8 weeks of endurance training) had lower level of circulatory HSP70 in compare to sedentary group (Isanejad et al. 2015). Intracellular HSP72 have anti-inflammatory effects by blocking the nuclear factor κB (NF-kB) activation (Jones et al. 2011). In contrast, eHSP70 have pro-inflammatory function by binding both Toll-like receptor 2 (TLR2) and TLR4 in a CD14-dependent fashion, consequently promoting innate immune activation (Asea et al. 2000b). The HSP as danger signal may regulate the immune response. In this regard, Garcia et al. (2013) investigated the effects of eHsp72-induced release of inflammatory cytokines by macrophages. They indicated that habitual exercise improved the release of pro-inflammatory cytokines such as IL-1β, IL-6, and TNFα. eHSP72 also can participated in stimulation of chemotaxis by acute exercise (Ortega et al. 2009). As we know, stress can stimulated immune system response to pathogens. Campisi and Fleshner (2003) demonstrated that both the stress status and physical fitness level of an organism have pronounced effects on the function of the immune system. They suggested that acute stress facilitate innate immunity and that physical activity status modulates the impact of stress on immunity. In addition, eHSP70 can suppress tumor growth through activation of natural killer cells (NK) (Moser et al. 2002).

Collectively, eHSP70 as chaperokine during exercise may enhance immune defense against pathogens or tumors. However there is needed to more inquiring to show the chaperokine activity of HSP70 in physical exercise trained subjects.

12.4.1 Exercise-Induced Cardio-Protection: The Role of HSP

Exercise training has been known to confer cardioprotection; however, the mechanisms of exercise-induced cardioprotection remain unclear (Gielen et al. 2010). One of these mechanisms is induction of myocardial heat shock proteins (Hamilton et al. 2003). Several HSP such as Hsp10, Hsp40, Hsp60, Hsp70 and Hsp90 are thought to play cardioprotective roles (Luo et al. 2005; Care et al. 2007). Some studies have been shown that overexpression of Hsp70 in cardiomyocytes protects against ischemic damage and is associated with increased cell survival (Martin et al. 1997).

Also, Hsp72 by chaperone activity can stabilizes the myocardium through mitochondrial preservation (Sammut et al. 2001) and membrane stabilization (Soti and Csermely 2007). Hsp72 overexpression is associated with myocardial preservation against necrotic and apoptotic cell death (Ravagnan et al. 2001; Tanonaka et al. 2004). After schemia-reperfusion (I/R), Hsp72 overexpression retards I/R-induced apoptosis through caspase-dependent and -independent pathways (Ravagnan et al. 2001). Specifically, Hsp72 binds with apoptosis protease-inducing factor, thereby inhibiting subsequent activation of both caspase-9 and caspase-3 (Ravagnan et al.

2001). Furthermore, caspase-independent inhibition by Hsp72 occurs through interference with the apoptotic effector molecule apoptosis-inducing factor (Baines and Molkentin 2005; Beere et al. 2000). Collectively, these findings demonstrated that increased cardiac HSP-72 levels are essential for protection against I/R-mediated cell death. Some studies have been shown that endurance exercise result in increase in cardiac HSP levels (Demirel et al. 2003; Hamilton et al. 2003; Taylor et al. 1999). Rinaldi et al. (2006) showed that Hsp70 and Hsp27 were elevated in trained old rats versus sedentary old rats, indicating improved resistance to ischemic insults (Rinaldi et al. 2006). Different stresses associated with exercise, including heat stress and hypoxia, reduced intracellular pH, reactive oxygen and nitrogen species (ROS and RNS) production, and depletion of glucose and glycogen stores, increase in cytosolic calcium levels and cardiomyocyte stretching can all contribute to HSP elevation in cardiac muscle (Powers et al. 2001). In addition to exercise training, antioxidants also could have cardioprotection effect. However, exercise and antioxidant supplementation each provided independent cardioprotection against I/R injury, and combination of these countermeasures did not enhance or inhibit the myocardial protection (Hamilton et al. 2003). In this respect, Hamilton et al. (2003) have been shown that antioxidant supplementation decreased the exercise-induced accumulation of myocardial Hsp72, and increased in cardiac MnSOD activity. In addition, this supplementation did not attenuate exercise-induced increases in cardiac MnSOD activity (Hamilton et al. 2003). Some studies have been indicated a sex-specific HSP response in cardiac (Paroo et al. 2002). Paroo et al. (2002) have been shown that in male rats compared with intact female rats, cardiac Hsp70 content was twofold greater after exercise. In this study also showed that although removal of the ovaries, the major endogenous source of estrogen, resulted in post exercise Hsp70 levels that were similar to those observed in male rats, but estrogen replacement in these animals reversed this effect. It has been seen that Hsp70 response to exercise is mediated by the female-specific hormone estrogen (Paroo et al. 2002). Furthermore, estrogen can directly protects cells from damage, thus reducing the need for HSP or this hormone interacts with HSP transcription and thus reducing the rate of HSP synthesis. Estrogen can affect the HSP72 expression in cardiac muscle. Hamilton et al. (2004) showed that there is a relationship between NFkB activation and decreased Hsp72 (Hamilton et al. 2004). Estrogen may attenuate cardiac damage and inflammation through downregulation of NFkB and with this attenuated damage, resulting in a blunted Hsp72 response (Yoshikawa and Yoshida 2000).

Although exercise induces cardiac HSP (which is cardioprotective), but some studies demonstrated that exercise-induced cardioprotective could be achievable through multiple mechanisms including increase in cardiac antioxidant capacity, expression of endoplasmic reticulum stress proteins, anatomical and physiological changes in the coronary arteries, changes in nitric oxide production, adaptation changes in cardiac mitochondria, increased autophagy, and improved function of sarcolemma and/or mitochondrial ATP-sensitive potassium channels (Golbidi and Laher 2011).

12.4.2 Exercise Induced HSP Expression in Aging

Aging is a multifactorial process resulting in damage of molecules, cells, and tissues, leading to a reduced efficacy of functions with different pathophysiological consequences and a variety of clinical phenotypes. It is characterized by altered regulation of many genes implicated in stress resistance and processes of tissue regeneration and repair (Westerheide et al. 2009; Rodgers et al. 2005). One of the first molecular markers of aging identified was the accumulation of abnormal proteins, specifically the accumulation of inactive enzymes in aging nematodes (Westerheide et al. 2009, Rodgers et al. 2005). Since then, denatured and aggregated proteins, typically associated with HSP, have been found to mark aging cells in virtually every species examined. In humans, numerous aging-related diseases involve the accumulation of abnormally processed and/or folded proteins (e.g., Alzheimer's disease, Parkinson's disease, Prion diseases), and these aggregates are typically associated with HSP (Aprahamian et al. 2013). The ability of the HSP to counter proteotoxicity might be one mechanism by which can promote longevity, consistent with the correlation between a robust stress response and long life. At the same time, during aging, chronic expression and accumulation of HSP might become toxic in some contexts, perhaps due to disruption of signaling pathways (Karvinen et al. 2016). It has been demonstrated that the expression and activity of HSP are modified in aging, with reduced cell ability to counteract the oxidant molecules, and consequent weak resistance to ROS accumulation. An important mechanism involved is represented by sirtuins, the activity of which is reduced by aging. Family of sirtuins, a cluster of proteins composed by seven homologues that regulate cellular biology and metabolism through deacetylation of histones and other cellular factors such as NFkB, HSF1, p53, FOXOs, and PGC-1 (Karvinen et al. 2016).

Exercises during the aging process have different effects on HSP. Murlasits et al. (2006) have shown that resistance training increased expression of some HSP in older rats (including hsp72 and hsp25 muscle tissue), but this expression was twice less than younger rats (Murlasits et al. 2006). It has been observed that in some studies that exercise (both resistance and aerobic) has no significant effect on HSP in aging (Siu and Alway 2005; Vasilaki et al. 2002, 2006; Kayani et al. 2008) in aging. It seems that oxidative stress is one of the possible mechanisms that can influence on exercise induced HSP during aging. According some evidences, aging is associated with the accumulation of oxidative stress (Wei and Lee 2002), and this accumulation can reduce the reaction of thermal shock (Demirel et al. 2003; Locke and Tanguay 1996). Physical activity increases the expression and the activity of HSP, with consequent reduction of ROS. Positive effects of physical exercise in terms of antioxidant activity could be ascribable to a greater expression and activity of SOD enzymes, HSP and SIRT1 activity (Karvinen et al. 2016).

During aging, apoptosis is affected. Apoptosis as programmed cell death is involved in diverse processes such as development and immune clearance. It has a crucial role in tissue normal development and homeostasis (Ko et al. 2013). Many factors such as HSP can regulate apoptosis. HSP can modulate apoptosis by interacting with components of the apoptotic machinery (Ko et al. 2013). In some condition, such as some cancer, expression of HSP increase, which lead to enhances tumor-genicity and resistance to cell death. On the other hand, increased expression of HSP associated with enhanced survival of post mitotic cells such as cardiomyocytes and neurons considered beneficial (Ko et al. 2013). Some studies have indicated that during aging, rate of apoptosis increases and Hsp70 expression decreases in rats (Soufi et al. 2008). In the aged myocardium, decreased expression of Hsp70 was linked with an increased expression of p38 (Taylor and Starnes 2003).

Hsp70 has been widely reported to be involved in the inhibition of p38 within the myocardium (Taylor and Starnes 2003). By activation of p38, apoptosis induced in the cardiomyocyte (Aceros et al. 2011; Sun et al. 2011). In general, exercise training can provide a cardioprotective effect on aging-induced apoptosis through the enhancement of Hsp70 expression. Hsp70 could suppress the expression of p38 in the heart (Moran et al. 2004; Rinaldi et al. 2006; Soufi et al. 2008).

Sarcopenia can occur during the aging process. Sarcopenia is 'a geriatric syndrome initially characterized by a decrease in muscle mass that will get worse causing a deterioration in strength and physical performance' (Muscaritoli et al. 2010; Cruz-Jentoft et al. 2010; Fielding et al. 2011; Morley et al. 2011). Muscle atrophy is a feature of the ageing process and has been suggested to be due to increased oxidative stress (Powers et al. 2005). Upregulation of HSP27 as a the cytoprotective marker, may ameliorate age-related skeletal muscle atrophy induced by inflammatory signaling (Lawler and Hindle 2011), and it seems that TNF-α contributes to a significant rate of phosphorylation of HSP27 in sarcopenia (Huey et al. 2007). It is known that elevated levels of HSP27 inhibits IKKB in non-muscle cells (Park et al. 2003); Moreover, recent work by Dodd et al. (2009) confirmed that HSP27 inhibits IKKB induced NF-kB activation also in skeletal muscle in vitro (Dodd et al. 2009). Ogawa et al. (2012) recently have shown that extracellular HSP72 can be a novel biomarker for sarcopenia (Ogawa et al. 2012). In this study, extracellular HSP72 had a negative correlation with muscle mass. Some studies found that the high levels of extracellular HSP72 associated with several pathological conditions and the low levels of HSP72 associated with the life span and quality of life in aging (Beyer et al. 2012; Macario and Conway de Macario 2005; Pockley 2002; Terry et al. 2006). In a study, extracellular HSP72 was significantly reduced after 12 weeks of low intensity resistance training in older women (Ogawa et al. 2010). In this study, it was also observed that in addition to a significant reduction of extracellular HSP72, circulating levels of TNF-α and CRP has also been decreased following low intensity resistance training (Ogawa et al. 2010). There was no relationship between extracellular HSP72 and muscle thickness (Ogawa et al. 2010). Therefore, it seems that in the relationship between extracellular HSP72 and muscle mass in aging, systemic inflammation, and some markers such as IL-6, CRP and TNF-a play an important role in muscle atrophy in aged people.

12.4.3 The Effects of Chaperokin Activity of HSP in Alzheimer's; Protective Effects of Exercise

Alzheimer's disease (AD) is a progressive, neurodegenerative disorder clinically characterized by the loss of memory and multiple cognitive dysfunctions (Aprahamian et al. 2013; Balin and Hudson 2014). In general, Alzheimer's disease is an example of a disorder related to protein folding (Muchowski and Wacker 2005, Selkoe 2003). In fact, the neurodegenerative disease such as Alzheimer's, with etiology of oxidative stress (Mariani et al. 2005), have been linked to CCT as chaperones down-regulation (McNaught and Jenner 2001). It has been proposed that CCT as chaperon maintains cytoarchitecture by correcting the three-dimensional structure of cytoskeletal proteins such as actin, tubulin (Sternlicht et al. 1993). CCT may also assist cell to tolerate the oxidative stress by promoting cellular hemostasis (Kane et al. 2004). HSP60 is a mitochondrial chaperone that may protect mitochondria from oxidative stress by facilitating the proper assembly of mitochondrial proteins (Boyd-Kimball et al. 2005). In addition, HSP 60 has anti-apoptotic functions (Lin et al. 2001). It is well known that the expression of HSP60 and HSP8 otherwise known as HSP71 are decreased in Alzheimer's disease (Yoo et al. 2001; Castegna et al. 2002). HSP are important chaperones that, in addition to participating in inhibit apoptotic, regulation of cell growth and cellular differentiation, are also involved in the proper folding of proteins, so they can be affected by Alzheimer's disease (Chen and Brown 2007). Heat shock proteins (HSP) in Alzheimer's disease, help refold or degrade mis-folded proteins, reduce abnormal protein accumulation and synaptic dysfunction, neuronal degeneration and cognitive decline and improves the associated symptoms (could increase the HSP70 protein in the brain of Alzheimer's rats (Koo et al. 2013; Gandy 2002; Lustbader et al. 2004; De la-Monte and Wands 2009; Jayaraman et al. 2008). According these data, physical exercise seems to reduce the damage caused by stress in neurons by improving the levels of HSP, especially HSP70 in the different tissues such as brain might play preventive role in of Alzheimer's (Najarzadegan et al. 2016). Ding et al. (2006) observed that voluntary exercise up-regulated the CCT, HSP60 and HSP60 in hippocampus tissue. They concluded that, exercise may protect neuron through upregulation of anti-apoptotic, cytoskeletal protective and anti-oxidative chaperones. Exercise is one of the interventions that can regulate HSP in Alzheimer's disease. However, little research has been done in this regard.

12.4.4 HSP and Cancer; The Role of Exercise

Heat shock response involved in tumor development usually by up-regulation of but also by down-regulation of specific HSP. Heat shock proteins can induce in response to a wide variety of physiological and environmental insults including anti-cancer chemotherapy, thus allowing the cell to survive to lethal conditions (Thomas et al.

2016). Different functions of HSP have been described to account for their cytoprotective function, including their role as molecular chaperones as these play central roles in the correct folding of misfolded proteins, and also their anti-apoptotic properties (Thomas et al. 2016). HSP are often overexpressed in cancer cells and this constitutive expression is necessary for cancer cells' survival. HSP may have oncogene-like functions and likewise mediate "non-oncogene addiction" of stressed tumor cells that must adapt to a hostile microenvironment, thereby becoming dependent for their survival on HSP. HSP-targeting drugs have therefore emerged as potential anti-cancer agents (Thomas et al. 2016). Some studies have shown that by adding HSP to cells through experimental methods, cellular resistance to anthracyclines has been increased (Saw et al. 1994). However, it is unclear whether exercise with the increase of intracellular HSP can deal with cells cancer.

Exercise is one of several lifestyle factors known to lower the risk of developing cancer and is associated with lower relapse rates and better survival. A few studies have examined the effects exercise-induced HSP on cancers. For example, Nilsen et al. (2016) showed that 16 weeks of resistance training in men with prostate cancer that were under Androgen deprivation therapy (ADT) resulted in a reduction HSP70 in vastuslateralis muscle (Nilsen et al. 2016). It probably indicates reduction cells stress in this muscle. Some studies have shown that increased HSP is a possible mechanism involving in exercise-protected heart, in patients with breast cancer who use anthracycline-based chemotherapy regimens (Lancaster et al. 2004; Kavazis et al. 2010; Scott 2011). Also, increased HSP is one of possible mechanism indicated the effect of exercise in reduce cognitive impairment during chemotherapy. This effect seems to be occurring by protecting from astrocytes and supporting cells within brain (Calabrese et al. 2003). Although cancer cells produce a lot of HSP, it is unlikely that changes in serum HSP levels after exercise will have an effect on the levels of the HSP intra-tumor (Scott 2011). This was recently supported in an experimental research. In this study, the response to Adriamycin in mice was better with exercise (Kavazis et al. 2010). In addition, we observed the increased serum level of HSP72 in breast cancer undergoing hormone therapy and exercise training (unpublished data). Because positive effects on HSP, physical exercise can consider as a non-pharmacologic methods for cancer prevention and treatment.

12.4.5 Type 2 Diabetes, Exercise and Chaperokine Activity of HSP

Diabetes is characterized by a chronic elevation of blood glucose (hyperglycaemia). The persistent hyperglycaemia can lead to further disorders, such as coronary artery disease, stroke and peripheral vascular disease. Also in type 2 diabetes, which is a risk factor for some diseases including cardiovascular disease, in whole body and peripheral tissues including liver, adipose and skeletal muscle tissue, insulin resistance occurs (Hooper et al. 2014). It is well known that physical inactivity and obesity are the main risk factors of type 2 diabetes. Chronic inflammation and mitochondrial dysfunction are the most important causing insulin resistance and type 2 diabetes. In particular, endoplasmic reticulum (ER), tumor necrosis factor alpha (TNF-α), and oxidative stress stimulate inflammatory kinases such as inhibitor of NF-kb kinase (IKK), c -jun amino terminal kinase (JNK), and protein kinase C (PKC), which inhibit insulin signaling. Mitochondrial dysfunction in skeletal muscle has also been proposed as a prominent factor in the pathogenesis of T2DM either by altering the cellular redox state and or by reducing the ability to oxidize fatty acids, leading to the accumulation of deleterious lipid species in peripheral tissues such as liver and skeletal muscle (Geiger and Gupte 2011; Hooper et al. 2014; Archer et al. 2018). Recent studies suggest that (HSP) play an important role in insulin resistance (Geiger and Gupte 2011) and improving the cellular stress response via HSP induction plays an important role in treating type 2 diabetes (Hooper et al. 2014). This improvement appears to be linked with a reduction in JNK1phosphorylation and/or an increase in oxidative capacity consequential of improvements in mitochondrial homeostasis. It has been accepted that exercise protects against or ameliorates insulin resistance; however, few studies have examined the exercise-induced HSP in improved insulin resistance. Tsuzuki et al. (2017) showed that attenuation of exercise-induced HSP72 expression partially blunts improvement in whole body insulin resistance and lipid metabolism in type 2 diabetic rats (Tsuzuki et al. 2017). Archer et al. (2018) suggest that exercise by inducing the secretion of extracellular HSP72, can stimulate the resume of HSP72 in insulin-resistant tissues which HSP are low, thereby contributing to metabolic hemostasis (Archer et al. 2018). Abubaker et al. (2013) showed that DNAJB3 expression was reduced in obese subjects (Abubaker et al. 2013). Human DNAJB3; also known as MSJ-1 in mouse (Berruti et al. 1998), is one of the 49 members of the HSP-40 family of proteins. It acts as a co-chaperone to regulate the activity and specificity of HSP-70 (Kampinga et al. 2009; Kampinga and Craig 2010; Hageman et al. 2011). Some studies suggest that exercise can restore DNAJB3 and return to its natural levels in obese subjects (Abubaker et al. 2013). Venojarvi et al. (2008) have been shown that the 2-year endurance training, combined with dietary counselling, significantly increase mitochondrial Hsp60 expressions and glucose-regulated protein 75 (GRP75) in middle-aged subjects with impaired glucose tolerance (Venojarvi et al. 2008). Endurance training could activate PGC-1α, which mainly mediates transcriptional regulation of nuclear genes encoding mitochondrial proteins (NGEMP). These nuclear-derived gene products have to import into mitochondria via the protein-import machinery. Imported NGEMPs interact with GRP75, and GRP75 drives translocation of these precursors across the mitochondrial matrix. In an ATP-dependent process precursors are released and bound to the Hsp60. Therefore, both GRP75 and Hsp60 are essential for mitochondrial function and biogenesis (Atalay and Laaksonen 2002; Venojarvi et al. 2008).

The adaptive changes in the expression of GRP75 and Hsp60 could support mitochondrial protein import, protein folding and enhance tissue protection in insulin-resistant muscle tissue. These adaptive changes of mitochondrial HSP and increased oxidative capacity are mainly due to increased contractile activity of skeletal muscle.

12.5 Conclusions

Heat Shock Proteins (HSP) as chaperone can play an important role in human health and physiology and are part of the body's protective system against stress. In addition to heat, environmental changes, exercise as a stressor cans stimuli HSP expression. On the other hand, regular physical exercise training by improving the levels of HSP consequently can protects cells and organelles from these stress and damaging factors. Factors such as oxidative stress, muscle damage, inflammation, metabolic stress, and temperature are involved in exercise-induced HSP adaptation. HSP as chaperokine can stimulate the immune response by interacted to cell receptors such as TLR-2. Furthermore, physical exercise by inducing the chaperokine activity of HSP might be positive in prevention/control of conditions such as cardiovascular diseases, diabetes, cancer and neurodegenerative disease such as Alzheimer. Despite these recent advances, many questions still remain about exercise-induced HSP. Further studies, by focusing on these questions; can give us a better understanding of the biology of the HSP response and its potential application.

Acknowledgment This work was supported by the Shahed University, Tehran, Iran.

References

Abubaker J, Tiss A, Abu-Farha M et al (2013) DNAJB3/HSP-40 cochaperone is downregulated in obese humans and is restored by physical exercise. PLoS One 8(7):e69217

Aceros H, Farah G, Cobos-Puc L (2011) Moxonidine improves cardiac structure and performance in SHR through inhibition of cytokines, p38 MAPK and Akt. Br J Pharmacol 164:946–957

Adrie C, Ritcher C, Bachelet M (2000) Contrasting effects of NO and peroxynitrites on HSP70 expression and apoptosis in human monocyctes. Am J Physiol 279:C452–C460

Akira S, Taga T, Kishimoto T (1993) Interleukin-6 in biology and medicine. Adv Immunol 54:1–78

Aprahamian I, Stella F, Forlenza OV (2013) New treatment strategies for Alzheimer's disease: is there a hope? Indian J Med Res 138:449–460

Archer EA, Von-Schulze TA, Geiger CP (2018) Exercise, heat shock proteins and insulin resistance. Phil Trans R Soc B 373(1738), pii20160529

Arrigo AP, Welch WJ (1987) Characterization and purification of the small 28,000-dalton mammalian heat shock protein. J Biol Chem 262:15359–15369

Asea A (2006) Initiation of the immune response by extracellular Hsp72: chaperokine activity of Hsp72. Curr Immunol Rev 2(3):209–215

Asea A, Kraeft SK, Kurt-Jones EA, Stevenson MA, Chen LB, Finberg RW et al (2000a) HSP70 stimulates cytokine production through a CD14-dependant pathway, demonstrating its dual role as a chaperone and cytokine. Nat Med 6(4):435

Asea A, Kabingu E, Ann Stevenson M, Calderwood SK (2000b) HSP70 peptide-bearing and peptide-negative preparations act as chaperokines. Cell Stress Chaperones 5(5):425–431

Asea A, Rehli M, Kabingu E, Boch JA, Baré O, Auron PE, Calderwood SK (2002) Novel signal transduction pathway utilized by extracellular HSP70 role of Toll-like receptor (TLR) 2 and TLR4. J Biol Chem 277(17):15028–15034

Atalay M, Laaksonen DE (2002) Diabetes, oxidative stress and physical exercise. J Sport Sci Med 1(1):1–14

Baines CP, Molkentin JD (2005) Stress signaling pathways that modulate cardiac myocyte apoptosis. J Mol Cell Cardiol 38:47–62
Balin BJ, Hudson AP (2014) Etiology and pathogenesis of late-onset Alzheimer's disease. Curr Allergy Asthma Rep 14(3):417
Barone R, Macaluso F, Sangiorgi C (2016) Skeletal muscle Heat shock protein 60 increases after endurance training and induces peroxisome proliferator-activated receptor gamma coactivator 1 α1 expression. Sci Rep 6:19781
Beckmann RP, Mizzen LE, Welch WJ (1990) Interaction of Hsp 70 with newly synthesized proteins: implications for protein folding and assembly. Science 248:850–854
Beere HM, Wolf BB, Cain K et al (2000) Heat-shock protein 70 inhibits apoptosis by preventing recruitment of procaspase-9 to the Apaf-1 apoptosome. Nat Cell Biol 2:469–475
Benini R, Prado-Nunes PR, Orsatti CL, Barcelos LC, Orsatti FL (2015) Influence of sex on cytokines, heat shock protein and oxidative stress markers in response to an acute total body resistance exercise protocol. J Exerc Sci Fit 55(4):337–344
Benjamin IJ, Horie S, Greenberg ML et al (1992) Induction of stress proteins in cultured myogenic cells: molecular signals for the activation of heat shock transcription factor during ischemia. J Clin Invest 89:1685–1689
Berruti G, Perego L, Borgonovo B, Martegani E (1998) MSJ-1, a new member of the DNAJ family of proteins, is a male germ cell-specific gene product. Exp Cell Res 239:430–441
Beyer I, Njemini R, Bautmans I et al (2012) Inflammation-related muscle weakness and fatigue in geriatric patients. Exp Gerontol 47:52–59
Boyd-Kimball D, Sultana R, Poon HF, Lynn BC, Casamenti F, Pepeu G, Klein JB, Butterfield DA (2005) Proteomic identification of proteins specifically oxidized by intracerebral injection of amyloid betapeptide (1–42) into rat brain: implications for Alzheimer's disease. Neuroscience 132:313–324
Bukau B, Weissman J, Horwich A (2006) Molecular chaperones and protein quality control. Cell 125(3):443–451
Calabrese V, Scapagnini G, Colombrita C et al (2003) Redox regulation of heat shock protein expression in aging and neurodegenerative disorders associated with oxidative stress: a nutritional approach. Amino Acids 25:437–444
Campisi J, Fleshner M (2003) Role of extracellular HSP72 in acute stress-induced potentiation of innate immunity in active rats. J Appl Physiol 94(1):43–52
Campisi j, Leem TH, Greenwood BN, Hansen MK, Moraska A, Higgins K, smith TP, Fleshner M (2003) Habitual physical activity facilitates stress induced HSP72 induction in brain, peripheral and immune tissues. Am J Physiol 284:R520–R530
Cappello F, Conway de Macario E, Marasa L, Zummo G, Macario AJ (2008) Hsp60 expression, new locations, functions and perspectives for cancer diagnosis and therapy. Cancer Biol Ther 7(6):801–809
Care A, Catalucci D, Felicetti F et al (2007) MicroRNA-133 controls. cardiac hypertrophy. Nat Med 13:613–618
Castegna A, Aksenov M, Thongboonkerd V, Klein JB, Pierce WM, Booze R, Markesbery WR, Butterfield DA (2002) Proteomic identification of oxidatively modified proteins in Alzheimer's disease brain. Part II: dihydropyrimidinase-related protein 2, alpha-enolase and heat shock cognate 71. J Neurochem 82:1524–1532
Castellani R, Hirai K, Aliev G, Drew KL, Nunomura A, Takeda A et al (2002) Role of mitochondrial dysfunction in Alzheimer's disease. J Neurosci Res 70(3):357–360
Chang J, Knowlton AA, Xu F, Wasser JS (2001) Activation of the heat shock response: relationship to energy metabolites. A (31)P NMR study in rat hearts. Am J Physiol Heart Circ Physiol 280(1):426–433
Chen S, Brown IR (2007) Neuronal expression of constitutive heat shock proteins: implications for neurodegenerative diseases. Cell Stress Chaperones 12:51–58
Chiang HL, Terlecky SR, Plant CP, Dice JF (1989) A role for a 70- kilodalton heat shock protein in lyosomal degradation of intracellular proteins. Science 246:382–385

Cruz-Jentoft AJ, Baeyens JP, Bauer JM et al (2010) Sarcopenia: European consensus on definition and diagnosis: report of the European working group on sarcopenia in older people. Age Ageing 39(4):412–423

De la-Monte SM, Wands JR (2009) Alzheimer's disease is type 3 diabetes-evidence reviewed. J Diabetes Sci Technol 6:1101–1113

Demirel HA, Hamilton KL, Shanely RA, Tumer N, Koroly MJ, Powers SK (2003) Age and attenuation of exercise-induced myocardial HSP72 accumulation. Am J Physiol Heart Circ Physiol 285(4):1609–1615

Dimauro I, Mercatelli N, Caporossi D (2016) Exercise induced ROS in heat shock proteins response. Free Radic Biol Med 98:46–55

Ding Q, Vaynman S, Souda P, Whitelegge JP, Gomez-Pinilla F (2006) Exercise affects energy metabolism and neural plasticity-related proteins in the hippocampus as revealed by proteomic analysis. Eur J Neurosci 24(5):1265–1276

Dodd SL, Hain B, Senf SM, Judge AR (2009) Hsp27 inhibits IKK beta-induced NF-kappa B activity and skeletal muscle atrophy. FASEB J 23:3415–3423

Escobedo J, Pucci AM, Koh TJ (2004) HSP25 protects skeletal muscle cells against oxidative stress. Free Radic Biol Med 37:1455–1462

Feasson L, Stockholm D, Freyssenet D et al (2002) Molecular adaptations of neuromuscular disease-associated proteins in response to eccentric exercise in human skeletal muscle. J Physiol 543:297–306

Febbraio MA (2000) Does muscle function and metabolism affect exercise performance in the heat? Exerc Sport Sci Rev 28:171–176

Febbraio MA, Koukoulas I (2000) HSP72 gene expression progressively increases in human skeletal muscle during prolonged, exhaustive exercise. J Appl Physiol 89(3):1055–1060

Febbraio MA, Steensberg A, Walsh R, Koukoulas I, van Hall G, Saltin B, Pedersen BK (2002) Reduced glycogen availability is associated with an elevation in HSP72 in contracting human skeletal muscle. J Physiol 538:911–917

Febbraio, M. A., Steensberg A, ., Keller, C., Starkie, R. L., Nielsen, H. B., Krustrup, P., Ott, P., Secher, N. H. Pedersen, B. K. (2003). Glucose ingestion attenuates interleukin-6 release from contracting skeletal muscle in humans. J Physiol, 549, 607–612

Fehrenbach E, Niess AM, Schlotz E, Passek F, Dickhuth HH, Northoff H (2000) Transcriptional and translational regulation of heat shock proteins in leukocytes of endurance runners. J Appl Physiol 89(2):704–710

Fehrenbach E, Niess AM, Voelker K, Northoff H, Mooren FC (2005) Exercise intensity and duration affect blood soluble HSP72. Int J Sports Med 26(07):552–557

Fehrenbach E, Northoff H (2001) Free radicals, exercise, apoptosis, and heat shock proteins. Exerc Immunol Rev 7:66–89

Fielding RA, Vellas B, Evans WJ et al (2011) Sarcopenia: an undiagnosed condition in older adults. Current consensus definition: prevalence, etiology, and consequences. International working group on sarcopenia. J Am Med Dir Assoc 12(4):249–256

Fischer CP, Hiscock NJ, Basu S et al (2006) Vitamin E isoform-specific inhibition of the exercise-induced heat shock protein 72 expression in humans. J Appl Physiol 100:1679–1687

Fittipaldi S, Dimauro I, Mercatelli N, Caporossi D (2014) Role of exercise-induced reactive oxygen species in the modulation of heat shock protein response. Free Radic Res 48:52–70

Fleshner M, Campisi J, Johnson JD (2003) Can exercise stress facilitate innate immunity? A functional role for stress-induced extracellular Hsp72. Exerc Immunol Rev 9:6–24

Folkesson M, Mackey AL, Holm L et al (2008) Immunohistochemical changes in the expression of HSP27 in exercised human vastus lateralis muscle. Acta Physiol 194:215–222

Fuqua SA, Oesterreich S, Hilsenbeck SG, Von Hoff DD, Eckardt J, Osborne CK (1994) Heat shock proteins and drug resistance. Breast Cancer Res Treat 32(1):67–71

Gandy S (2002) Molecular basis for anti-amyloid therapy in the prevention and treatment of Alzheimer's disease. Neurobiol Aging 23:1009–1016

Garcia JJ, Martin-Cordero L, Hinchado MD, Bote ME, Ortega E (2013) Effects of habitual exercise on the eHsp72-induced release of inflammatory cytokines by macrophages from obese Zucker rats. Int J Sports Med 34(06):559–564

Gielen S, Schuler G, Adams V (2010) Cardiovascular effects of exercise training: molecular mechanisms. Circulation 122(12):1221–1238

Geiger PC, Gupte AA (2011) Heat shock proteins are important mediators of skeletal muscle insulin sensitivity. Exerc Sport Sci Rev 39:34–42

Gjovaag TF, Dahl HA (2006) Effect of training and detraining on the expression of heat shock proteins in m. triceps brachii of untrained males and females. Eur J Appl Physiol 98:310–322

Golbidi S, Laher I (2011) Molecular mechanisms in exercise-induced cardioprotection. Cardiol Res Pract 2011:972807

Gonzalez B, Hernando R, Manso R (2000) Stress proteins of 70kDa in chronically exercised skeletal muscle. Eur J Phys 440:42–49

Gorman AM, Heavey B, Creagh E, Cotter TG, Samali A (1999) Antioxidant-mediated inhibition of the heat shock response leads to apoptosis. FEBS Lett 445:98–102

Grivennikov SI, Greten FR, Karin M (2010) Immunity, inflammation, and cancer. Cell 140(6):883–899

Guttman SD, Glover CVC, Allis CD, Gorovsky MA (1980) Heat shock, deciliation and release from anoxia induce the synthesis of the same polypeptides in starved *T. pyriformis*. Cell 22:299–307

Hageman J, Van Waarde MA, Zylicz A, Walerych D, Kampinga HH (2011) The diverse members of the mammalian HSP70 machine show distinct chaperone like activities. Biochem J 435:127–142

Hamilton KL, Staib JL, Phillips T, Hess A, Lennon SL, SK P (2003) Exercise, antioxidants, and HSP72: protection against myocardial ischemia/reperfusion. Free Radic Biol Med 34:800–809

Hamilton KL, Gupta S, Knowlton AA (2004) Estrogen and regulation of heat shock protein expression in female cardiomyocytes: cross-talk with NF kappa B signaling. J Mol Cell Cardiol 36(4):577–584

Harris MB, Starnes JW (2001) Effects of body temperature during exercise on myocardial adaptations. Am J Phys 280:2271–2280

Hashimoto M, Rockenstein E, Crews L, Masliah E (2003) Role of protein aggregation in mitochondrial dysfunction and neurodegeneration in Alzheimer's and Parkinson's diseases. NeuroMolecular Med 4(1–2):21–35

Hawley JA, Hargreaves M, Joyner MJ, Zierath JR (2014) Integrative biology of exercise. Cell 159:738–749

Henstridge CD, Febbraio AM, Hargreaves M (2016) Heat shock proteins and exercise adaptations. Our knowledge thus far and the road still ahead. J Appl Physiol 120:683–691

Hernando R, Manso R (1997) Muscle fibre stress in response to exercise: synthesis, accumulation and isoform transitions of 70-kDa heat-shock proteins. Eur J Biochem 243(1–2):460–467

Hood DA (2001) Contractile activity induced mitochondrial biogenesis in skeletal muscle. J Appl Physiol 90:1031–1035

Hooper PL, Balogh G, Rivas E (2014) The importance of the cellular stress response in the pathogenesis and treatment of type 2 diabetes. Cell Stress Chaperones 19(4):447–464

Hopkins EM, Davis CF, Van-Tieghem RM, Whalen JP, Bucci JD (2012) Differential effects of acute and regular physical exercise on cognition and affect. Neuroscience 26(215):59–68

Huey KA, Mc-Call GE, Zhong H, Roy RR (2007) Modulation of HSP25 and TNF-alpha during the early stages of functional overload of a rat slow and fast muscle. J Appl Physiol 102:2307–2314

Isanejad A, Saraf HZ, Mahdavi M, Gharakhanlou R, Shamsi MM, Paulsen G (2015) The effect of endurance training and downhill running on the expression of IL-1b, IL-6, and TNF-α and HSP72 in rat skeletal muscle. Cytokine 73:302–308

Jackson MJ (2005) Reactive oxygen species and redox-reulation of skeletal muscle adaptations to exercise. Philos Trans R Soc Lond Biol Sci 360:2285–2291

Jackson MJ, Khassaf M, Vasilaki A, McArdle F, McArdle A (2004) Vitamin E and the oxidative stress of exercise. Ann N Y Acad Sci 1031:158–168

Jakob U, Gaestel M, Engel K, Buchner J (1993) Small heat shock proteins are molecular chaperones. J Biol Chem 268:1517–1520

Jayaraman M, Kannayiram G, Rajadas J (2008) Amyloid toxicity in skeletal myoblasts: Implications for inclusion-body myositis. Arch Biochem Biophys 474(1):15–21

Jones Q, S Voegeli T, Li G, Chen Y, William Currie R (2011) Heat shock proteins protect against ischemia and inflammation through multiple mechanisms. Inflamm Allergy-Drug Targets (Formerly Curr Drug Targets-Inflamm Allergy) 10(4):247–259

Kampinga HH, Craig EA (2010) The HSP70 chaperone machinery: J proteins as drivers of functional specificity. Nat Rev Mol Cell Biol 11:579–592

Kampinga HH, Hageman J, Vos MJ et al (2009) Guidelines for the nomenclature of the human heat shock proteins. Cell Stress Chaperones 14:105–111

Kane JK, Konu O, Ma JZ, Li MD (2004) Nicotine coregulates multiple pathways involved in protein modification/degradation in rat brain. Brain Res Mol Brain Res 132:181–191

Karvinen S, Silvennoinen M, Vainio P et al (2016) Effects of intrinsic aerobic capacity, aging and voluntary running on skeletal muscle sirtuins and heat shock proteins. Exp Gerontol 79:46–54

Katschinski DM (2004) On heat and cells and proteins. News Physiol Sci 19:11–15

Kavazis AN, Smuder AJ, Min K et al (2010) Short-term exercise training protects against doxorubicin induced cardiac mitochondrial damage independent of HSP72. Am J Physiol Heart Circ Physiol 299:1515–1524

Kayani AC, Close GL, Jackson MJ, McArdle A (2008) Prolonged treadmill training increases HSP70 in skeletal muscle but does not affect age-related functional deficits. Am J Physiol Regul Integr Comp Physiol 294:568–576

Kregel KC (2002) Invited review: heat shock proteins: modifying factors in physiological stress responses and acquired thermotolerance. J Appl Phys 92(5):2177–2186

Khassaf M, McArdle A, Esanu C et al (2003) Effect of vitamin C supplements on antioxidant defence and stress proteins in human lymphocytes and skeletal muscle. J Physiol 549:645–652

Kim KB, Kim MH, Lee DJ (2004) The effect of exercise in cool, control and hot environments on cardioprotective HSP70 induction. J Physiol Anthropol Appl Hum Sci 23:225–230

Ko IG, Kim SE, Kim CJ, Jee YS (2013) Treadmill exercise alleviates aging-induced apoptosis in rat cardiac myocytes. Int J Gerontol 7:152–157

Koo HJ, Kwon SN, Kang BE et al (2013) Neuroprotective effects of treadmill exercise on BDNF and PI3-K/Akt signaling pathway in the cortex of transgenic mice model of Alzheimer's disease. J Exerc Nutr Biochem 17(4):151–160

Lancaster GI, Febbraio MA (2005) Mechanisms of stress-induced cellular HSP72 release: implications for exercise-induced increases in extracellular HSP72. Exerc Immunol Rev 11:46–52

Lancaster GI, Møller K, Nielsen B, Secher NH, Febbraio MA, Nybo L (2004) Exercise induces the release of heat shock protein 72 from the human brain in vivo. Cell Stress Chaperones 9(3):276–280

Lawler JM, Hindle A (2011) Living in a box or call of the wild? Revisiting lifetime inactivity and sarcopenia. Antioxid Redox Signal 15:2529–2541

Lin MT, Beal MF (2006) Mitochondrial dysfunction and oxidative stress in neurodegenerative diseases. Nature 443(7113):787

Lin KM, Lin B, Lian IY, Mestril R, Scheffler IE, Dillmann WH (2001) Combined and individual mitochondrial HSP60 and HSP10 expression in cardiac myocytes protects mitochondrial function and prevents apoptotic cell deaths induced by simulated ischemia-reoxygenation. Circulation 103:1787–1792

Liu Y, Steinacker JM (2001) Changes in skeletal muscle heat shock proteins: pathological significance. Front Biosci 6:12–25

Liu Y, Mayr S, Opitz-Gress A et al (1999) Human skeletal muscle HSP70 response to training in highly trained rowers. J Appl Physiol 86:101–104

Liu Y, Lormes W, Baur C et al (2000) Human skeletal muscle HSP70 response to physical training depends on exercise intensity. Int J Sports Med 21:351–355
Liu Y, Lormes W, Wang L, Reissnecker S, Steinacker JM (2004) Different skeletal muscle HSP70 responses to high-intensity strength training and low-intensity endurance training. Eur J Appl Physiol 91:330–335
Liu Y, Gampert L, Nething K, Steinacker JM (2006) Response and function of skeletal muscle heat shock protein 70. Front Biosci 11:2802–2827
Locke M (1997) The cellular stress response to exercise: role of stress proteins. Exerc Sports Sci Rev 25:105–136
Locke M, Noble E (2002) Overview of the stress response. In: Exercise and stress response. CRC Press, Boca Raton, pp 8–19
Locke M, Noble EG (2002) Exercise and stress response: the role of stress proteins. CRC Press LLC, Boca Raton
Locke M, Tanguay RM (1996) Diminished heat shock response in the aged myocardium. Cell Stress Chaperones 1(4):251–260
Locke M, Noble EG, Atkinson BG (1990) Exercising mammals synthesise stress proteins. Am J Physiol 258:C723–C729
Luo J, McMullen JR, Sobkiw CL et al (2005) Class IA phosphoinositide 3-kinase regulates heart size and physiological cardiac hypertrophy. Mol Cell Biol 25:9491–9502
Lustbader JW, Cirilli M, Lin C et al (2004) ABAD directly links Abeta to mitochondrial toxicity in Alzheimer's disease. Science 304(5669):448–452
Lyashko VN, Chernicov VG, Ivanov VI, Ulmasov KA, Zatsepina OG, Evgenev MB (1994) Comparison of the heat shock response in ethnically and ecologically different human populations. Proc Ntl Acad Sci USA 91:2492–2495
Macario AJ, Conway de Macario E (2005) Sick chaperones, cellular stress, and disease. New Engl J Med 353:1489–1501
Maglara AA, Vasilaki A, Jackson MJ et al (2003) Damage to developing mouse skeletal muscle myotubes in culture: protective role of heat shock proteins. J Physiol 548:837–846
Malm C (2001) Exercise-induced muscle damage and inflammation: fact or fiction? Acta Physiol 171(3):233–239
Marber MS, Meistrel R, Chi SR, Sayen MR, Yellon DM, Dillman WH (1995) Overexpression of the rat inducible 70 k-Da heat stress protein in a transgenic mouse increases the resistance of the heart to ischemic injury. J Clin Investig 95:1446–1456
Mariani E, Polidori MC, Cherubini A, Mecocci P (2005) Oxidative stress in brain aging, neurodegenerative and vascular diseases: an overview. J Chromatogr B Anal Technol Biomed Life Sci 827:65–75
Markuns J, Wojtaszewski JFP, Goodyear JL (1999) Insulin and exercise decrease glycogen synthase kinase-3 activity by different mechanisms in rat skeletal muscle. J Biol Chem 274(35):24896–24900
Marshall HC, Ferguson RA, Nimmo MA (2006) Human resting extracellular heat shock protein 72 concentration decreases during the initial adaptation to exercise in a hot, humid environment. Cell Stress Chaperones 11(2):129–134
Martin JL, Mestril R, Hilal-Dandan R, Brunton LL, Dillmann WH (1997) Small heat shock proteins and protection against ischemic injury in cardiac myocytes. Circulation 96:4343–4348
Mattson JP, Ross CR, Kilgore JL, Musch TI (2000) Induction of mitochondrial stress proteins following treadmill running. Med Sci Sports Exerc 32(2):365–369
McArdle A, Pattwell D, Vasilaki A et al (2001) Contractile activity- induced oxidative stress: cellular origin and adaptive responses. Am J Phys 280:C621–C627
McArdle F, Spiers S, Aldemir H, Vasilaki A, Beaver A, Iwanejko I, McArdle A, Jackson MJ (2004) Preconditioning of skeletal muscle against contractioninduced damage: the role of adaptations to oxidants in mice. J Physiol 561:233–244

McNaught KS, Jenner P (2001) Proteasomal function is impaired in substantia nigra in Parkinson's disease. Neurosci Lett 297:191–194

Melkani GC, Cammarato A, Bernstein SI (2006) aB-crystallin maintains skeletal muscle myosin enzymatic activity and prevents its aggregation under heat shock stress. J Mol Biol 358:635–645

Melling CJ, Krause MP, Noble EG (2006) PKA-mediated ERK1/2 inactivation and hsp70 gene expression following exercise. J Mol Cell Cardiol 41(5):816–822

Melling CW, Thorp DT, Milne JK, Krause MP, Noble EG (2007) Exercise-mediated regulation of Hsp70 expression following aerobic exercise training. Am J Physiol Heart Circ Physiol 293:3692–3698

Millarski KL, Morimoto RI (1989) Mutational analysis of the human HSP70 protein: distinct domains for nuclear localization and ATP binding. J Cell Biol 109:1947–1962

Milne KJ, Noble EG (2002) Exercise-induced elevation of HSP70 is intensity dependent. J Appl Physiol 92:561–568

Moran M, Delgado J, Gonzalez B (2004) Responses of rat myocardial antioxidant defences and heat shock protein HSP72 induced by 12 and 24-week treadmill training. Acta Physiol Scand 180:157–166

Morley JE, Abbatecola AM, Argiles JM et al (2011) Sarcopenia with limited mobility: an international consensus. J Am Med Dir Assoc 12(6):403–409

Morton JP, Mac Laren DPM, Cable NT et al (2007) Elevated core and muscle temperature to levels comparable to exercise do not increase heat shock protein content of physically active men. Acta Physiol 190:319–327

Morton JP, Croft L, Bartlett J et al (2009a) Reduced carbohydrate availability does not modulate training-induced heat shock protein adaptations but does up-regulate oxidative enzyme activity in human skeletal muscle. J Appl Physiol 106:1513–1521

Morton JP, Kayani AC, McArdle A, Drust B (2009b) The exercise-induced stress response of skeletal muscle, with specific emphasison humans. Sports Med 39(8):643–662

Moser C, Schmidbauer C, Gürtler U, Gross C, Gehrmann M, Thonigs G et al (2002) Inhibition of tumor growth in mice with severe combined immunodeficiency is mediated by heat shock protein 70 (Hsp70)-peptide–activated, CD94 positive natural killer cells. Cell Stress Chaperones 7(4):365–373

Muchowski PJ, Wacker JL (2005) Modulation of neurodegeneration by molecular chaperones. Nat Rev Neurosci 6:11–22

Multhoff G, Botzler C, Jennen L, Schmidt J, Ellwart J, Issels R (1997) Heat shock protein 72 on tumor cells: a recognition structure for natural killer cells. J Immunol 158(9):4341–4350

Murlasits Z, Cutlip RG, Geronilla KB, Rao KM, Wonderlin WF, Alway SE (2006) Resistance training increases heat shock protein levels in skeletal muscle of young and old rats. Exp Gerontol 41:398–406

Muscaritoli M, Anker SD, Argiles J et al (2010) Consensus definition of sarcopenia, cachexia and pre-cachexia: joint document elaborated by Special Interest Groups (SIG) "cachexia-anorexia in chronic wasting diseases" and "nutrition in geriatrics". Clin Nutr (Edinb Scot) 29(2):154–159

Naito H, Powers SK, Demirel HA et al (2000) Heat stress attenuates skeletal muscle atrophy in hindlimb-unweighted rats. J Appl Physiol 88:359–363

Najarzadegan MR, Ataei E, Akbarzadeh F et al (2016) The role of heat shock proteins in Alzheimer disease: a systematic review. J Syndr 3(1):6

Nething K, Wang L, Liu Y, Lormes W, Steinacker JM (2004) Blunted HSP70 response to acute exercise in well trained skeletal muscle. Med Sci Sports Exerc 36:S318

Nielsen JN, Wojtaszewski JF (2004) Regulation of glycogen synthase activity and phosphorylation by exercise. Proc Nutr Soc 63(2):233–237

Niess AM, Dickhuth, Northoff H, Fehrenbach E (1999) Free radicals and oxidative stress in exercise-immunological aspects. Exerc Immunol Rev 5:22–56

Niess AM, Fehrenbach E, Schlotz E, Sommer M, Angres C, Tschositsch K (2002) Effects of RRR-alpha-tocopherolon leukocyte expression of HSP72 in response to exhaustive treadmill exercise. Int J Sports Med 23:445–452

Nilsen ST, Thorsen L, Kirkegaard C, Ugelstad U, Fossa SD, Raastad S (2016) The effect of strength training on muscle cellular stress in prostate cancer patients on ADT. Endocr Connect 5(2):74–82

Noble EG (2002) Heat shock proteins and their induction with exercise. In: Exercise and stress response. CRC Press, Boca Raton, pp 49–84

Noble EG, Milne KJ, Melling J (2008) Heat shock proteins and exercise: a primer. Appl Physiol Nutr Metab 33:1050–1065

O'Neill DET, Aubrey FK, Zeldin DA, Michel RN, Noble EG (2006) Slower skeletal muscle phenotypes are critical for constitutive expression of HSP70 in overloaded rat plantaris muscle. J Appl Physiol 100:981–987

Obinata T, Maruyama K, Sugita H, Kohama K, Ebashi S (1981) Dynamic aspects of structural proteins in vertebrate skeletal muscle. Muscle Nerve 4:456–488

Ogawa K, Sanada K, Machida S et al (2010) Resistance exercise training-induced muscle hypertrophy was associated with reduction of inflammatory markers in elderly women. Mediat Inflamm 2010:171023. https://doi.org/10.1155/2010/171023

Ogawa K, Kim HK, Shimizu T et al (2012) Plasma heat shock protein 72 as a biomarker of sarcopenia in elderly people. Cell Stress Chaperones 17:349–359

Ogura Y, Naito H, Kurosaka M, Sugiura T, Aoki J, Katamoto S (2006) Sprint-interval training induces heat shock protein 72 in rat skeletal muscles. J Sports Sci Med 5:194–201

Oishi Y, Taniguchi K, Matsumoto H et al (2002) Muscle type specific response of HSP60, HSP72 and HSC73 during recovery after elevation of muscle temperature. J Appl Physiol 92:1097–1103

Oishi Y, Taniguchi K, Matsumoto H et al (2003) Differential responses of HSP to heat stress in slow and fast regions of rat gastrocnemius muscle. Muscle Nerve 28:587–594

Ortega E, Hinchado MD, Martin-Cordero L, Asea A (2009) The effect of stress-inducible extracellular Hsp72 on human neutrophil chemotaxis: a role during acute intense exercise. Stress 12(3):240–249

Park KJ, Gaynor RB, Kwak YT (2003) Heat shock protein 27 association with the I kappa B kinase complex regulates tumor necrosis factor alpha-induced NF-kappa B activation. J Biol Chem 278:35272–35278

Paroo Z, Haist JV, Karmazyn M, Noble EG (2002) Exercise improves postischemic cardiac function in males but not females: consequences of a novel sex-specific heat shock protein 70 response. Circ Res 90(8):911–917

Paroo Z, Tiidus PM, Noble EG (1999) Estrogen attenuates HSP 72 expression in acutely exercised male rodents. Eur J Appl Physiol Occup Physiol 80(3):180–184

Paulsen G, Hanssen KE, Ronnestad BR et al (2012) Strength training elevates HSP27, HSP70 and alphaB-crystallin levels in musculivastuslateralis and trapezius. Eur J Appl Physiol 112:1773–1782

Pelham HR (1986) Speculations on the functions of the major heat shock and glucose related proteins. Cell 46:959–961

Petriz BA, Gomes CP, Almeida JA, de Oliveira GP, Ribeiro FM, Pereira RW, Franco OL (2017) The effects of acute and chronic exercise on skeletal muscle proteome. J Cell Physiol 232(2):257–269

Pockley AG (2002) Heat shock proteins, inflammation, and cardiovascular disease. Circulation 105:1012–1017

Powers SK, Hamilton KL (2002) Heat shock proteins and reactive oxygen species. In: Exercise and stress response. CRC Press, Boca Raton, pp 128–140

Powers SK, Kavazis AN, Deruisseau KC (2005) Mechanisms of disuse muscle atrophy: role of oxidative stress. Am J Physiol Regul Integr Comp Physiol 288(2):R337–R344

Powers SK, Locke M, Demirel HA (2001) Exercise, heat shock proteins, and myocardial protection from I-R injury. Med Sci Sports Exerc 33(3):386–392

Powers SK, Talbert EE, Adhihetty PJ (2011) Reactive oxygen and nitrogen species as intracellular signals in skeletal muscle. J Physiol 589:2129–2138
Puntschart A, Vogt M, Widmer HR, Hoppeler H, Billeter R (1996) HSP70 expression in human skeletal muscle after exercise. Acta Physiol Scand 157:411–417
Ravagnan L, Gurbuxani S, Susin SA et al (2001) Heat-shock protein 70 antagonizes apoptosis-inducing factor. Nat Cell Biol 3:839–843
Richter K, Haslbeck M, Buchner J (2010) The heat shock response: life on the verge of death. Mol Cell 22:253–266
Rinaldi B, Corbi G, Boccuti S et al (2006) Exercise training affects age-induced changes in SOD and heat shock protein expression in rat heart. Exp Gerontol 41:764–770
Rodgers JT, Lerin C, Haas W, Gygi SP, Spiegelman BM, Puigserver P (2005) Nutrient control of glucose homeostasis through a complex of PGC-1alpha and SIRT1. Nature 434(7029):113–118
Salo DC, Donovan CM, Davies KJA (1991) HSP70 and other possible heat shock or oxidative stress proteins are induced in skeletal muscle, heart and liver during exercise. Free Radic Biol Med 11:239–246
Samali A, Orrenius S (1998) Heat shock proteins: regulators of stress response and apoptosis. Cell Stress Chaperone 3:228–236
Sammut IA, Jayakumar J, Latif N et al (2001) Heat stress contributes to the enhancement of cardiac mitochondrial complex activity. Am J Pathol 158:1821–1831
Sciandra JJ, Subjeck JR (1983) The effect of glucose on protein synthesis and thermosensitvity in Chinese hamster ovary cells. J Biol Chem 258:12091–12093
Scott JM (2011) Modulation of anthracycline-induced cardiotoxicity by aerobic exercise in breast cancer—current evidence and underlying mechanisms. Circulation 124:642–650
Selkoe DJ (2003) Folding proteins in fatal ways. Nature 426:900–904
Shamsi MM, Mahdavi M, Quinn SL, Gharakhanlou R, Isanegad A (2016) Effect of resistance exercise training on expression of Hsp70 and inflammatory cytokines in skeletal muscle and adipose tissue of STZ-induced diabetic rats. Cell Stress Chaperone 21:783–791
Silver JT, Kowalchuk H, Noble EG (2012) hsp70 mRNA temporal localization in rat skeletal myofibers and blood vessels post-exercise. Cell Stress Chaperone 17:109–120
Siu PM, Alway SE (2005) Age-related apoptotic responses to stretch-induced hypertrophy in quail slow-tonic skeletal muscle. Am J Physiol Cell Physiol 289:1105–1113
Skidmore R, Gutierrez JA, Guerriero VJ, Kregel KC (1995) HSP70 induction during exercise and heat stress in rats: role of internal temperature. Am J Phys 268:92–97
Smolka MB, Zoppi CC, Alves AA et al (2000) HSP72 as a complementary protection against oxidative stress induced by exercise in the soleus muscle of rats. Am J Physiol Regul Integr Comp Physiol 279:1539–1545
Sondermann H, Becker T, Mayhew M, Wieland F, Hartl FU (2000) Characterization of a receptor for heat shock protein 70 on macrophages and monocytes. Biol Chem 381(12):1165–1174
Soti C, Csermely P (2007) Aging cellular networks: chaperones as major participants. Exp Gerontol 42:113–119
Soufi FG, Farajnia S, Aslanabadi N et al (2008) Long-term exercise training affects age induced changes in HSP70 and apoptosis in rat heart. Gen Physiol Biophys 27:263–270
Staib JL, Quindry JK, French JP, Criswell DS, Powers SK (2007) Increased temperature, not cardiac load, activates heat shock transcription factor 1 and heat shock protein 72 expression in the heart. Am J Physiol Regul Integr Comp Physiol 292:432–439
Starkie RL, Angus DJ, Rolland J, Hargreaves M, Febbraio MA (2000) Effect of prolonged submaximal exercise and carbohydrate ingestion on monocyte intracellular cytokine production in humans. J Physiol 528:647–655
Steensberg A, van Hall G, Osada T, Sacchetti M, Saltin B, Pedersen BK (2000) Production of interleukin-6 in contracting human skeletal muscles can account for the exercise-induced increase in plasma interleukin-6. J Physiol 529:237–242
Steensberg A, Dalsgaard MK, Secher NH, Pedersen BK (2006) Cerebrospinal fluid IL-6, HSP72, and TNF-alpha in exercising humans. Brain Behav Immun 20(6):585–589

Stephanou A, Isenberg DA, Akira S, Kishimoto T, Latchman DS (1998) The nuclear factor interleukin-6 (NF-IL6) and signal transducer and activator of transcription 3 (STAT3) signalling pathways co-operate to mediate the activation of hsp90 gene by interleukin-6 but have opposite effects on its inducibility by heat shock. Biochem J 330:189–195

Sternlicht H, Farr GW, Sternlicht ML, Driscoll JK, Willison K, Yaffe MB (1993) The t-complex polypeptide 1 complex is a chaperonin for tubulin and actin in vivo. Proc Natl Acad Sci U S A 90:9422–9426

Sullivan GW, Sarembock IJ, Linden J (2000) The role of inflammation in vascular diseases. J Leukoc Biol 67(5):591–602

Sun Z, Zhang J, Zhang J (2011) Rad GTPase induces cardiomyocyte apoptosis through the activation of p38 mitogen-activated protein kinase. Biochem Biophys Res Commun 409:52–57

Suzuki K, Smolenski RT, Jayakumar J, Murtuza B, Brand NY, Yacoub MH (2000) Heat shock treatment enhances graft cell survival in skeletal myoblast transplantation to the heart. Circulation 102(Suppl 3):111216–111221

Tamura Y, Matsunaga Y, Masuda H et al (2014) Postexercise whole body heat stress additively enhances endurance training-induced mitochondrial adaptations in mouse skeletal muscle. Am J Physiol Regul Integr Comp Physiol 307:931–943

Tanonaka K, Toga W, Takahashi M, Yoshida H, Oikawa R, Takeo S (2004) Induction of heat shock protein 72 in the failing heart is attenuated after an exposure to heat shock. Mol Cell Biochem 259:211–215

Taylor RP, Starnes JW (2003) Age, cell signalling and cardioprotection. Acta Physiol Scand 178:107–116

Taylor RP, Harris MB, Starnes JW (1999) Acute exercise can improve cardioprotection without increasing heat shock protein content. Am J Physiol Heart Circ Physiol 276:1098–1102

Terry DF, Wyszynski DF, Nolan VG et al (2006) Serum heat shock protein 70 level as a biomarker of exceptional longevity. Mech Ageing Dev 127:862–868

Thomas JR, Kenfield AS, Jimenez A (2016) Exercise-induced biochemical changes and their potential influence on cancer: a scientific review. Br J Sports Med 0:1–8

Thompson HS, Maynard EB, Morales ER, Scordilis SP (2003) Exercise-induced HSP27, HSP70 and MAPK responses in human skeletal muscle. Acta Physiol Scand 178(1):61–72

Tissieres A, Mitchell HK, Tracy UM (1974) Protein synthesis in salivary glands of *Drosophila melanogaster*: relation to chromosone puffs. J Mol Biol 84:389–398

Tsuzuki T, Kobayashi T, Yoshihara H et al (2017) Attenuation of exercise-induced heat shock protein 72 expression blunts improvements in whole-body insulin resistance in rats with type 2 diabetes. Cell Stress Chaperones 22(2):263–269

Tupling AR, Bombardier E, Stewart RD et al (2007) Muscle fibre specific response of HSP70 expression in human quadriceps following acute isometric exercise. J Appl Physiol 103:2105–2111

Tuppo EE, Arias HR (2005) The role of inflammation in Alzheimer's disease. Int J Biochem Cell Biol 37(2):289–305

Tytell M (2005) Release of heat shock proteins (Hsps) and the effects of extracellular Hsps on neural cells and tissues. Int J Hyperth 21(5):445–455

Vasilaki A, Jackson MJ, McArdle A (2002) Attenuated HSP70 response in skeletal muscle of aged rats following contractile activity. Muscle Nerve 25:902–905

Vasilaki A, McArdle F, Iwanejko LM, McArdle A (2006) Adaptive responses of mouse skeletal muscle to contractile activity: the effect of age. Mech Ageing Dev 127:830–839

Venojarvi M, Aunola S, Puhke R (2008) Exercise training with dietary counselling increases mitochondrial chaperone expression in middle-aged subjects with impaired glucose tolerance. BMC Endocr Disord 8:3

Walsh RC, Koukoulas I, Garnham A, Moseley PL, Hargreaves M, Febbraio MA (2001) Exercise increases serum Hsp72 in humans. Cell Stress Chaperones 6(4):386–393

Walters G, Ryan KL, Tehrany MR, Jones MB, Paulus LA, Mason PA (1998) HSP70 expression in the CNS in response to exercise and heat stress in rats. J Appl Physiol 84:1269–1277

Wei YH, Lee HC (2002) Oxidative stress, mitochondrial DNA mutation, and impairment of antioxidant enzymes in aging. Exp Biol Med 227(9):671–682
Weitzel G, Pilatus U, Rensing L (1985) Similar dose response of heat shock protein synthesis and intracellular pH change in yeast. Exp Cell Res 159:252–256
Welch WJ (1992) Mammalian stress response: cell physiology, structure/function of stress proteins, and implications for medicine and disease. Physiol Rev 72:1063–1081
Welch WJ, Garrells JI, Thomas GP, Lin JJ, Feramisco JR (1983) Biochemical characterization of the mammalian stress proteins and identification of two stress proteins as glucose and Ca2+ ionophore regulated proteins. J Biol Chem 258:7102–7111
Westerheide SD, Anckar J, Stevens JR, Sistonen L, Morimoto RI (2009) Stress inducible regulation of heat shock factor 1 by the deacetylase SIRT1. Science 323(5917):1063–1066
Yoo BC, Kim SH, Cairns N, Fountoulakis M, Lubec G (2001) Deranged expression of molecular chaperones in brains of patients with Alzheimer's disease. Biochem Biophys Res Commun 280:249–258
Yoshikawa T, Yoshida N (2000) Vitamin E and leukocyte-endothelial cell interactions. Antioxid Redox Signal 2(4):821–825
Zuo LI, Christofi FL, Wright VP, Liu CY, Merola AJ, Berliner LJ, Clanton TL (2000) Intra-and extracellular measurement of reactive oxygen species produced during heat stress in diaphragm muscle. Am J Physiol Cell Physiol 279(4):C1058–C1066

Chapter 13
Extracellular Hsp90 Governs Spatio-Temporal Regulation of Biological Responses

Yasuaki Tamura, Akihiro Yoneda, Norio Takei, and Kaori Sawada

Abstract Heat shock protein (HSP) plays a pivotal role in the maintenance of cellular homeostasis within cells. However, recent studies have unveiled unexpected roles of extracellular HSP in various types of biological responses. For example, extracellular Hsp90 facilitates wound healing via recruiting skin cells. In the field of immunology, it has been shown that the historically important and widely studied extracellular HSP-antigenic peptide complex (HSP-PC) activates adaptive immunity through antigen cross-presentation by professional antigen-presenting cells (pAPCs). In addition, extracellular HSP activates innate immunity as a chaperokine and play a role in the etiology of autoimmune diseases including systemic lupus erythematosus. Therefore, the role of extracellular HSP in immunity is now gaining much attention. The most prominent feature of extracellular HSP in immunity is that they function both on their own and as an HSP-ligand complex such as HSP-antigen peptide complex and HSP-nucleic acid complex. Herein, recapitulating the history, we introduce a unique feature of extracellular HSP as a chaperokine: targeting chaperoned molecules into a particular compartment of dendritic cells and thereby inducing innate immune responses via spatiotemporal regulation. Furthermore, we discuss how such chaperokine activity of extracellular HSP impacts the pathogenesis of autoimmune disease and point out that targeting HSP might be an effective therapeutic approach.

Keywords Autoimmune disease · Chaperokine · Dendritic cell · Endosome · Heat shock protein · Toll-like receptor · Tumor immunity · Wound healing

Y. Tamura (✉) · A. Yoneda · N. Takei · K. Sawada
Department of Molecular Therapeutics, Center for Food & Medical Innovation, Institute for Innovation and Business Promotion, Hokkaido University, Sapporo, Japan

A. A. A. Asea, P. Kaur (eds.), *Chaperokine Activity of Heat Shock Proteins*,
Heat Shock Proteins 16, https://doi.org/10.1007/978-3-030-02254-9_13

Abbreviations

APC	Antigen-presenting cell
CpG	Cytosine-phosphate- guanine
DC	Dendritic cell
HIF	Hypoxia-inducible factor
HSP	Heat shock protein
Hsp72	Heat shock protein 72
Hsp90	Heat shock protein 90
IFN	Interferon
MVB	Multivesicular body
SLE	Systemic lupus erythematosus
TGF-β	Tumor growth factor-β
TLR	Toll-like receptor
TNF	Tumor necrosis factor

13.1 Introduction

A link between HSP and immune responses was first demonstrated by Srivastava (Srivastava 2002). Much attention has been paid to the significance of his discoveries for the field of tumor immunity. No one could imagine that intracellular HSP acted as a tumor antigen, or more correctly speaking as a chaperone for tumor antigen. He found that the Hsp90 family member gp96 bound tumor antigen peptide(s) and that immunization of tumor-derived gp96 induced tumor-specific CD8^{+} T cell responses as well as an antitumor effect. Researchers in Srivastava's group demonstrated one after another that immunization of tumor-derived Hsp70, Hsp90 and calreticulin elicited tumor-specific cytotoxic T-cell responses and antitumor responses (Basu et al. 2001; Udono and Srivastava 1994). They also revealed the mechanisms by which antitumor immunity is induced. HSP chaperones various types of tumor antigen peptides generated within tumor cells. Immunized HSP-antigen peptide complexes are taken-up by pAPCs such as dendritic cells, resulting in cross presentation of an antigen peptide to specific CD8^{+} T-cells (Suto and Srivastava 1995). In fact, gp96 and Hsp70 isolated from tumor cells showed drastic immuotherapeutic effects in various murine tumor models (Tamura et al. 1997). Very interestingly, they also showed that not only CD8^{+} T cells but also NK cells play an important role in HSP-mediated antitumor immunity. Their following studies revealed that most of the HSP including Hsp70, Hsp90, gp96 and calreticulin stimulated innate immunity (Basu et al. 2000). Administration of HSP elicited cytokine production through Toll-like receptors (TLRs) expressed on APCs (Asea et al. 2002). A landmark discovery was also made by Srivastava's group. It was found that CD91 (LDL receptor-related protein 1, LRP1) acted as a receptor for antigen cross-presentation by APCs (Basu et al. 2001; Binder et al. 2000). Following this discovery, various types of HSP receptor expressed on APC have been demonstrated. Scavenger receptor that were identified include SR-A for gp96 (Berwin et al. 2003) and grp170 (ORP150) (Wang et al. 2007), and LOX-1 for Hsp70 (Delneste et al. 2002). In the late of 1990s, another important finding attracted much attention. The Calderwood's group demonstrated that Hsp70 stimulated the cytokine production from APCs and they therefore proposed this activity as "chaperokine" activity (Asea et al. 2000). The chaperokine activities are mediated through HSP receptors expressed on APC (Asea 2005)s.

In contrast, we demonstrated that extracellular Hsp90 chaperoned immune ligands but not Hsp90 of its own regulated innate immunity (Okuya et al. 2010; Tanaka et al. 2015). Importantly, such extracellular HSP-mediated immune activation such as production of IFN-α by dendritic cells (DCs) was regulated via spatiotemporal targeting of chaperoned ligands. Moreover, we showed that such activation of innate immunity by an HSP-ligand complex may be a pathogenesis of autoimmune diseases (Okuya et al. 2010; Saito et al. 2015). These characteristic functions can also be regarded as chaperokine activities. Thus, extracellular HSP as a chaperokine plays a very important role in the regulation of innate and adaptive immunity. In this chapter, we will introduce such unique feature of extracellular HSP in innate immunity and autoimmune diseases.

13.2 A (Sad) History of Extracellular Hsp90

In the late 1990s, a number of laboratories showed that Hsp90 is expressed on the cell surface, either as a tumor antigen or a protein that assists antigen presentation to antigen-presenting cells (Csermely et al. 1998; Tsutsumi and Neckers 2007). In addition, the secreted form of Hsp90 was reported to induce activation of ERK1/2 in rat vascular smooth muscle cells and stimulation of lymphoid cells (Liao et al. 2000). However, results of these earlier studies on extracellular Hsp90 were largely viewed as experimental artefacts such as a pathophysiological process or a leak from a small number of dead cells in culture. Multhoff and Hightower tried to address this critical issue in studies in which various pharmacological approaches were used, and they argued that extracellular Hsp90 was not an artefact (Multhoff and Hightower 1996; Palladino et al. 1987). The primary reason for this skepticism is that Hsp90 does not fit into any of conventional categories of actively secreted proteins. First, Hsp90 has neither any signal peptide for secretion via the endoplasmic reticulum/Golgi protein secretory pathway nor a recognizable transmembrane domain for membrane anchoring. Second, it had already been reported that Hsp90 could be released to the extracellular microenvironment following cell necrosis (Multhoff and Hightower 1996). Therefore, it was arguable that extracellular Hsp90 found in conditioned media of cultured cells may come from a small portion of dead cells. The turning point of secreted HSP is a wound healing story.

13.3 Secreted HSP Grabs the Spotlight Again!

It was striking that Eustace et al. demonstrated that cancer cells secreted Hsp90α but not Hsp90β, which in turn associated with the matrix metalloproteinase MMP2, causing cancer cell invasion. Inhibition of extracellular Hsp90 by using the Hsp90 inhibitor geldanamycin decreased both MMP2 activity and invasiveness (Eustace et al. 2004). These results indicate that cell-impermeant anti-Hsp90 drugs might decrease invasiveness without the concerns inherent in inhibiting intracellular Hsp90.

Furthermore, recent studies by independent groups have provided stronger arguments that cells secrete Hsp90 for a purpose. First, quiescent normal cells do not secrete Hsp90 (Cheng et al. 2008; Li et al. 2007; Liao et al. 2000). However, normal cells are induced to secrete Hsp90 by many pathological and stress cues including reactive oxygen species (ROS) (Tsutsumi and Neckers 2007), heat (Clayton et al. 2005; Hightower and Guidon Jr. 1989), hypoxia (Li et al. 2007), γ-irradiation and tissue injury-related cytokines such as TGF-α (Cheng et al. 2008). The level of TGF-α is low or undetectable in intact skin, and TGF-α only appears in a wound following skin injury. TGF-α stimulation causes rapid membrane translocation and secretion of Hsp90α (Cheng et al. 2011). Moreover, Tsutumi et al. demonstrated that a conserved hydrophobic motif in a β-strand at the boundary between the N-terminal domain and charged linker of Hsp90 is required for Hsp90 secretion (Tsutsumi et al. 2009). In addition, Wang et al. showed that a C-terminal EEVD motif and Thr-90 phosphorylation both play a regulatory role in Hsp90 secretion (Wang et al. 2009). Thus, these findings strongly support the notion that Hsp90 secretion is a regulated process. However, it should be clarified how multiple distinct sequence motifs across the entire protein regulates its secretion. In particular, it is important to demonstrate how these motifs plays a role in interaction with most of the reported exosome trafficking pathways for Hsp90 secretion as described later. Furthermore, Li et al. reported that hypoxia-inducible factor-1α (HIF-1α) regulates Hsp90α secretion in dermal fibroblasts and keratinocytes (Li et al. 2007; Woodley et al. 2009). Dominant negative HIF-1α blocks the secretion, whereas a constitutively active mutant HIF-1α makes the cells secrete Hsp90 even under a condition of normoxia. The same mechanism appears to operate in tumor cells. Depletion of HIF-1α or HIF-1β from the breast cancer cell line MDA-MB-231 completely blocked the secretion of Hsp90α, which could be rescued by exogenously re-introducing the constitutively active mutant HIF-1α, but not dominant negative HIF-1α (Sahu et al. 2012). These findings indicate that HIF-1α is a crucial and direct upstream regulator of Hsp90α secretion. Since overexpression of HIF-1α has been shown to occur in 40% of human tumors (Semenza 2007), it is thought that extracellular Hsp90α plays an important role in tumors. In fact, Eutace et al. demonstrated that Hsp90α, but not Hsp90β was secreted from HT-1080 human colon cancer cells (Eustace et al. 2004). It was also reported that Hsp90α is secreted from MCF-7 human breast cancer cell (Wang et al. 2009) and HCT-8 colon cancer cells (Suzuki and Kulkarni 2010). Furthermore, the human breast cancer cell lines

MDA-MB-231 and MDA-MB-468 overexpress HIF-1α which causes constitutive secretion of Hsp90α in an HIF-1-dependent manner (Sahu et al. 2012). However, studies on the detailed routes of its translocation and secretion in response to environmental cues continue to be a matter of debate. Moreover, it has still not been determined how HIF-1α-mediated signals are connected to this protein secretory pathway and what the roles of these reported secretion motifs in Hsp90 are. Further studies are needed to demonstrate the secretion machinery of Hsp90 in response to various stresses by living cells.

13.4 Secretion of Hsp90 via an Exosomal Protein Scretory Pathway

What is the mechanism of Hsp90 secretion from live cells? In general, there are several cellular protein trafficking machineries. First, the classical ER/Golgi protein secretory pathway requires a signal peptide at the amino terminus of protein. Another pathway is an exoxomal pathway, which is mediated by secreted nanovesicles, called exoxomes This protein secretory pathway is used for secretory proteins without signal peptide sequences. Exosomes are internal vesicles of multivesicular bodies (MVB) released into the extracelluar milieu upon fusion of MVB with the cell surface. Asea's group showed that whereas some extracellular Hsp72 could be found as free Hsp72, a proportion of extracellular Hsp72 was released within exosomes (Bausero et al. 2005; Gastpar et al. 2005). Hsp90 has also been identified in exosomes (Hegmans et al. 2004).

13.5 Wound Healing Story Put an End to the Sad History

In 2007, Li's group found that secreted Hsp90 promotes wound healing in mice far more strongly than the growth factor becaplermin gel (recombinant human PDGF-BB), which is approved by the FDA (Cheng et al. 2008, 2011; Li et al. 2007). They found that Hsp90 was secreted from stressed skin cells such as keratinocytes and dermal fibroblasts. Furthermore, topical application of recombinant Hsp90α proteins to diabetic mouse wounds dramatically shortened the time of wound closure from 35 days to 18 days, a significantly stronger effect than becaplermin gel treatment. The authors proposed that Hsp90 drives inward migration of dermal cells into the wound. Their findings shed light again on the extraxcellular Hsp90. They also demonstrated that a 115-aa fragment of secreted Hsp90α, called F-5 (aa-236 to aa-350), acts as a wound healing agent in mice. F-5 is located between the linker region (LR) and the middle (M) domain of human Hsp90α. F-5 promotes skin cell migration in vitro and wound healing in vivo as effectively as does full-length Hsp90. Topical application of an F-5 peptide promoted acute and diabetic wound

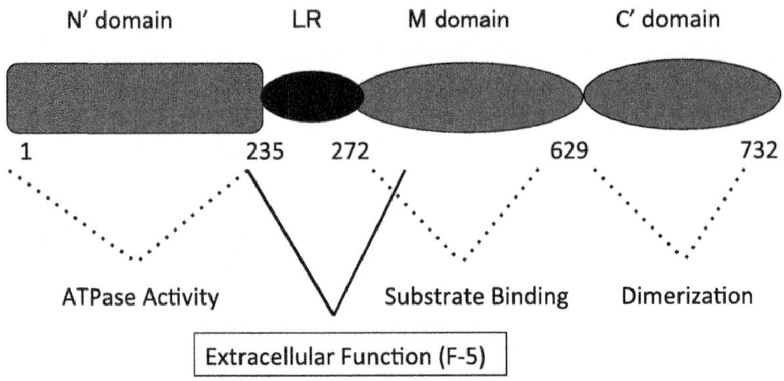

Fig. 13.1 (**a**) Functional elements of the pro-motility function of Hsp90α (F-5). Functional elements of intracellular Hsp90α and extracellular Hsp90α. Hsp90 has an amino terminal (N′) domain for ATPase activity, a carboxyl terminal (C′) domain for dimerization, and a middle (M) domain for chaperone activity. F-5, a 115-aminoacid-fragment, F-5 at the boundary between the linker region (LR) and the M domain, is responsible for the pro-motility function of extracellular Hsp90α

closure in mice far more effectively than did PDGF-BB. Interestingly, they found that F-5 has three properties: it has the ability to recruit both epidermal and dermal cells, it has the ability to promote dermal cell migration, which was not inhibited by TGF-β, and it has the ability to override the inhibitory effects of hyperglycemia on cell migration in diabetes. Very interestingly, Cheng et al. demonstrated that several Hsp90α mutants that have decreased or no ATPase activity retained a pro-motility activity similar to that of Hsp90α-wt (Cheng et al. 2011). These results indicate that the N-terminal ATPase domain and the C-terminal dimer-forming and co-factor binding domain are dispensable for extracellular Hsp90α to promote cell migration. A schematic representation of the structure and function requirements for intracellular Hsp90 and F-5 is shown in Fig. 13.1.

13.6 LRP1 (CD91) is an Essential Receptor for Extracellular Hsp90

Binder et al. first identified LDL receptor-related protein 1 (LRP1), also known as CD91, as a receptor responsible for cross-presentation of Gp96 expressed on APCs (Binder et al. 2000). Further examination revealed that CD91 was the common receptor for the extracellular HSP including Gp96, Hsp70, Hsp90, and calreticulin (Basu et al. 2001). LRP1 is a widely expressed cell surface receptor. Interestingly, independent research revealed a beautiful encounter between tumor immunity and biological reactions such as wound healing. Cheng et al. found that extracellular Hsp90 transduced a signal to promote skin cell migration for wound healing through LRP1 (Cheng et al. 2008). These marvelous findings provided overwhelming evidence that LRP1 is an essential receptor for extracellular Hsp90. However, since

LRP1 shows large ligand binding repertoire, LRP1 signal transduction is complicated and redundant. For instance, as described above, besides extracellular Hsp90 and gp96, other extracellular HSP such as Hsp60, Hsp70 and calretuculin also bind to LRP1, however, extracellular HSP other than Hsp90, including extracellular Hsp70, gp96 and calreticulin, did not exhibit cell migration ability. Therefore, to understand its mechanism of action, the specific binding site in LRP1 for extracellular Hsp90α must be identified.

13.7 Targeting Extracellular Hsp90 is Better Than Targeting Intracellular Hsp90 for Treatment of Tumors

N-terminal ATP-binding pocket of Hsp90 is a target site for Hsp90 inhibitors such as geldanamycin and its derivatives. These drugs binds to the ATP-binding pocket with higher affinity than ATP/ADP, and consequently direct Hsp90 client proteins to proteasomal degradation (Whitesell and Lindquist 2005). Therefore, Hsp90 inhibitors for cancer therapy have been extensively investigated and tested in many clinical trials. One hurdle of for the use of these drugs is how to minimize their potential interference with the physiological chaperone function of Hsp90 in surrounding normal tissue and cells and to selectively harm cancer Hsp90 in tumor cells embedded next to normal cells. The secretion of Hsp90 from normal cells under normal conditions has not been reported. Moreover, extracellular Hsp90α does not require the ATPase region, which is the target for geldanamycin-derived inhibitors and the reason for the drug-induced toxicity. Thus, the F-5 epitope in extracellular Hsp90α can be an excellent target for the design of safer, more effective and more specific inhibitors for treatment of Hsp90α-secreting tumors such as HIF-1α-positive tumors. Therefore, selective inhibitors against extracellular Hsp90α but not intracellular Hsp90α, or more specifically speaking, drugs targeting the F-5 region, should have higher efficacy and exert less toxicity in cancer patients.

13.8 Hsp90-Mediated Spatio-Temporal Regulation in Innate Immunity

In contrast to the idea that an HSP itself acts as a chaperokine, we have shown that HSP facilitates chaperoned innate ligands to activate an innate immune response via spatiotemporal regulation (Okuya et al. 2010). Bacterial and viral DNAs rich in cytosine-phosphate-guanine (CpG) motifs or small synthetic oligodeoxynucleotides (ODNs) containing CpG motifs are recognized by Toll-like receptor 9 (TLR9), which is expressed primarily by plasmacytoid DCs (pDCs) and B cells, leading to a large amount of IFN-α production (Krieg 2002; Wagner 1999). TLR9 plays a central role in viral immunity as well as in certain types of autoimmune diseases.

Compartmental control of cell signaling allows for specific, dynamic, and tightly regulated responses to various stimuli. In the case of immunostimulatory nucleic acid sensing, to avoid unnecessary stimulation, TLR9 localizes in the ER in a quiescent state and translocates to early endosomes only upon ligand recognition. Moreover, retention of the CpG-TLR9 complex in endosomes is the primary determinant of TLR9 signaling (Guiducci et al. 2006; Honda et al. 2005). Previous studies revealed that longer retention of CpG-A or various DNAs in the endosomes of pDCs is prerequisite for sufficient activation of TLR9 signaling (Lande et al. 2007; Yasuda et al. 2005). The manner of CpG internalization and the retention time of CpG in endosomes differ between CpG-A and CpG-B. CpG-A ODNs form large multimeric aggregates with diameters ~50 μm, but CpG-B ODNs are monomeric and do not form such higher-order structures. These TLR9 ligands exhibit distinct trafficking properties. Multimeric CpG-A ODNs are retained for longer periods of time in transferrin receptor-positive early endosomes, resulting in the extended activation of the TLR9-MyD88-IRF7 signal-transducing complex, which leads to robust IFN-α production. In contrast, CpG-B ODNs rapidly traffic through early endosomes into late endosomes or lysosomes of pDCs, resulting in the production of TNF-α and IL-6. In conventional DCs (cDCs), CpG-A is not retained in early endosomes and therefore does not stimulate production of IFN-α. However, if CpG-A is complexed with the cationic lipid DOTAP, cDCs retain the CpG-A-DOTAP complex in the early endosomes and produce IFN-α. Furthermore, Guiducci et al. demonstrated that TLR9-ligand interaction in a compartment of $pH > 6.0$, (i.e., early endosome) was required to trigger IFN-α production. In contrast, under the condition of pH below 5.75 (i.e., late endosome/lysosome), the ligand is released from TLR9 and supports pDC maturation but not IFN-α production (Guiducci et al. 2006). Thus, CpG-ODNs-mediated TLR9 activation is spatiotemporally controlled with TLR9 being activated in DCs.

Previously, we showed that translocation of Hsp90-Ag complex into the "static" early endosome after endocytosis was crucial for efficient cross-presentation (Kurotaki et al. 2007; Tanaka et al. 2015). Lakadamyali et al. demonstrated that early endosomes are comprised of two distinct populations: a population of dynamic early endosomes that are highly mobile on microtubules and mature rapidly into late endosomes, and a population of a static early endosomes that mature much more slowly (Lakadamyali et al. 2006). Cargoes destined for degradation, including LDL, EGF and influenza virus, are internalized and targeted to the $Rab5^+$, $EEA1^-$-dynamic early endosomes and thereafter traffic to $Rab7^+$-late endosomes. In contrast, the recycling ligand transferrin is delivered to $Rab5^+$, $EEA1^+$-static early endosomes, followed by translocation to $Rab11^+$-recycling endosomes. Based on our findings of the intriguing ability of Hsp90 to target and retain chaperoned ligands within "static" early endosomes, we examined whether Hsp90 can target chaperoned CpG-DNA within static early endosomes (Okuya et al. 2010). We found that murine pDCs pulsed with an Hsp90-CpG-A DNA complex produced a larger amount of IFN-α than that in the case of CpG-A alone (Fig. 13.2). Unlike human DCs, since murine cDCs express both TLR7 and TLR9, we examined the ability of Hsp90 to target and retain chaperoned CpG-A in static early endosomes of murine cDCs, leading to

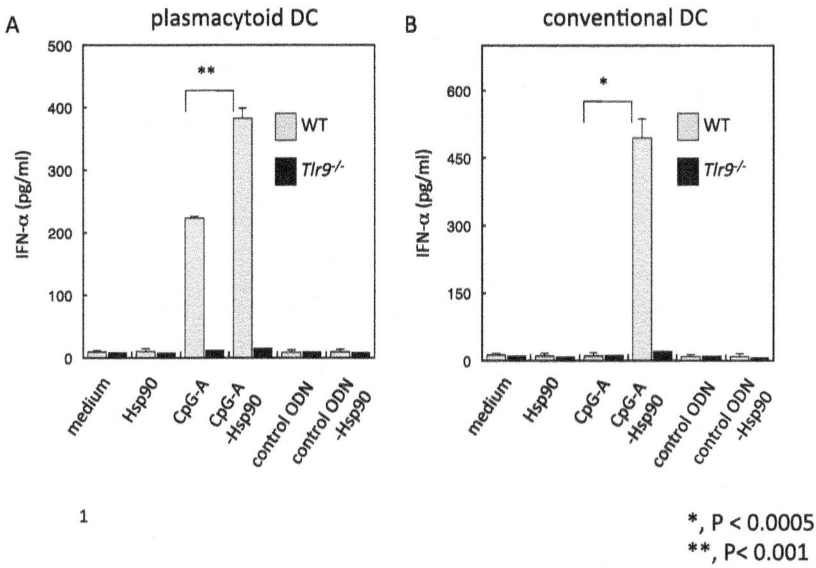

Fig. 13.2 CpG-A-Hsp90 complex enhances production of IFN-α by plasmacytoid DCs and conventional DCs. Murine plasmacytoid DCs (**a**) or conventional DCs (**b**) from WT or Tlr9$^{-/-}$ mice were loaded with CpG-A or CpG-A-Hsp90 complex. IFN-α production was determined by ELISA. Figures modified from Okuya et al. (2010)

IFN-α production. We found that Hsp90 targeted chaperoned CpG-A into static early endosomes and stimulated the production of IFN-α. The observed IFN-α production was TLR9-dependent because DCs isolated from TLR9 knockout mice did not do so. We found that Hsp90-chaperoned CpG-A was retained within static early endosomes for longer periods in cDCs (at least 2 h), thereby leading to sustained activation of cDCs and eliciting TLR9 signaling for IFN-α production. In contrast, CpG-A alone trafficked to late endosomes and lysosomes within cDCs. Another class of CpG-ODN, CpG-B, does not stimulate IFN-α production but stimulates the production of TNF-α and IL-6 by pDCs. Interestingly, not only CpG-A but also CpG-B when chaperoned by Hsp90 could stimulate TLR9 signaling via targeting and longer retention of CpG-B within static early endosomes, resulting in the production of IFN-α. Thus, extracellular Hsp90 has the extraordinary ability to direct associated CpGs into static early endosomes, leading to interferon regulatory factor (IRF)7 activation and IFN-α production (Fig. 13.3). Thus, an extracellular Hsp90-ligand complex also acts as a chaperokine. These results suggest that the use of extracellular Hsp90 would augment the effect of cancer vaccine in combination with CpG. Thus, Hsp90 regulates TLR9-mediated responses through spatiotemporal regulation of Hsp90-chaperoned ligand complexes. More importantly, extracellular Hsp90 might play a pivotal role in the pathogenesis of nucleic acid-mediated autoimmune disease such as systemic lupus erythematosus (SLE) as described in the following.

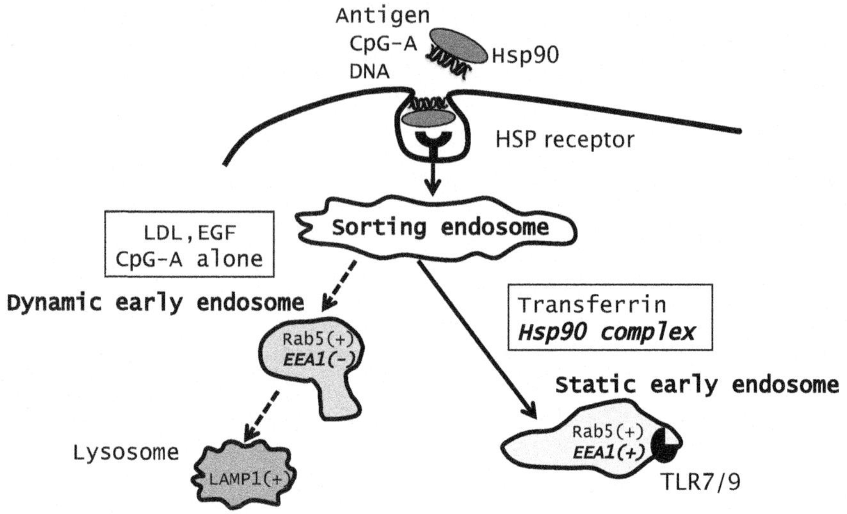

Fig. 13.3 Extracellular Hsp90 targets chaperoned molecules into static early endosome. Hsp90-ligand complexes are preferntilally targeted and retained in static early endosomes (Rab5$^+$ and EEA1$^+$) but not in dynamic early endosomes (Rab5$^+$ and EEA1$^-$)

13.9 IFN-α, Self-DNA and Autoimmune Diseases

Plasmacytoid DCs normally do not respond to self-DNA. This is due to the fact that viral/bacterial DNA sequences contain multiple CpG nucleotides that bind and activate TLR9, whereas mammalian self-DNA contains fewer such motifs, which are most likely masked by methylation. Moreover, it has been thought that self-DNA fails to access the TLR9-containing endolysosomal compartments and is thereby unable to stimulate TLR9. DNase easily and rapidly breaks down extracellular DNA, thereby preventing self-DNA from being targeted into endocytic compartments. The importance of this mechanism in inhibition of autoimmune responses has been shown by the fact that mice deficient in DNase II develop systemic lupus erythematosus (SLE)-like syndrome (Yoshida et al. 2005). Recent evidence, however, suggests that self-DNA has the potential to activate TLR9 when self-DNA is engaged TLR9. For instance, it was shown in an in vitro study that mammalian DNA in the form of chromatin-containing immune complexes could stimulate TLR9 in association with B-cell receptors (Viglianti et al. 2003). A breakdown of innate tolerance to self-nucleic acids occurs when tissue injury or necrosis causes the release of some endogenous molecules, including antimicrobial peptides such as LL37 (Lande et al. 2007) and the nuclear protein HMGB-1, that help to promote stabilization and delivery of associated innate ligands including nucleic acids into early endosomes (Yanai et al. 2009). These molecules have been shown to play a pivotal role in the initiation of autoimmune diseases through IFN-α production.

SLE is a prototype of autoimmune disease characterized by the production of autoantibodies specific for self-nuclear antigens such as dsDNA and RNA-containing antigens including Sm and RNP. These autoantibodies complexed with their cognate self-antigens are deposited within capillaries of various organs and subsequently induce systemic disorders (Ripley et al. 2001). The commonly targeted organs include the skin, heart, kidneys, lungs, joints, and central nervous system. Recently, it has been demonstrated that a prominent feature of SLE is increased expression of type I IFN-regulated genes in blood cells, often being associated with an increase of IFN-α in serum (Bennett et al. 2003; Ronnblom et al. 2009; Trinchieri 2010). In fact, it has been shown that the levels of serum IFN-α are significantly elevated in SLE patients compared with the level in healthy subjects (Crow 2014). Moreover, it has been shown that elevated IFN-α in serum of SLE patients accelerates the disease severity of SLE and is associated with disease activity (Crow 2014; Crow and Kirou 2004; Elkon and Stone 2011). The major producer of IFN-α appears to be pDCs, which are decreased in number in blood but are abundant in the skin and lymph nodes (Ronnblom et al. 2009). In SLE patients, immune complexes consisting of autoantibodies bound to self-DNA and RNA can stimulate production of IFN-α through TLR9 and TLR7 after uptake via Fc receptors. The pathogenic role of IFN-α in SLE is mediated partly by its ability to induce immune activation, including a positive feedback loop that induces plasma cell maturation and increases autoantibody production (Jego et al. 2003). The role of IFN-α in this disease has been confirmed in lupus-prone mouse models (Agrawal et al. 2009; Triantafyllopoulou et al. 2010). Thus, the accumulated data indicate that IFN-α plays a key role in the development and disease activity of SLE.

13.10 Extracellular Hsp90 as a Possible Accelerator for Autoimmune Diseases

As described above, since Hsp90 can bind DNA, Hsp90-DNA complex may activate the production of IFN-α via targeting DNA to static early endosomes. Therefore, we examined whether Hsp90 targets self-DNA into static early endosomes, resulting in IFN-α production by human pDCs (Okuya et al. 2010). We showed that upon Hsp90-mediated endosomal translocation as well as longer retention, human self-DNA could activate pDCs via TLR9 to produce IFN-α. These results indicated that Hsp90 could inactivate innate tolerance to self-DNA by forming a complex with self-DNA. Through this mechanism, pDCs may drive autoimmunity in autoimmune diseases. Thus, extracellular Hsp90-self nucleic acid complex may accelerate the disease activity of autoimmune diseases such as SLE.

13.11 Extracellular Hsp90 Plays a Pivotal Role in the Pathological Condition of SLE

Autoreactivity to HSP is often associated with autoimmune diseases. Previous studies have demonstrated the presence of autoantibodies to Hsp90 (Conroy et al. 1994; Minota et al. 1988) and enhanced expression of Hsp90 in peripheral blood mononuclear cells (PBMC) of patients with active SLE (Ripley et al. 2001; Twomey et al. 1993), suggesting a role of Hsp90 in the pathogenesis of SLE. Furthermore, under stressful conditions, cytosolic Hsp90 translocates to the nucleus (Akner et al. 1992). This suggests that Hsp90 may bind self-DNA within the nucleus of damaged cells. When cells undergo necrosis, self-DNA associated with endogenous Hsp90 could be released into the extracellular milieu and might trigger IFN-α production by pDCs as described above. In fact, we found that serum levels of Hsp90 were significantly increased in patients with active SLE compared with the levels in patients with inactive SLE or other autoimmune diseases (Saito et al. 2015) (Fig. 13.4). Of interest, extracellular self-Hsp90 found in SLE sera could stimulate IFN-α production by pDCs, and we found that extracellular self-Hsp90 associates with self-DNA or an anti-DNA antibody-self DNA complex. This Hsp90-self DNA complex or

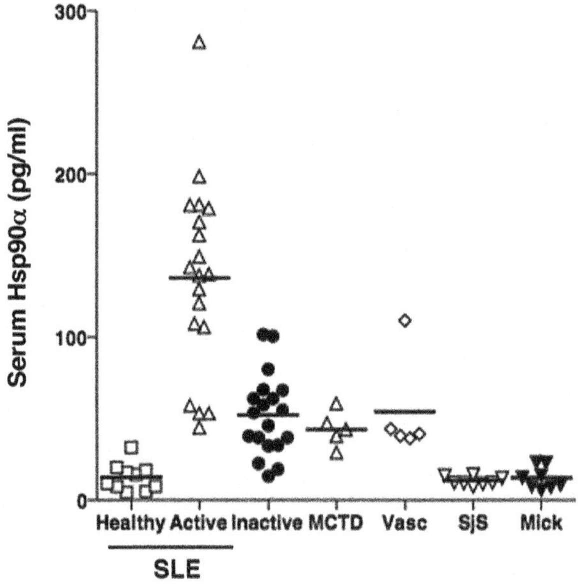

Fig. 13.4 Serum Hsp90α level is increased in SLE patients and is associated with disease activity. SLE patients were divided into subgroups according to SLEDAI scores of 0–5 (inactive) and >6 (active). Serum Hsp90α levels were increased in active SLE patients. In contrast, serum Hsp90α levels were significantly lower in patients with other autoimmune diseases. Healthy, healthy controls, *MCTD* mixed connective tissue disease, *Vasc* autoimmune vasculitis, *SjS* Sjögren's syndrome, *Mck* Micklickz disease (Adapted from Saito et al. (2015))

Hsp90-anti-DNA autoantibody-self DNA complex might be efficiently endocytosed and targeted to early endosomes via the action of Hsp90, leading to the robust IFN-α production observed in SLE sera (Okuya et al. 2010). Moreover, we showed that immunodepletion of extracellular Hsp90 from SLE serum decreased IFN-α production by pDCs, confirming that extracellular Hsp90 played a pivotal role in the production of IFN-α. These findings indicated that Hsp90 absorption using a column during hemodialysis might reduce disease activity and might induce remission and prevent end-organ damage. Thus, control of deregulated pDC activation and IFN-α production would provide an alternative treatment strategy for SLE. Collectively, our results indicate that extracellular Hsp90 plays a crucial role in the disease activity of SLE and that Hsp90 inhibitors have promise for treatment of IFN-α-mediated autoimmune diseases including SLE.

13.12 Intracellular Hsp90 Chaperones TLR7/9 to Sense Self Nucleic Acids

The involvement of IFN-α in the pathogenesis of SLE indicates the possibility of therapeutically targeting either IFN-α or the mechanisms leading to IFN-α production. As described above, pDCs play a pathogenic role in development of the autoimmune process by their ability to produce robust IFN-α through sensing nucleic acids by TLR7 and TLR9. The localization of TLR7/9 is unique, because they reside in the ER at the quiescent stage and traffic to endosomes only upon ligand recognition. Therefore, the underlying mechanism for the adequate transportation of these TLRs has been investigated. Recently, Unc93B1, a multitransmembrane ER-resident protein, has been shown to associate with and deliver TLR7/9 from the ER to endosomes, where TLR7/9 recognize their ligands (Fukui et al. 2009). In addition, gp96, an ER-resident HSP, has been shown to be a master immune chaperone for both cell surface and intracellular TLRs including TLR1, 2, 4, 5, 7, and 9 (Yang et al. 2007). Furthermore, the ER luminal protein PRAT4A (also known as CNPY3) also translocates TLR1, 2, 4, 5, 7, and 9 to either cell surfaces or endolysosomes (Takahashi et al. 2007). Recently, it has been shown that gp96 cooperates with PRAT4A for the trafficking of theses TLRs (Liu et al. 2010). More recently, AP-3 was identified as a regulator of TLR9 trafficking to an IFN-signaling compartment (the authors named as IRF7 endosome) by using a murine macrophage cell line (Sasai et al. 2010). AP-3 is associated with clathrin coats on sorting endosomes and directs cargo to functionally distinct lysosome-related organelles. The authors reported that AP-3 sorts TLR9 from a VAMP3$^+$ sorting endosome to a LAMP2$^+$ endosome. However, this model is different from the prevailing finding that early endosomal retention of TLR9 ligands promotes production of type I IFN, whereas lysosomal translocation of ligands elicits production of inflammatory cytokines (Guiducci et al. 2006; Honda et al. 2005).

We have recently shown that endogenous Hsp90 also has a crucial function in the TLR7 and TLR 9-mediated induction of IFN-α in pDCs (Saito et al. 2015). We first examined the effect of the Hsp90 inhibitor 17-AAG on the production of IFN-α in response to CpG-A by human pDCs. Treatment with 17-AAG dramatically inhibited the production of IFN-α. Using several independent methods to investigate how Hsp90 regulates IFN-α production, we found that Hsp90 physically associated with TLR7/9 and, more importantly, that Hsp90 chaperoned TLR7/9 from the ER to the early endosome. Inhibition of Hsp90 by 17-AAG resulted in disruption of the interaction of Hsp90 with TLR7/9, leading to inhibition of IRF7 phosphorylation and nuclear localization, which impaired the production of IFN-α. Moreover, laser confocal analysis revealed that treatment of pDCs with an Hsp90 inhibitor and depletion of Hsp90 using siRNA to Hsp90 resulted in inhibition of the endosomal translocation of TLR7/9, indicating that Hsp90 is required to deliver TLR7/9 from the ER to the early endosome, where TLR7 or TLR9 recognizes their ligands. Thus, Hsp90 binds directly to TLR7 and TLR9, indicating that Hsp90 is a cytosolic chaperone for TLR7 and TLR9 and is essential for TLR7/9-mediated innate immune responses (Saito et al. 2015).

13.13 Possible Therapeutic Use of an Hsp90 Inhibitor in SLE

Since pDC activation in blood by self-nucleic acids is regarded as a pathogenic trigger of the autoimmune process and since deregulated IFN-α elevation in serum by continuous pDC activation amplifies the pathogenic spiral in SLE, we examined the effect of an Hsp90 inhibitor in SLE-prone MRL/lpr mice (Saito et al. 2015). MRL/lpr mice develops an autoimmune disease that reflects pathologies of human SLE including lymph node enlargement, increased IgG levels, anti-nuclear antibodiy production, proteinuria, and kidney damage caused by inflammation of the glomeruli (Perry et al. 2011). We showed that treatment with the Hsp90 inhibitor 17-DMAG significantly ameliorated disease activity and improved histopathological findings in diseased mice. Administration of the Hsp90 inhibitor led to a significant reduction in serum levels of nucleic acid-specific autoantibodies (anti-dsDNA, anti-Sm, and anti-RNP), amelioration of proteinuria, and kidney damage, and an increased in survival. Shimp et al. also demonstrated that MRL/lpr mice treated with17-DMAG showed decreased proteinuria and reduced serum anti-dsDNA antibody levels (Shimp et al. 2012). They showed that 17-DMAG treatment led to an increase in the number of $CD8^+$ T cells, reduction in the number of double-negative T cells, decrease in the CD4/CD8 ratio, and reduction in the number of follicular B cells. They suggested that Hsp90 may play a pivotal role in the regulation of T cell differentiation and activation. However, the precise mechanism of Hsp90-mediated T cell differentiation in SLE has not been determined. Han et al. showed that stimulation of the T-cell receptor leading to T-cell activation requires Hsp90 to stabilize lymphocyte-specific protein tyrosine kinase (Lck) to initiate activation (Liu and Davidson 2012), which may be a mechanism behind the change in T-cell subtypes. Trepel et al. have indicated that Hsp90 inhibitors led to degradation of Hsp90-dependent substrate proteins such as NF-κB, which regulate inflammation

(Trepel et al. 2010). Moreover, it has been demonstrated that anti-inflammatory effects of Hsp90 inhibitors are due to the release of HSF-1 from Hsp90, resulting in the induction of a number of genes, including IL-10 and Hsp70, both of which are known to suppress pro-inflammatory and activate anti-inflammatory genes (Collins et al. 2013; Tukaj et al. 2014; Zhang et al. 2012). The immunosuppressive action of Hsp70 is mediated through (i) inactivation of DCs, (ii) expansion of regulatory T cells and (iii) blockade of NF-κB activity (Stocki and Dickinson 2012). It is possible that the dramatic decrease in severity of SLE is due to simultaneous blocking of TLR7 and TLR9 signaling by the Hsp90 inhibitor. Interestingly, similar to that in patients with SLE, serum Hsp90 levels in mice that developed SLE were clearly decreased in the 17-DMAG-treated group compared with the levels in the untreated group. Thus, our results indicate that both intracellular Hsp90 and extracellular Hsp90 play a crucial role in the pathogenesis of SLE and that Hsp90 inhibitors have promise for treatment of IFN-α-mediated autoimmune diseases including SLE.

Conclusions It is now understood that Hsp90 has two fundamental roles. One role is as an intracellular chaperone and the other role is as an extracellular tissue-repairing factor as well as immune system modulator, both of which seem to be designed for cells to cope with environmental stresses such as tissue injury. Both functions are taken advantage of by tumor cells during invasion and metastasis. More importantly, recognition of the existence of extracellular Hsp90 and its role in tissue repair and cancer invasion has revealed a new line of therapeutic intervention. However, important issues such as the mechanisms of Hsp90 secretion and extracellular Hsp90 action remain to be clarified. Not only Hsp90 but also other HSP have important roles in many aspects of life homeostasis. Therefore, the major unsolved issue, i.e., the mechanism for the trafficking of HSP into the extracellular milieu, must be clarified. By doing so, researchers will inevitably spark enthusiasm for the fascinating world of HSP biology.

Acknowledgements This research was partially supported by the Center of Innovation Program from MEXT and JST.

References

Agrawal H, Jacob N, Carreras E, Bajana S, Putterman C, Turner S, Neas B, Mathian A, Koss MN, Stohl W, Kovats S, Jacob CO (2009) Deficiency of type I IFN receptor in lupus-prone New Zealand mixed 2328 mice decreases dendritic cell numbers and activation and protects from disease. J Immunol 183:6021–6029

Akner G, Mossberg K, Sundqvist KG, Gustafsson JA, Wikstrom AC (1992) Evidence for reversible, non-microtubule and non-microfilament-dependent nuclear translocation of hsp90 after heat shock in human fibroblasts. Eur J Cell Biol 58:356–364

Asea A (2005) Stress proteins and initiation of immune response: chaperokine activity of hsp72. Exerc Immunol Rev 11:34–45

Asea A, Kabingu E, Stevenson MA, Calderwood SK (2000) HSP70 peptidembearing and peptide-negative preparations act as chaperokines. Cell Stress Chaperones 5:425–431

Asea A, Rehli M, Kabingu E, Boch JA, Bare O, Auron PE, Stevenson MA, Calderwood SK (2002) Novel signal transduction pathway utilized by extracellular HSP70: role of toll-like receptor (TLR) 2 and TLR4. J Biol Chem 277:15028–15034

Basu S, Binder RJ, Suto R, Anderson KM, Srivastava PK (2000) Necrotic but not apoptotic cell death releases heat shock proteins, which deliver a partial maturation signal to dendritic cells and activate the NF-kappa B pathway. Int Immunol 12:1539–1546

Basu S, Binder RJ, Ramalingam T, Srivastava PK (2001) CD91 is a common receptor for heat shock proteins gp96, hsp90, hsp70, and calreticulin. Immunity 14:303–313

Bausero MA, Gastpar R, Multhoff G, Asea A (2005) Alternative mechanism by which IFN-gamma enhances tumor recognition: active release of heat shock protein 72. J Immunol 175:2900–2912

Bennett L, Palucka AK, Arce E, Cantrell V, Borvak J, Banchereau J, Pascual V (2003) Interferon and granulopoiesis signatures in systemic lupus erythematosus blood. J Exp Med 197:711–723

Berwin B, Hart JP, Rice S, Gass C, Pizzo SV, Post SR, Nicchitta CV (2003) Scavenger receptor-A mediates gp96/GRP94 and calreticulin internalization by antigen-presenting cells. EMBO J 22:6127–6136

Binder R, Han D, Srivastava P (2000) CD91: a receptor for heat shock protein gp96. Nat Immun 1:151–155

Cheng CF, Fan J, Fedesco M, Guan S, Li Y, Bandyopadhyay B, Bright AM, Yerushalmi D, Liang M, Chen M, Han YP, Woodley DT, Li W (2008) Transforming growth factor alpha (TGFalpha)-stimulated secretion of HSP90alpha: using the receptor LRP-1/CD91 to promote human skin cell migration against a TGFbeta-rich environment during wound healing. Mol Cell Biol 28:3344–3358

Cheng CF, Sahu D, Tsen F, Zhao Z, Fan J, Kim R, Wang X, O'Brien K, Li Y, Kuang Y, Chen M, Woodley DT, Li W (2011) A fragment of secreted Hsp90alpha carries properties that enable it to accelerate effectively both acute and diabetic wound healing in mice. J Clin Invest 121:4348–4361

Clayton A, Turkes A, Navabi H, Mason MD, Tabi Z (2005) Induction of heat shock proteins in B-cell exosomes. J Cell Sci 118:3631–3638

Collins CB, Aherne CM, Yeckes A, Pound K, Eltzschig HK, Jedlicka P, de Zoeten EF (2013) Inhibition of N-terminal ATPase on HSP90 attenuates colitis through enhanced Treg function. Mucosal Immunol 6:960–971

Conroy SE, Faulds GB, Williams W, Latchman DS, Isenberg DA (1994) Detection of autoantibodies to the 90 kDa heat shock protein in systemic lupus erythematosus and other autoimmune diseases. Br J Rheumatol 33:923–926

Crow MK (2014) Type I interferon in the pathogenesis of lupus. J Immunol 192:5459–5468

Crow MK, Kirou KA (2004) Interferon-alpha in systemic lupus erythematosus. Curr Opin Rheumatol 16:541–547

Csermely P, Schnaider T, Soti C, Prohaszka Z, Nardai G (1998) The 90-kDa molecular chaperone family: structure, function, and clinical applications. A comprehensive review. Pharmacol Ther 79:129–168

Delneste Y, Magistrelli G, Gauchat J, Haeuw J, Aubry J, Nakamura K, Kawakami-Honda N, Goetsch L, Sawamura T, Bonnefoy J, Jeannin P (2002) Involvement of LOX-1 in dendritic cell-mediated antigen cross-presentation. Immunity 17:353–362

Elkon KB, Stone VV (2011) Type I interferon and systemic lupus erythematosus. J Interferon Cytokine Res: Off J Int Soc Interferon Cytokine Res 31:803–812

Eustace BK, Sakurai T, Stewart JK, Yimlamai D, Unger C, Zehetmeier C, Lain B, Torella C, Henning SW, Beste G, Scroggins BT, Neckers L, Ilag LL, Jay DG (2004) Functional proteomic screens reveal an essential extracellular role for hsp90 alpha in cancer cell invasiveness. Nat Cell Biol 6:507–514

Fukui R, Saitoh S, Matsumoto F, Kozuka-Hata H, Oyama M, Tabeta K, Beutler B, Miyake K (2009) Unc93B1 biases Toll-like receptor responses to nucleic acid in dendritic cells toward DNA- but against RNA-sensing. J Exp Med 206:1339–1350

Gastpar R, Gehrmann M, Bausero MA, Asea A, Gross C, Schroeder JA, Multhoff G (2005) Heat shock protein 70 surface-positive tumor exosomes stimulate migratory and cytolytic activity of natural killer cells. Cancer Res 65:5238–5247

Guiducci C, Ott G, Chan JH, Damon E, Calacsan C, Matray T, Lee KD, Coffman RL, Barrat FJ (2006) Properties regulating the nature of the plasmacytoid dendritic cell response to Toll-like receptor 9 activation. J Exp Med 203:1999–2008

Hegmans JP, Bard MP, Hemmes A, Luider TM, Kleijmeer MJ, Prins JB, Zitvogel L, Burgers SA, Hoogsteden HC, Lambrecht BN (2004) Proteomic analysis of exosomes secreted by human mesothelioma cells. Am J Pathol 164:1807–1815

Hightower LE, Guidon PT Jr (1989) Selective release from cultured mammalian cells of heat-shock (stress) proteins that resemble glia-axon transfer proteins. J Cell Physiol 138:257–266

Honda K, Ohba Y, Yanai H, Negishi H, Mizutani T, Takaoka A, Taya C, Taniguchi T (2005) Spatiotemporal regulation of MyD88-IRF-7 signalling for robust type-I interferon induction. Nature 434:1035–1040

Jego G, Palucka AK, Blanck JP, Chalouni C, Pascual V, Banchereau J (2003) Plasmacytoid dendritic cells induce plasma cell differentiation through type I interferon and interleukin 6. Immunity 19:225–234

Krieg A (2002) CpG motifs in bacterial DNA and their immune effects. Annu Rev Immunol 20:709–760

Kurotaki T, Tamura Y, Ueda G, Oura J, Kutomi G, Hirohashi Y, Sahara H, Torigoe T, Hiratsuka H, Sunakawa H, Hirata K, Sato N (2007) Efficient cross-presentation by heat shock protein 90-peptide complex-loaded dendritic cells via an endosomal pathway. J Immunol 179:1803–1813

Lakadamyali M, Rust MJ, Zhuang X (2006) Ligands for clathrin-mediated endocytosis are differentially sorted into distinct populations of early endosomes. Cell 124:997–1009

Lande R, Gregorio J, Facchinetti V, Chatterjee B, Wang YH, Homey B, Cao W, Su B, Nestle FO, Zal T, Mellman I, Schroder JM, Liu YJ, Gilliet M (2007) Plasmacytoid dendritic cells sense self-DNA coupled with antimicrobial peptide. Nature 449:564–569

Li W, Li Y, Guan S, Fan J, Cheng CF, Bright AM, Chinn C, Chen M, Woodley DT (2007) Extracellular heat shock protein-90alpha: linking hypoxia to skin cell motility and wound healing. EMBO J 26:1221–1233

Liao DF, Jin ZG, Baas AS, Daum G, Gygi SP, Aebersold R, Berk BC (2000) Purification and identification of secreted oxidative stress-induced factors from vascular smooth muscle cells. J Biol Chem 275:189–196

Liu Z, Davidson A (2012) Taming lupus-a new understanding of pathogenesis is leading to clinical advances. Nat Med 18:871–882

Liu B, Yang Y, Qiu Z, Staron M, Hong F, Li Y, Wu S, Hao B, Bona R, Han D, Li Z (2010) Folding of Toll-like receptors by the HSP90 paralogue gp96 requires a substrate-specific cochaperone. Nat Commun 1:79

Minota S, Koyasu S, Yahara I, Winfield J (1988) Autoantibodies to the heat-shock protein hsp90 in systemic lupus erythematosus. J Clin Invest 81:106–109

Multhoff G, Hightower LE (1996) Cell surface expression of heat shock proteins and the immune response. Cell Stress Chaperones 1:167–176

Okuya K, Tamura Y, Saito K, Kutomi G, Torigoe T, Hirata K, Sato N (2010) Spatiotemporal regulation of heat shock protein 90-chaperoned self-DNA and CpG-oligodeoxynucleotide for type I IFN induction via targeting to static early endosome. J Immunol 184:7092–7099

Palladino MA Jr, Srivastava PK, Oettgen HF, DeLeo AB (1987) Expression of a shared tumor-specific antigen by two chemically induced BALB/c sarcomas. Cancer Res 47:5074–5079

Perry D, Sang A, Yin Y, Zheng YY, Morel L (2011) Murine models of systemic lupus erythematosus. J Biomed Biotechnol 2011:271694

Ripley BJ, Isenberg DA, Latchman DS (2001) Elevated levels of the 90 kDa heat shock protein (hsp90) in SLE correlate with levels of IL-6 and autoantibodies to hsp90. J Autoimmun 17:341–346

Ronnblom L, Alm GV, Eloranta ML (2009) Type I interferon and lupus. Curr Opin Rheumatol 21:471–477

Sahu D, Zhao Z, Tsen F, Cheng CF, Park R, Situ AJ, Dai J, Eginli A, Shams S, Chen M, Ulmer TS, Conti P, Woodley DT, Li W (2012) A potentially common peptide target in secreted heat shock protein-90alpha for hypoxia-inducible factor-1alpha-positive tumors. Mol Biol Cell 23:602–613

Saito K, Kukita K, Kutomi G, Okuya K, Asanuma H, Tabeya T, Naishiro Y, Yamamoto M, Takahashi H, Torigoe T, Nakai A, Shinomura Y, Hirata K, Sato N, Tamura Y (2015) Heat shock protein 90 associates with Toll-like receptors 7/9 and mediates self-nucleic acid recognition in SLE. Eur J Immunol 45:2028–2041

Sasai M, Linehan MM, Iwasaki A (2010) Bifurcation of Toll-like receptor 9 signaling by adaptor protein 3. Science 329:1530–1534

Semenza GL (2007) Evaluation of HIF-1 inhibitors as anticancer agents. Drug Discov Today 12:853–859

Shimp SK 3rd, Chafin CB, Regna NL, Hammond SE, Read MA, Caudell DL, Rylander M, Reilly CM (2012) Heat shock protein 90 inhibition by 17-DMAG lessens disease in the MRL/lpr mouse model of systemic lupus erythematosus. Cell Mol Immunol 9:255–266

Srivastava P (2002) Interaction of heat shock proteins with peptides and antigen presenting cells: chaperoning of the innate and adaptive immune responses. Annu Rev Immunol 20:395–425

Stocki P, Dickinson AM (2012) The immunosuppressive activity of heat shock protein 70. Autoimmune Dis 2012:617213

Suto R, Srivastava PK (1995) A mechanism for the specific immunogenicity of heat shock protein-chaperoned peptides. Science 269:1585–1588

Suzuki S, Kulkarni AB (2010) Extracellular heat shock protein HSP90beta secreted by MG63 osteosarcoma cells inhibits activation of latent TGF-beta1. Biochem Biophys Res Commun 398:525–531

Takahashi K, Shibata T, Akashi-Takamura S, Kiyokawa T, Wakabayashi Y, Tanimura N, Kobayashi T, Matsumoto F, Fukui R, Kouro T, Nagai Y, Takatsu K, Saitoh S, Miyake K (2007) A protein associated with Toll-like receptor (TLR) 4 (PRAT4A) is required for TLR-dependent immune responses. J Exp Med 204:2963–2976

Tamura Y, Peng P, Liu K, Daou M, Srivastava PK (1997) Immunotherapy of tumors with autologous tumor-derived heat shock protein preparations. Science 278:117–120

Tanaka T, Okuya K, Kutomi G, Takaya A, Kajiwara T, Kanaseki T, Tsukahara T, Hirohashi Y, Torigoe T, Hirata K, Okamoto Y, Sato N, Tamura Y (2015) Heat shock protein 90 targets a chaperoned peptide to the static early endosome for efficient cross-presentation by human dendritic cells. Cancer Sci 106:18–24

Trepel J, Mollapour M, Giaccone G, Neckers L (2010) Targeting the dynamic HSP90 complex in cancer. Nat Rev Cancer 10:537–549

Triantafyllopoulou A, Franzke CW, Seshan SV, Perino G, Kalliolias GD, Ramanujam M, van Rooijen N, Davidson A, Ivashkiv LB (2010) Proliferative lesions and metalloproteinase activity in murine lupus nephritis mediated by type I interferons and macrophages. Proc Natl Acad Sci U S A 107:3012–3017

Trinchieri G (2010) Type I interferon: friend or foe? J Exp Med 207:2053–2063

Tsutsumi S, Neckers L (2007) Extracellular heat shock protein 90: a role for a molecular chaperone in cell motility and cancer metastasis. Cancer Sci 98:1536–1539

Tsutsumi S, Mollapour M, Graf C, Lee CT, Scroggins BT, Xu W, Haslerova L, Hessling M, Konstantinova AA, Trepel JB, Panaretou B, Buchner J, Mayer MP, Prodromou C, Neckers L (2009) Hsp90 charged-linker truncation reverses the functional consequences of weakened hydrophobic contacts in the N domain. Nat Struct Mol Biol 16:1141–1147

Tukaj S, Tiburzy B, Manz R, de Castro Marques A, Orosz A, Ludwig RJ, Zillikens D, Kasperkiewicz M (2014) Immunomodulatory effects of heat shock protein 90 inhibition on humoral immune responses. Exp Dermatol 23:585–590

Twomey BM, Dhillon VB, McCallum S, Isenberg DA, Latchman DS (1993) Elevated levels of the 90 kD heat shock protein in patients with systemic lupus erythematosus are dependent upon enhanced transcription of the hsp90 beta gene. J Autoimmun 6:495–506

Udono H, Srivastava PK (1994) Comparison of tumor-specific immunogenicities of stress-induced proteins gp96, hsp90, and hsp70. J Immunol 152:5398–5403

Viglianti GA, Lau CM, Hanley TM, Miko BA, Shlomchik MJ, Marshak-Rothstein A (2003) Activation of autoreactive B cells by CpG dsDNA. Immunity 19:837–847

Wagner H (1999) Bacterial CpG DNA activates immune cells to signal infectious danger. Adv Immunol 73:329–368

Wang XY, Facciponte J, Chen X, Subjeck JR, Repasky EA (2007) Scavenger receptor-A negatively regulates antitumor immunity. Cancer Res 67:4996–5002

Wang X, Song X, Zhuo W, Fu Y, Shi H, Liang Y, Tong M, Chang G, Luo Y (2009) The regulatory mechanism of Hsp90alpha secretion and its function in tumor malignancy. Proc Natl Acad Sci U S A 106:21288–21293

Whitesell L, Lindquist SL (2005) HSP90 and the chaperoning of cancer. Nat Rev Cancer 5:761–772

Woodley DT, Fan J, Cheng CF, Li Y, Chen M, Bu G, Li W (2009) Participation of the lipoprotein receptor LRP1 in hypoxia-HSP90alpha autocrine signaling to promote keratinocyte migration. J Cell Sci 122:1495–1498

Yanai H, Ban T, Wang Z, Choi MK, Kawamura T, Negishi H, Nakasato M, Lu Y, Hangai S, Koshiba R, Savitsky D, Ronfani L, Akira S, Bianchi ME, Honda K, Tamura T, Kodama T, Taniguchi T (2009) HMGB proteins function as universal sentinels for nucleic-acid-mediated innate immune responses. Nature 462:99–103

Yang Y, Liu B, Dai J, Srivastava PK, Zammit DJ, Lefrancois L, Li Z (2007) Heat shock protein gp96 is a master chaperone for toll-like receptors and is important in the innate function of macrophages. Immunity 26:215–226

Yasuda K, Yu P, Kirschning CJ, Schlatter B, Schmitz F, Heit A, Bauer S, Hochrein H, Wagner H (2005) Endosomal translocation of vertebrate DNA activates dendritic cells via TLR9-dependent and -independent pathways. J Immunol 174:6129–6136

Yoshida H, Okabe Y, Kawane K, Fukuyama H, Nagata S (2005) Lethal anemia caused by interferon-beta produced in mouse embryos carrying undigested DNA. Nat Immunol 6:49–56

Zhang H, Zhang L, Yu F, Liu Y, Liang Q, Deng G, Chen G, Liu M, Xiao X (2012) HSF1 is a transcriptional activator of IL-10 gene expression in RAW264.7 macrophages. Inflammation 35:1558–1566

Chapter 14
HSP47 and Its Involvement in Fibrotic Disorders

Haiyan Chu, Li Jin, and Jiucun Wang

Abstract Heat shock protein 47 (HSP47) is a collagen specific chaperone that is critical for the synthesis and maturation of collagen. It is encoded by the gene of *SERPINH1* and belongs to the serpin superfamily, while it lacks the active site essential for proteinase inhibition. HSP47 is an endoplasmic reticulum (ER) resident protein having the RDEL retention signal at C terminus. As a collagen chaperone, HSP47 preferentially recognizes the Gly-X-Y on procollagen triple helix. It binds to procollagen transiently once the nascent procollagen enters the ER to prevent the misfolding and aggregation of the procollagen, afterwards it dissociates from the triple helix at the cis-Golgi and cycles back to ER under the lead of RDEL signal. HSP47 expresses strictly in parallel with collagen proteins by collagen-producing cells, and it has been demonstrated to be involved in various fibrotic disorders including liver, lung and skin fibrosis. The specialty and necessity of HSP47 for collagen synthesis make it a good candidate for the treatment of fibrotic diseases. Several approaches designed targeting HSP47 has shown promising therapeutic effect as investigated in experimental animal models. This chapter

H. Chu
State Key Laboratory of Genetic Engineering, Collaborative Innovation Center for Genetics and Development, Department of Anthropology and Human Genetics, School of Life Sciences, Fudan University, Shanghai, China

L. Jin
State Key Laboratory of Genetic Engineering, Collaborative Innovation Center for Genetics and Development, Department of Anthropology and Human Genetics, School of Life Sciences, Fudan University, Shanghai, China

Human Phenome Institute, Fudan University, Shanghai, China

J. Wang (✉)
State Key Laboratory of Genetic Engineering, Collaborative Innovation Center for Genetics and Development, Department of Anthropology and Human Genetics, School of Life Sciences, Fudan University, Shanghai, China

Human Phenome Institute, Fudan University, Shanghai, China

Institute of Rheumatology, Immunology and Allergy, Fudan University, Shanghai, China
e-mail: jcwang@fudan.edu.cn

A. A. A. Asea, P. Kaur (eds.), *Chaperokine Activity of Heat Shock Proteins*, Heat Shock Proteins 16, https://doi.org/10.1007/978-3-030-02254-9_14

briefly reviews the recent advances of HSP47 as collagen chaperone and its participant in fibrotic disorders.

Keywords Molecular chaperone · Chaperokine · Collagen synthesis · Fibrosis · Fibrotic disorders · Heat shock protein 47

Abbreviations

BiP	Immunoglobulin heavy chain-binding protein
CCI4	Carbon tetrachloride
CTGF	Connective tissue growth factor
ECM	Extracellular matrix
ER	Endoplasmic reticulum
ERQC	ER quality control
Grp94	94-kDa glucose-regulated protein
HSCs	Hepatic stellate cells
HSE	Heat shock element
HSF1	Heat shock transcription factor I
HSP	Heat shock proteins
IL	Interleukin
IPF	Idiopathic pulmonary fibrosis
KLF	Kruppel-like factor
NASH	Nonalcoholic steatohepatitis
P4H	Prolyl 4-hydroxylase
PBMC	Peripheral blood mononuclear cell
PDI	Protein disulphide isomerase
SLE	Systemic lupus erythematosus
SSc	Systemic sclerosis
TGF-β	Transform growth factor β

14.1 Introduction

Heat shock proteins (HSP) are highly conserved and constitutively expressed in cells. HSP could not only be induced by heat shock as suggested by the name, but other stressful conditions such as hypoxia or endoplasmic reticulum (ER) stress can also increase the production of HSP (Ito and Nagata 2017; Kiang and Tsokos 1998). HSP play a fundamental role in cellular processes, such as protein assembly, export, turn over and regulation. In cells, the three dimensional structure of a protein is predetermined by its sequence, while its maturation involves the assistance of other proteins (Gething and Sambrook 1992). In other instances, a certain biogenesis requires the protein to display as a temporarily unfolded or partially folded conformation, as a consequence, the region normally buried within this protein are exposed and are tended to get aggregation (Becker and Craig 1994). To meet the needs described above, cells have their own chaperone proteins to help with this process. Many of HSP are known as molecular chaperones that facilitate the synthesis and maturation of other proteins, including stabilizing the nascent peptides, preventing the misfolding and aggregation, as well as assisting certain modifications of these peptides (Ito and Nagata 2017). Recently, it was demonstrated that some HSP has the activity of cytokine, such as HSP72. The intracellular HSP72 could act as chaperone, in contrast, the extracellular HSP72 released from the necrotic cell or produced in response to stress interacts with immune cells, such as natural killer (NK) cells and antigen present cells (APC) (including dendritic cells, macrophages, peripheral blood monocytes and B lymphocytes) through the toll-like receptor (TLR) 2/4. Upon the interaction, extracellular HSP72 exerts its immunoregulatory effects and specifically activates the intracellular signaling cascades, which is known as the chaperokine of the HSP72 (Asea 2003, 2005).

HSP47, a 47-kDa heat-shock protein, is initially characterized as a cell surface-associated protein, and later identified as a specific collagen binding protein (Kurkinen et al. 1984). HSP47 is encoded by *SERPINH1* gene, and belongs to the serpin family, but it lacks the active site essential for the inhibition of proteases (Nagata 2003). Although other HSP proteins may have broad substrate specificity, HSP47 is a collagen-specific molecular chaperone localized in the ER, where it transiently binds to the nascent pro-collagen proteins upon their synthesis to facilitate the assembly/export of collagens (Chu et al. 2015; Ishida and Nagata 2011). The expression pattern of HSP47 was identified to correlate closely with collagen expression (Nagata 1996). Thereby, HSP47 is a critical mediator in the regulation of collagen related disorders.

Fibrotic disorders are main causes of morbidity and mortality worldwide. Excessive extracellular matrix (ECM) such as collagen and fibronectin deposition because of aberrant injury repair leads to fibrosis, although it is initially beneficial for tissue repair. Fibrosis can affect almost every organs in the body and can be the results of many stimulus, for example, pulmonary fibrosis secondary to autoimmune diseases (Chu et al. 2017; Young et al. 2007) or many other triggers in the absence or presence of chronic inflammation (Kolahian et al. 2016), liver fibrosis following infection or alcohol abuse (Bataller and Brenner 2005), and systemic sclerosis resulting from chronic autoimmune inflammation or genetic influences (Pattanaik et al. 2015). It is commonly accepted that recurrent infection or other triggers induce the damage of epithelial or endothelial cells, and the release of inflammatory mediators from these damaged cells initiate the repair cascade, which includes leukocytes and fibroblast as well as the pro-inflammatory and pro-fibrotic cytokines produced by these cells to make the repair (Wynn 2004). However, when injury occurs recurrently or the repair goes uncontrolled, fibrosis develops subsequently.

As a collagen chaperone, HSP47 has been found to be involved in various fibrotic processes (Taguchi and Razzaque 2007), such as liver cirrhosis (Masuda et al. 1994), pulmonary fibrosis (Hagiwara et al. 2007c), systemic sclerosis (Chu et al.

2015), and peritoneal fibrosis (Nishino et al. 2003). Since abnormal accumulations of collagen is the main feature of fibrotic disorders, the understanding of the maturation of collagen and the regulation of HSP47 production as well as its specific function during the collagen biogenesis is critical for better understanding of the fibrosis pathogenesis. This chapter summaries the findings on HSP47 so far and its involvement in fibrotic disorders.

14.2 Heat Shock Protein 47

HSP47 was first discovered as a cell surface-associated protein "colligin" with a molecular weight of 47 kDa in mouse embryo cells as well as other cell lines producing type IV collagen (Kurkinen et al. 1984). Lately, this 47 kDa protein was identified to bind to gelatin and collagen in chicken embryo (Nagata et al. 1986) and then it was also identified as "gp46" in rat (Cates et al. 1987), and the binding ability of this protein to collagen was regulated by pH, while the binding was abolished when the pH is 6.3 (Saga et al. 1987).

ER is a major protein synthesis organelle and has an elaborate surveillance system called ER quality control (ERQC) system (Araki and Nagata 2011). Various ER-resident chaperones and several enzymes are involved in the ERQC system to guarantee and aid the folding and post-translational modification of proteins (Araki and Nagata 2011). HSP47 is an ER-resident protein, but its expression is regulated by heat-shock rather than ER stress, which is different from other HSP (Ishida and Nagata 2011; Saga et al. 1987). Acquisition of the RDEL retention signal at C terminus features HSP47 as an ER-resident protein (Hirayoshi et al. 1991). Although HSP47 belongs to the serpin family, a superfamily of plasma serine protease inhibitors, but it lacks the active site essential for inhibition and cannot bind to intracellular proteases (Hirayoshi et al. 1991).

HSP47 is identified to be synthesized as a molecule of 42 KDa, and then decreases to 41 KDa after processing, but finally gains to 47 KDa after glycosylation as a mature molecule (Nagata et al. 1988). Expression of HSP47 is closely correlated with that of collagen (Masuda et al. 1994; Takechi et al. 1992) and only in collagen-producing cells (Nagata 2003). It is reported that HSP47 as well as type IV collagen expression could be significantly induced during the differentiation of mouse teratocarcinoma F9 cell (Takechi et al. 1992). Also, during the progression of carbon tetrachloride (CCl4)-induced liver fibrosis, HSP47 transcription was markedly enhanced in parallel with type I and III collagen mRNAs (Masuda et al. 1994). Moreover, an increase in HSP47 expression was observed in murine belomycin-induced pneumopathy and the relative amounts of HSP47 were significantly correlated with the lung hydroxyproline content, an indicator of collagen deposition (Ishii et al. 2003). In this regards, expression of HSP47 also seems to be regulated by other factors other than heat stress. As a heat shock protein, the expression of HSP47 is mediated by its heat shock element (HSE) in its promoter region (Hirayoshi et al. 1991; Takechi et al. 1992). Both TGF-β and IL-1β have been shown to be involved in the regulation of HSP47 expression via enhancement of the binding activity between HSE and heat shock transcription factor I (HSF1) (Sasaki et al. 2002). Afterwards, further study demonstrated another transcriptional regulatory element exists within a 280-bp promoter region of HSP47, which was identified as a putative Sp1-binding site (Hirata et al. 1999). Moreover, HSP47 harbors two more not yet identified elements, BS5-B and Kruppel-like factor (KLF), in the first intron which is necessary for the tissue-specific expression of HSP47 (Ishida and Nagata 2011; Nagata 2003).

14.3 Interaction of HSP47 and Collagen

Collagen is a kind of abundant protein that represents 30% of protein in human body (Gonzalez Santander et al. 1999), and 28 different collagens have been described so far. Collagen biosynthesis is a complex process including intra- and extracellular steps that implicates various modification enzymes and maturation chaperones. Once the α-chain of collagen is synthesized, it then proceeds to the procollagen assembly, one or more distinct procollagen α-chains are translocated to the ER where procollagen undergoes triple helix assembly and further modifications (Widmer et al. 2012). The protein chaperones such as HSP47, protein disulphide isomerase (PDI) and prolyl 4-hydroxylase (P4H) are involved in these processes (Lamande and Bateman 1999). After α-chain enters into ER, PDI binds subsequently to the nascent α-chain to facilitate the trimer formation at the C-terminus by catalyzing disulfide-bond. Meanwhile, P4H binds to the α-chain and hydroxylates the proline residues. With the assistance of these two chaperones, the triple-helix forms from the C-terminus to N-terminus. During the progression of the triple-helix formation, HSP47 binds to the procollagen to assist correct folding and stabilize the triple-helix procollagen molecules (Myllyharju 2003; Nagata 2003).

A study which investigated the interaction of HSP47 and collagen by using a synthetic peptide model demonstrated that recognition of HSP47 required a certain collagen sequence and length, and the typical collagen motif Pro-Pro-Gly repeats need to be more than seven (Koide et al. 1999). Then using the yeast two-hybrid techniques, HSP47 was suggested to preferentially recognize collagen Gly-X-Y repeats in triple-helical conformation (Koide et al. 2000). Gly-X-Y serviced as the dominant binding sites for HSP47 when Arg occupied the Y site (Koide et al. 2002). Furthermore, when the synthetic collagen peptides linked with each other as a triple helix by disulfide bond at the C-termini, the binding between HSP47 and these peptides was markedly enhanced, indicating the triple helix of collagen-like peptides has the superiority for the interaction with HSP47 (Koide et al. 2006; Nagata 2003). Therefore, the triple helix form (Gly-X-Arg) n, when n more than seven is the consensus motif for HSP47 binding (Tasab et al. 2000, 2002; Widmer et al. 2012).

The dynamic interaction between HSP47 and collagen has been explored by pulse label-chase experiments accompanied with chemical crosslinking and immunoprecipitation analysis (Nagata 1996; Nakai et al. 1992; Satoh et al. 1996).

Normally, HSP47 was observed to dissociate from procollagen within 10–15 min after pulse labeling, while the total procollagen decreases after 20–30 min as determined by precipitation, suggesting the whole intra-cellular events process completes within 30 min under normal condition. However, no dissociation of HSP47 from procollagen was detected even after 2 h when the cell was treated with an iron chelate, α-α-dipyridyl, which can inhibit procollagen triple helix formation or underwent heat shock stress. Whereas, the co-localization of HSP47 and procollagen was then found in the ER under these conditions. Afterwards, HSP47 and procollagen proceeded along the secretory path into Golgi, as revealed by pulse-chase experiments, when treated cells with brefeldin A or low temperature to inhibit the transport from ER to Golgi, co-localization of HSP47 and procollagen in the intermediate compartment occurred. If the cells were treated with monensin or bafiromycin A1 to inhibit the secretion at the post-*cis*-Golgi and post-trans-Golgi, respectively, there is no complex of HSP47 and procollagen under both conditions. Meanwhile, the accumulation of procollagen in a compartment could be detected. Collectively, HSP47 binds to the nascent procollagen chain in ER, where it prevents the degradation and aggregation of procollagen to ensure correct helix formation, and then procollagen triple-helix accompanied by HSP47 entries into *cis*-Golgi and dissociates from HSP47 at this site, then it goes through to the trans-Golgi network to be secreted eventually. At the same time, HSP47 recycles to the ER under the lead of the RDEL retention sequence. HSP47's functions to prevent the unfolding/aggregation of procollagen, is an essential convoy for the successful secretion and maturation of collagen.

14.4 HSP47$^{-/-}$ Phenotype

Although the function of HSP47 in the maturation of collagen was relative clear, the exact role of HSP47 in collagen biosynthesis in vitro or in vivo had not been elucidated until one decade ago. To address this question, Nagai and colleagues established *Hsp47* knockout mouse by homologous recombination (Nagai et al. 2000). The deficiency of *Hsp47* in mouse caused severe abnormality in collagen maturation in the embryo, leading to embryonic lethality before 11.5 days postcoitus (dpc) in homozygous mice. *Hsp47*$^{-/-}$ embryonic collagen was more sensitive to protease digestion, suggesting the failure in the formation of stable triple-helix without HSP47 assistance. Compared to *Hsp47*$^{+/+}$, only a limited number of collagen fibrils were observed in the lamina fibroreticularis and stroma of *hsp47*$^{-/-}$ embryos. The aberrance in collagen formation results in the fracture of endothelial cell surrounding blood vessels and abnormally orientated epithelial tissues. These observations suggest HSP47 is essential for mouse normal development.

Hsp47$^{-/-}$ fibroblasts showed retarded type I collagen secretion and abnormally accumulation of type I collagen in ER, as a results, ER of *Hsp47*$^{-/-}$ fibroblasts was markedly dilated (Ishida et al. 2006). Meanwhile, the type I collagen with unprocessed N-terminus pro-peptide was observed to be secreted from *Hsp47*$^{-/-}$ cells (Ishida et al. 2006). As revealed by immunostaining, type IV collagen in the basement membrane was absent due to the lack of HSP47 of *Hsp47*$^{-/-}$ embryos, but laminin and nidogen-1 were unaffected (Marutani et al. 2004). Thus, it suggests that HSP47 is critical for both type I and type IV collagen folding and processing and *Hsp47* disruption specifically alters procollagen maturation.

The consequence of *Hsp47* deficiency is not just limited to the collagen formation aberration. Due to the accumulation of unfolded procollagen ER stress, apoptosis in *Hsp47*$^{-/-}$ embryos was likely induced by increased transcription of CHOP, an ER-stress-induced transcription factor responsible for apoptosis (Marutani et al. 2004). This phenomenon was also observed in *Hsp47*$^{-/-}$ hepatic stellate cells (HSCs). In this cell, accumulation of immature procollagen caused the expression of ER stress-inducible proteins BiP (immunoglobulin heavy chain-binding protein) and Grp94 (94-kDa glucose-regulated protein) as well as CHOP (Kawasaki et al. 2015).

Hsp47 conditional knockout mouse was also established. This mouse strain was created using mouse with LoxP-Hsp47 gene to cross the mouse with Cre gene under type II collagen-promoter (Masago et al. 2012). This *hsp47* conditional null mouse died just before or shortly after birth with lower levels of type II and type XI collagen as well as misaligned type I collagen molecules. Whole-mount skeletal staining of this mouse showed severe cartilage and bone malformation, confirming the pivotal role of HSP47 during the development of mouse.

14.5 HSP47 and Fibrotic Disorders

HSP47 is a collagen-specific chaperone that is necessary to the maturation of collagen. Accordingly, HSP47 is associated with collagen-related disorders such as fibrosis and keloid. Many cytokines have been shown to be involved in the development of fibrosis, including transforming growth factor (TGF)-β (Chu et al. 2015), interleukin (IL)-1β (Ning et al. 2017) as well as IL-17A (Chi et al. 2017). Recent studies showed that these cytokines regulate HSP47 expression. Both TGF-β and IL-1β induce HSP47 expression via enhancement of the binding activity of HSFI (heat shock transcription factor I) to HSE (Heat shock element) (Sasaki et al. 2002). Analysis of IL-17A signaling pathways revealed IL-17A could induce HSP47 expression through c-Jun N-terminal kinase (Honzawa et al. 2014). Fibrosis affects almost every tissue in the body, including liver, lung, kidney, skin fibrosis, and is characterized by excessive accumulation of collagens in the tissues. Currently, the available treatment strategies for fibrosis are limited. If highly progressive, fibrosis eventually leads to the organ failure and death. Given that HSP47 is collagen specific and critical synthesis chaperone and besides, the principal fibrogenesis cytokines are commonly involved in HSP47 induction, HSP47 could be a good candidate target for fibrosis controlling.

14.5.1 HSP47 and Liver Fibrosis

Liver fibrosis is the results of excessive accumulation of ECM, especially collagen, resulting from chronic damage to the liver. Among different kinds of stimuli, chronic HCV infection, alcohol abuse and nonalcoholic steatohepatitis (NASH) are the main causes of liver fibrosis (Bataller and Brenner 2005). HSCs are the main cell responsible for the production of ECM in the development of liver fibrosis. Under normal condition, HSCs reside in the space of Disse as the storage sites for vitamin A (Bataller and Brenner 2005). Following chronic injury, the HSCs are activated and differentiate into myofibrobalst-like cells that acquiring pro-inflammatory and fibrogenic properties (Marra 1999). In an experimental rat liver fibrosis model induced by CCI4, *Hsp47* mRNA was markedly increased in parallel with type I and III collagen mRNA. The distribution of HSP47 was similar to that of type I collagen as revealed by in situ hybridization and HSCs were the sources of HSP47 as identified by immunohistochemistry and dual fluorescence (Brown et al. 2005; Masuda et al. 1994).

HSCs play a crucial role in the progression of liver fibrosis and can be the target for therapy design. Taken advantage of the vitamin A up taking and storing ability of the HSCs, a specific strategy for HSP47 targeting drug delivery system was established by Sato and his colleagues in rat liver fibrosis model (Sato et al. 2008). They encapsulated the siRNA targeting HSP47 in vitamin A-coupled liposomes to ensure this siRNA target preferentially to HSCs. Their results demonstrated that this siRNA-bearing vitamin A-coupled system almost completely prevented rat liver fibrosis and prolonged the survival of rat models (Sato et al. 2008), suggesting a reasonable approach for liver treatment by targeting HSP47. This result was confirmed in *Hsp47*-KO HSCs (Kawasaki et al. 2015). HSCs were isolated from the liver of HSP47 floxed mice. The deficiency in *Hsp47* of these cells was induced by infecting cells with an adenovirus harboring the Cre gene. In *Hsp47*-KO HSCs, there is no mature collagen secretion, and as a result, the aggregated procollagen induced ER stress. Because autophagy degrades accumulated procollagen, the accumulation and ER-stress were augmented when autophagy was inhibited. The inhibition of autophagy also resulted in the increase of CHOP that responsible for apoptosis. Thus, disruption of *Hsp47* in HSCs leads to the blockade of collagen synthesis with the additional effect of inducing the apoptosis of HSCs when autophagy is inhibited. Therefore, HSP47 could be a candidate target for the treatment of liver fibrosis.

14.5.2 HSP47 and Scleroderma

Scleroderma, also known as systemic sclerosis (SSc), is an autoimmune disease with the characteristics of alteration of the microcirculation and fibrosis in skin and internal organs involving lung, heart, gastrointestine, and kidney (Chu et al. 2015). In the US, the prevalence and incidence of SSc has continued to rise and the annual prevalence ranged from 13.5 in 2003 to 18.4 (per 100,000) in 2008 (Furst et al. 2012). SSc causes severe mortality around the world. Visceral organ fibrosis is the leading cause of death in SSc, especially SSc-related interstitial lung disease (Chu et al. 2017). Although certain immunosuppressive or anti-fibrotic therapies may slow disease progression, current treatment approaches are not curative.

HSP47 was found highly expressed in cultured SSc fibroblasts together with type I collagen, which was correlated with shorter disease duration (Kuroda et al. 1998). As to other type of dermal fibrotic disease such as keloid and cicatricial pemphigoid, highly regulated HSP47 together with the increase of both type I collagen and type III was also observed, besides, HSP47 could be induced by exogenous TGF-β administration in vitro (Naitoh et al. 2001). Anti-HSP47 autoantibodies were present in SSc patients with a frequency of 26% compared with healthy control or patients with other skin diseases or systemic lupus erythematosus (SLE), and this autoantibody was also positive in 79% of TSK/+ mice, a genetic SSc mouse model (Fujimoto et al. 2004). We then investigated the potential role HSP47 in SSc systemically. We found that *Hsp47* was elevated in the peripheral blood mononuclear cell (PBMC) of SSc patients, and the enhanced level of HSP47 in PBMC was positively correlated with the presence of anti-centromere antibody. In addition, increased *Hsp47* level was also detected in the skin biopsy and plasma of SSc patients and in the skin of SSc mouse model. Further in vitro studies demonstrated that *Hsp47* knockdown could block the intracellular and extracellular collagen overproductions induced by exogenous TGF-β (Chu et al. 2015). These observations implicate HSP47 might be involved in the pathogenesis of SSc and could be the potential treatment target for SSc. Wound healing is a collagen active production process, and HSP47 was an important determinant of scar formation during wound healing (Wang et al. 2002). Blunted the expression of *Hsp47* by its antisense oligonucleotide in rat wound skin has been shown a therapeutic potential to suppress the scar formation of skin (Wang et al. 2003; Wang and Li 2008), confirming the possibility of HSP47 as a therapy target for skin fibrosis including SSc.

14.5.3 HSP47 and Pulmonary Fibrosis

Pulmonary fibrosis, including idiopathic pulmonary fibrosis (IPF) and interstitial lung fibrosis (secondary to conditions such as SSc and rheumatoid arthritis) are particularly severe lung diseases characterized by epithelial injury, and impaired wound healing are caused by the accumulation of fibroblasts and ECM in the lung (Herzog et al. 2014; Noble et al. 2012; Toonkel et al. 2013; Veraldi et al. 2010; Wolters et al. 2014). Pulmonary fibrosis is a devastating kind of lung diseases, with IPF having the median survival of 2.5–3.5 years (Ryu et al. 2017). Pulmonary fibrosis is recognized to result from various factors (Ueno et al. 2011). It is considered that injury arising from various etiologies leads to activation of inflammatory cytokines (e.g. IL-1β, IL-6) that contributing to the progression of lung fibrosis in the early stages (Di Paola et al. 2011; Vaillant et al. 1996). In later stages of lung injury, a variety of pro-fibrotic mediators, such as TGF-β and connective tissue growth factor (CTGF) interact to maintain fibrosis (Wynn 2011).

Considering the pivotal function of HSP47 in the production of collagen, the role of HSP47 in pulmonary fibrosis was investigated recently. To identify the cell that expresses HSP47, Iwashita *et al* examined the expression level of HSP47 and type I procollagen in patients with usual interstitial pneumonia. The results demonstrated that both HSP47 and type I procollagen were localized predominantly in α-smooth muscle actin (SMA)-positive myofibroblasts and surfactant protein-A-positive type II pneumocytes in active fibrotic areas, and the positive cells were increased in this areas compared to the samples from excised lung cancer (Iwashita et al. 2000). This was also observed in mice with lung fibrosis induced by bleomycin administration. HSP47 was found to localize in α-SMA-positive myofibroblasts, surfactant-protein-A-positive type II pneumocytes, and F4/80-positive macrophages in the active fibrotic areas, meanwhile, the *Hsp47* mRNA was markedly increased in bleomycin-treated lungs (Kakugawa et al. 2010). Moreover, serum HSP47 level is highly elevated in acute exacerbation of IPF than stable IPF (Kakugawa et al. 2013). These observations suggest HSP47 might be involved in the fibrogenesis of pulmonary fibrosis and it contributes to the active fibrogenetic progression. Other studies also showed that lung fibrosis could be improved by disruption of *Hsp47* with its antisense oligonucleotides (Hagiwara et al. 2007a, b) or with pirfenidone, an antifibrotic agent that has been shown effective in clinical studies (Kakugawa et al. 2004). Therefore, HSP47 has been suggested a promising therapeutic target for these collagen-related diseases, particularly fibrotic disorders.

14.6 Conclusions

HSP47, an ER-resident molecular chaperone, is essential for collagen synthesis and secretion. The company of HSP47 ensures procollagen proceeding successfully from ER to Golgi along the secretory folding and modification ways. Induced increase of HSP47 by cytokines (e.g. TGF-β, IL-1β) or drug (bleomycin) leads to the excessive deposition of collagen that features fibrotic diseases. Manipulated disruption of HSP47 results in the aggregation of procollagen that induced ER stress as well as apoptosis that may help control fibrogenesis progression. On the other hand, HSP47 has been shown to be related with other collagen related diseases such as osteogenesis imperfecta (OI). Most cases of this disease are caused by mutations in type I collagen genes that associated with defective bone collagen assembly, whereas recent findings showed the mutations in HSP47 could be another reason for this disease (Ito and Nagata 2017). Similar to HSP72, there are antibodies of HSP47 in the circulation (Fujimoto et al. 2004), and thus it is possible that HSP47 could be released from cells and function as a chaperokine. Therefore, in addition to the chaperone ability, HSP47 might has the ability of chaperokine to activate the pro-inflammatory signaling, which are documented involved in the development of fibrosis. Thus, as a chaperokine, HSP47 would enhance the fibrotic ability of itself, and we will include this in our future study of HSP47. Overall, HSP47 could be the excellent therapeutic target for collagen-related diseases, particularly for fibrotic disorders.

Acknowledgements This study was partially supported by the grants from the National Science Foundation of China (81470254, 31521003, 81770066), Shanghai Municipal Science and Technology Major Project (2017SHZDZX01), and the International S & T Cooperation Program of China (2013DFA30870).

References

Araki K, Nagata K (2011) Protein folding and quality control in the ER. Cold Spring Harb Perspect Biol 3:a007526

Asea A (2003) Chaperokine-induced signal transduction pathways. Exerc Immunol Rev 9:25–33

Asea A (2005) Stress proteins and initiation of immune response: chaperokine activity of hsp72. Exerc Immunol Rev 11:34–45

Bataller R, Brenner DA (2005) Liver fibrosis. J Clin Invest 115:209–218

Becker J, Craig EA (1994) Heat-shock proteins as molecular chaperones. Eur J Biochem 219:11–23

Brown KE, Broadhurst KA, Mathahs MM, Brunt EM, Schmidt WN (2005) Expression of HSP47, a collagen-specific chaperone, in normal and diseased human liver. Lab Inv; J Tech Methods Pathol 85:789–797

Cates GA, Nandan D, Brickenden AM, Sanwal BD (1987) Differentiation defective mutants of skeletal myoblasts altered in a gelatin-binding glycoprotein. Biochem Cell Biol Biochim Biol Cell 65:767–775

Chi HH, Hua KF, Lin YC et al (2017) IL-36 signaling facilitates activation of the NLRP3 Inflammasome and IL-23/IL-17 Axis in renal inflammation and fibrosis. J Am Soc Nephrol 28:2022–2037

Chu H, Wu T, Wu W et al (2015) Involvement of collagen-binding heat shock protein 47 in scleroderma-associated fibrosis. Protein Cell 6:589–598

Chu H, Shi Y, Jiang S et al (2017) Treatment effects of the traditional Chinese medicine Shenks in bleomycin-induced lung fibrosis through regulation of TGF-beta/Smad3 signaling and oxidative stress. Sci Rep 7:2252

Di Paola R, Talero E, Galuppo M et al (2011) Adrenomedullin in inflammatory process associated with experimental pulmonary fibrosis. Respir Res 12:41

Fujimoto M, Hamaguchi Y, Yazawa N, Komura K, Takehara K, Sato S (2004) Autoantibodies to a collagen-specific molecular chaperone, heat-shock protein 47, in systemic sclerosis. Clin Exp Immunol 138:534–539

Furst DE, Fernandes AW, Iorga SR, Greth W, Bancroft T (2012) Epidemiology of systemic sclerosis in a large US managed care population. J Rheumatol 39:784–786

Gething MJ, Sambrook J (1992) Protein folding in the cell. Nature 355:33–45

Gonzalez Santander R, Plasencia Arriba MA, Martinez Cuadrado G et al (1999) Intracellular biogenesis of collagen fibrils in 'activated fibroblasts' of tendo Achillis. An ultrastructural study in the New Zealand rabbit. J Bone Jt Surg Br 81:522–530

Hagiwara S, Iwasaka H, Matsumoto S, Noguchi T (2007a) Antisense oligonucleotide inhibition of heat shock protein (HSP) 47 improves bleomycin-induced pulmonary fibrosis in rats. Respir Res 8:37

Hagiwara S, Iwasaka H, Matsumoto S, Noguchi T (2007b) An antisense oligonucleotide to HSP47 inhibits paraquat-induced pulmonary fibrosis in rats. Toxicology 236:199–207

Hagiwara S, Iwasaka H, Matsumoto S, Noguchi T, Yoshioka H (2007c) Coexpression of HSP47 gene and type I and type III collagen genes in LPS-induced pulmonary fibrosis in rats. Lung 185:31–37

Herzog EL, Mathur A, Tager AM, Feghali-Bostwick C, Schneider F, Varga J (2014) Review: interstitial lung disease associated with systemic sclerosis and idiopathic pulmonary fibrosis: how similar and distinct? Arthritis Rheumatol 66:1967–1978

Hirata H, Yamamura I, Yasuda K et al (1999) Separate cis-acting DNA elements control cell type- and tissue-specific expression of collagen binding molecular chaperone HSP47. J Biol Chem 274:35703–35710
Hirayoshi K, Kudo H, Takechi H et al (1991) HSP47: a tissue-specific, transformation-sensitive, collagen-binding heat shock protein of chicken embryo fibroblasts. Mol Cell Biol 11:4036–4044
Honzawa Y, Nakase H, Shiokawa M et al (2014) Involvement of interleukin-17A-induced expression of heat shock protein 47 in intestinal fibrosis in Crohn's disease. Gut 63:1902–1912
Ishida Y, Nagata K (2011) Hsp47 as a collagen-specific molecular chaperone. Methods Enzymol 499:167–182
Ishida Y, Kubota H, Yamamoto A, Kitamura A, Bachinger HP, Nagata K (2006) Type I collagen in Hsp47-null cells is aggregated in endoplasmic reticulum and deficient in N-propeptide processing and fibrillogenesis. Mol Biol Cell 17:2346–2355
Ishii H, Mukae H, Kakugawa T et al (2003) Increased expression of collagen-binding heat shock protein 47 in murine bleomycin-induced pneumopathy. Am J Physiol Lung Cell Mol Physiol 285:L957–L963
Ito S, Nagata K (2017) Biology of Hsp47 (Serpin H1), a collagen-specific molecular chaperone. Semin Cell Dev Biol 62:142–151
Iwashita T, Kadota J, Naito S et al (2000) Involvement of collagen-binding heat shock protein 47 and procollagen type I synthesis in idiopathic pulmonary fibrosis: contribution of type II pneumocytes to fibrosis. Hum Pathol 31:1498–1505
Kakugawa T, Mukae H, Hayashi T et al (2004) Pirfenidone attenuates expression of HSP47 in murine bleomycin-induced pulmonary fibrosis. Eur Respir J 24:57–65
Kakugawa T, Mukae H, Hishikawa Y et al (2010) Localization of HSP47 mRNA in murine bleomycin-induced pulmonary fibrosis. Virchows Archiv: Int J Pathol 456:309–315
Kakugawa T, Yokota S, Ishimatsu Y et al (2013) Serum heat shock protein 47 levels are elevated in acute exacerbation of idiopathic pulmonary fibrosis. Cell Stress Chaperones 18:581–590
Kawasaki K, Ushioda R, Ito S, Ikeda K, Masago Y, Nagata K (2015) Deletion of the collagen-specific molecular chaperone Hsp47 causes endoplasmic reticulum stress-mediated apoptosis of hepatic stellate cells. J Biol Chem 290:3639–3646
Kiang JG, Tsokos GC (1998) Heat shock protein 70 kDa: molecular biology, biochemistry, and physiology. Pharmacol Ther 80:183–201
Koide T, Asada S, Nagata K (1999) Substrate recognition of collagen-specific molecular chaperone HSP47. Structural requirements and binding regulation. J Biol Chem 274:34523–34526
Koide T, Aso A, Yorihuzi T, Nagata K (2000) Conformational requirements of collagenous peptides for recognition by the chaperone protein HSP47. J Biol Chem 275:27957–27963
Koide T, Takahara Y, Asada S, Nagata K (2002) Xaa-Arg-Gly triplets in the collagen triple helix are dominant binding sites for the molecular chaperone HSP47. J Biol Chem 277:6178–6182
Koide T, Nishikawa Y, Asada S et al (2006) Specific recognition of the collagen triple helix by chaperone HSP47. II. The HSP47-binding structural motif in collagens and related proteins. J Biol Chem 281:11177–11185
Kolahian S, Fernandez IE, Eickelberg O, Hartl D (2016) Immune mechanisms in pulmonary fibrosis. Am J Respir Cell Mol Biol 55:309–322
Kurkinen M, Taylor A, Garrels JI, Hogan BL (1984) Cell surface-associated proteins which bind native type IV collagen or gelatin. J Biol Chem 259:5915–5922
Kuroda K, Tsukifuji R, Shinkai H (1998) Increased expression of heat-shock protein 47 is associated with overproduction of type I procollagen in systemic sclerosis skin fibroblasts. J Invest Dermatol 111:1023–1028
Lamande SR, Bateman JF (1999) Procollagen folding and assembly: the role of endoplasmic reticulum enzymes and molecular chaperones. Semin Cell Dev Biol 10:455–464
Marra F (1999) Hepatic stellate cells and the regulation of liver inflammation. J Hepatol 31:1120–1130
Marutani T, Yamamoto A, Nagai N, Kubota H, Nagata K (2004) Accumulation of type IV collagen in dilated ER leads to apoptosis in Hsp47-knockout mouse embryos via induction of CHOP. J Cell Sci 117:5913–5922

Masago Y, Hosoya A, Kawasaki K et al (2012) The molecular chaperone Hsp47 is essential for cartilage and endochondral bone formation. J Cell Sci 125:1118–1128

Masuda H, Fukumoto M, Hirayoshi K, Nagata K (1994) Coexpression of the collagen-binding stress protein HSP47 gene and the alpha 1(I) and alpha 1(III) collagen genes in carbon tetrachloride-induced rat liver fibrosis. J Clin Invest 94:2481–2488

Myllyharju J (2003) Prolyl 4-hydroxylases, the key enzymes of collagen biosynthesis. Matrix Biol: J Int Soc Matrix Biol 22:15–24

Nagai N, Hosokawa M, Itohara S et al (2000) Embryonic lethality of molecular chaperone hsp47 knockout mice is associated with defects in collagen biosynthesis. J Cell Biol 150:1499–1506

Nagata K (1996) Hsp47: a collagen-specific molecular chaperone. Trends Biochem Sci 21:22–26

Nagata K (2003) HSP47 as a collagen-specific molecular chaperone: function and expression in normal mouse development. Semin Cell Dev Biol 14:275–282

Nagata K, Saga S, Yamada KM (1986) A major collagen-binding protein of chick embryo fibroblasts is a novel heat shock protein. J Cell Biol 103:223–229

Nagata K, Hirayoshi K, Obara M, Saga S, Yamada KM (1988) Biosynthesis of a novel transformation-sensitive heat-shock protein that binds to collagen. Regulation by mRNA levels and in vitro synthesis of a functional precursor. J Biol Chem 263:8344–8349

Naitoh M, Hosokawa N, Kubota H et al (2001) Upregulation of HSP47 and collagen type III in the dermal fibrotic disease, keloid. Biochem Biophys Res Commun 280:1316–1322

Nakai A, Satoh M, Hirayoshi K, Nagata K (1992) Involvement of the stress protein HSP47 in procollagen processing in the endoplasmic reticulum. J Cell Biol 117:903–914

Ning ZW, Luo XY, Wang GZ et al (2017) MicroRNA-21 mediates angiotensin II-induced liver fibrosis by activating NLRP3 Inflammasome/IL-1beta Axis via targeting Smad7 and Spry1. Antioxid Redox Signal 27:1–20

Nishino T, Miyazaki M, Abe K et al (2003) Antisense oligonucleotides against collagen-binding stress protein HSP47 suppress peritoneal fibrosis in rats. Kidney Int 64:887–896

Noble PW, Barkauskas CE, Jiang D (2012) Pulmonary fibrosis: patterns and perpetrators. J Clin Invest 122:2756–2762

Pattanaik D, Brown M, Postlethwaite BC, Postlethwaite AE (2015) Pathogenesis of systemic sclerosis. Front Immunol 6:272

Ryu C, Sun H, Gulati M et al (2017) Extracellular mitochondrial DNA is generated by fibroblasts and predicts death in idiopathic pulmonary fibrosis. Am J Respir Crit Care Med 196:1571–1581

Saga S, Nagata K, Chen WT, Yamada KM (1987) pH-dependent function, purification, and intracellular location of a major collagen-binding glycoprotein. J Cell Biol 105:517–527

Sasaki H, Sato T, Yamauchi N et al (2002) Induction of heat shock protein 47 synthesis by TGF-beta and IL-1 beta via enhancement of the heat shock element binding activity of heat shock transcription factor 1. J Immunol 168:5178–5183

Sato Y, Murase K, Kato J et al (2008) Resolution of liver cirrhosis using vitamin A-coupled liposomes to deliver siRNA against a collagen-specific chaperone. Nat Biotechnol 26:431–442

Satoh M, Hirayoshi K, Yokota S, Hosokawa N, Nagata K (1996) Intracellular interaction of collagen-specific stress protein HSP47 with newly synthesized procollagen. J Cell Biol 133:469–483

Taguchi T, Razzaque MS (2007) The collagen-specific molecular chaperone HSP47: is there a role in fibrosis? Trends Mol Med 13:45–53

Takechi H, Hirayoshi K, Nakai A, Kudo H, Saga S, Nagata K (1992) Molecular cloning of a mouse 47-kDa heat-shock protein (HSP47), a collagen-binding stress protein, and its expression during the differentiation of F9 teratocarcinoma cells. Eur J Biochem 206:323–329

Tasab M, Batten MR, Bulleid NJ (2000) Hsp47: a molecular chaperone that interacts with and stabilizes correctly-folded procollagen. EMBO J 19:2204–2211

Tasab M, Jenkinson L, Bulleid NJ (2002) Sequence-specific recognition of collagen triple helices by the collagen-specific molecular chaperone HSP47. J Biol Chem 277:35007–35012

Toonkel RL, Hare JM, Matthay MA, Glassberg MK (2013) Mesenchymal stem cells and idiopathic pulmonary fibrosis. Potential for clinical testing. Am J Respir Crit Care Med 188:133–140

Ueno M, Maeno T, Nomura M et al (2011) Hypoxia-inducible factor-1alpha mediates TGF-beta-induced PAI-1 production in alveolar macrophages in pulmonary fibrosis. Am J Physiol Lung Cell Mol Physiol 300:L740–L752

Vaillant P, Menard O, Vignaud JM, Martinet N, Martinet Y (1996) The role of cytokines in human lung fibrosis. Monaldi Arch Chest Dis Arch Monaldi mal torace/Fondazione clin lav, IRCCS [and] Ist clin tisiologica mal apparato respiratorio, Univ Napoli, Secondo ateneo 51:145–152

Veraldi KL, Hsu E, Feghali-Bostwick CA (2010) Pathogenesis of pulmonary fibrosis in systemic sclerosis: lessons from interstitial lung disease. Curr Rheumatol Rep 12:19–25

Wang Z, Li L (2008) Adenovirus-mediated RNA interference against collagen-specific molecular chaperone 47-KDa heat shock protein suppresses scar formation on mouse wounds. Cell Biol Int 32:484–493

Wang ZL, Inokuchi T, Ikeda H et al (2002) Collagen-binding heat shock protein HSP47 expression during healing of fetal skin wounds. Int J Oral Maxillofac Surg 31:179–184

Wang Z, Inokuchi T, Nemoto TK, Uehara M, Baba TT (2003) Antisense oligonucleotide against collagen-specific molecular chaperone 47-kDa heat shock protein suppresses scar formation in rat wounds. Plast Reconstr Surg 111:1980–1987

Widmer C, Gebauer JM, Brunstein E et al (2012) Molecular basis for the action of the collagen-specific chaperone Hsp47/SERPINH1 and its structure-specific client recognition. Proc Natl Acad Sci U S A 109:13243–13247

Wolters PJ, Collard HR, Jones KD (2014) Pathogenesis of idiopathic pulmonary fibrosis. Annu Rev Pathol 9:157–179

Wynn TA (2004) Fibrotic disease and the T(H)1/T(H)2 paradigm. Nat Rev Immunol 4:583–594

Wynn TA (2011) Integrating mechanisms of pulmonary fibrosis. J Exp Med 208:1339–1350

Young A, Koduri G, Batley M et al (2007) Mortality in rheumatoid arthritis. Increased in the early course of disease, in ischaemic heart disease and in pulmonary fibrosis. Rheumatology 46:350–357

Index

A. A. A. Asea, P. Kaur (eds.), *Chaperokine Activity of Heat Shock Proteins*, Heat Shock Proteins 16, https://doi.org/10.1007/978-3-030-02254-9

The manufacturer's authorised representative in the EU is Springer Nature Customer Service Centre GmbH, Europaplatz 3, 69115 Heidelberg, Germany. If you have any concerns regarding our products, please contact ProductSafety@springernature.com

Printed and bound by CPI Group (UK) Ltd, Croydon, CR0 4YY
28/07/2026
02179208-0007